D0509442

Planning for Sustainability

Planning for Sustainability presents a straightforward, accessible introduction to the concept of planning for more sustainable and livable communities. The author explores topics such as how more compact, walkable cities and towns might be created, how local ecosystems can be restored, how social inequalities might be reduced, and how more sustainable forms of economic development can be brought about. Many photographs, graphics, and examples help illustrate key points.

Building upon previous schools of planning theory, *Planning for Sustainability* lays out a sustainability planning framework that pays special attention to the rapidly evolving institutions and power structures of a globalizing world. By considering in turn each scale of planning – international, national, regional, municipal, neighborhood, and site and building – the book illustrates how sustainability initiatives at different levels can inter-relate. The author argues that only by weaving together planning initiatives at different scales, and by integrating efforts across disciplines, can long-term human and ecological well-being be assured.

In addressing this important topic in a highly readable manner, *Planning for Sustainability* will appeal to students, practitioners, and all those interested in the future of our communities.

Stephen M. Wheeler is Assistant Professor of Community and Regional Planning at the University of New Mexico. He formerly served as editor of *The Urban Ecologist* journal, and is co-editor (with Timothy Beatley) of *The Sustainable Urban Development Reader* (Routledge, 2004).

Planning for Sustainability

Creating livable, equitable, and ecological communities

Stephen M. Wheeler

Routledge
Taylor & Francis Group

LONDON AND NEW YORK

To Mimi

First published 2004
by Routledge
2 Park Square, Milton Park, Abingdon, Oxon OX14 4RN

Simultaneously published in the USA and Canada
by Routledge
270 Madison Ave, New York NY 10016

Routledge is an imprint of the Taylor & Francis Group

© 2004 Stephen M. Wheeler

Typeset in Times by Keystroke, Jacaranda Lodge, Wolverhampton
Printed and bound in Great Britain by TJ International Ltd, Padstow, Cornwall

British Library Cataloguing in Publication Data
A catalogue record for this book is available from the British Library

Library of Congress Cataloging in Publication Data
Wheeler, Stephen (Stephen Maxwell)
    Planning for sustainability : creating livable, equitable, and ecological
communities / Stephen M. Wheeler.
        p. cm.
    Includes bibliographical references and index.
    1. City planning—Environmental aspects. 2. City planning—Social aspects.
    3. Urbanization—Environmental aspects. 4. Community development, Urban.
    5. Regional planning—Environmental aspects. I. Title.
    HT166.W484 2004
    307.1′216—dc22   2004000361

ISBN 0–415–32285–5 (hbk)
ISBN 0–415–32286–3 (pbk)

Contents

Acknowledgments

Many thanks to all those who have inspired me to work in the field of urban sustainability, especially former staff and board members of Urban Ecology, and some years before that colleagues at Friends of the Earth. Thanks especially to the inimitable Dave Brower, who had an inordinate faith in the printed word and image, and never tired of asking "Steve, where's the book?" Many years later, here it is.

Thanks also to four anonymous reviewers, who provided very useful comments on a previous draft, and to my colleagues at the University of California, Berkeley and the University of New Mexico, who have been very helpful sounding boards for various ideas.

The excellent staff at Routledge shepherded this book through its development and production. Thanks especially to Andrew Mould, Melanie Attridge, Belinda Dearbergh, Richard Willis and Lesley Edwards.

I am deeply grateful to my parents for their constant love and encouragement during years of sometimes quixotic pursuits. Most of all, thanks to my wife, Mimi, for her love, humor, and insight, from personal experience as an editor, into the long process of book production.

The author and publishers would like to thank the following for granting permission to reproduce copyright material:

Metro Resource Data Center for Figure 5.1; Tom Jones for Figure 5.3; Classic Communities for Figure 5.4; Contra Costa County Redevelopment Agency for Figures 6.1 and 6.2; Sustainable Seattle for Box 6.1; Princeton Architectural Press for Figure 10.1d; The Brookings Institution for Figure 10.2; James Taylor Chair in Landscape and Liveable Environments for Figures 12.2 and 12.3; Town of Markham for Figure 12.7; City of Albany, California and Freedman Tung & Bottomley for Figure 12.9; Calthorpe Associates for Figures 12.11, 12.12, and 12.13; Steve Price, Urban Advantage (www.urbanadvantage.com) for Figures 12.14 and 12.15; University of California Press for Figure 12.16.

1 Introduction

The past century has seen the rise of an urban and suburban landscape that is profoundly different from anything created before. From the office parks, malls, freeways, residential tracts, and abandoned inner cities of many affluent nations to enormous Third World megacities, human communities have taken on dramatically new forms and characteristics. Cities that once occupied a few square miles now cover thousands; populations that once walked most places are now utterly dependent on the automobile. Although recent patterns of urbanization have brought many benefits, they have also created enormous problems and are unsustainable in that they cannot be continued in the same ways in the long run. Today's development practices – both economic and physical forms of development – consume enormous amounts of land and natural resources, damage ecosystems, produce a wide variety of pollutants and toxic chemicals, create ever-growing inequities between groups of people, fuel global warming, and undermine local community, economies, and quality of life. Since the changes are incremental, it is hard to appreciate how rapidly our world is being transformed and how fundamentally these processes affect our lives and the choices available to us.

One of the main challenges of the twenty-first century will be to bring about more sustainable human communities. The broad and diverse field that has come to be known as planning can play a central role in meeting this goal, in that it deals with the nuts-and-bolts of how communities, regions, and nations are built and run, including how they relate to natural ecosystems. Since the origins of formal urban and regional planning activities about 100 years ago, results have been decidedly mixed. Much good has been done in terms of improving human welfare, but unfortunately planners themselves have led many unsustainable development practices. They have issued the building permits for suburban sprawl, programmed the monies for ever-expanding freeway systems, set up urban renewal programs that at times have bulldozed vibrant neighborhoods, assisted with the rise of an economy run by global corporations, and most important, failed to be as creative as they might at developing alternative visions. Planning can do better. It can and should reorient itself in the twenty-first century to focus on the challenge of creating more sustainable communities. This role will acknowledge existing politics and traditions but seek every opportunity to bring about creative change. Though it will not be easy, such sustainability planning can be an exhilarating and meaningful path for new generations of planners, architects, landscape architects, engineers, political leaders, progressive developers, and community activists.

The purpose of this book is to provide a systematic background to the subject of sustainability planning as it cuts across many different specialties and scales. The following chapters examine how the sustainability concept has developed and assumed center stage in global debates, its general implications for planning, how it relates to planning theory, how it relates to a range of specific issues and planning tools, and then how it might be pursued at various scales of planning. To develop this latter point, Chapters 7 through 13 examine international, national, state and provincial, regional, municipal, neighborhood, and site- and building-scale planning in turn. Sustainability efforts of different types are possible at each level, and one main point of this book is that action at all of these scales is necessary, and must become better integrated into a coherent, mutually reinforcing framework if sustainable development is to come about.

Although much of the focus of the book is on North American planning, it will consider as well planning in Britain, the European Union, Asia, Latin America, and elsewhere. Similar urban problems exist in many parts of the world, not least because the same technologies and modes of development are spreading everywhere. Many of the potential solutions to current problems are similar as well. Every part of the world can learn from one other, and the most rapid and creative change will come about if we do learn in this way.

Though many existing books discuss sustainable development in one way or another, the intent here is to more fully develop sustainability planning theory in a way that is broadly accessible and useful to students, practitioners, and lay readers alike. I would also like to invite readers personally to become more actively involved in planning for sustainability, whether through projects in their homes, yards, neighborhoods, local ecosystems, or within larger-scale political, economic, or social systems. All of us can play some role in bringing about a more sustainable society, and doing so may be one of the most satisfying and meaningful activities we can undertake.

The need for change

To start with, we need to take note of the nature and variety of current sustainability problems. We have around us an ongoing disaster that – partly because it takes place in slow and incremental fashion, and at scales and in places seemingly removed from our daily lives – is little discussed by the media, politicians, or most citizens. In part this crisis is one of misguided physical development, leaving us with the fragmented landscape of subdivisions, freeways, malls, commercial strips, and office parks that covers vast quantities of land and forms the daily reality for many of us. In part it is one of transportation, having to do with rising traffic congestion and over-dependency on the automobile. In part it is a problem of economic development, having to do with the growing power of global corporations and the decline of smaller-scale businesses that could form a more locally-oriented, socially responsible economy. In part we have a challenge of housing, which is frequently scarce, unaffordable, or inappropriately designed and located. In part we face a crisis of growing poverty and inequality, which leaves enormous numbers of people worldwide without access to decent-paying work, good schools, health care, or other necessities of life. And of course in substantial part our development crisis is one of environmental damage, leading to phenomena such as global warming, resource depletion, and the loss of species that are difficult or impossible to reverse.

All of these problems are interrelated. Though much effort is underway to address them, they are not being acknowledged adequately by current political and professional leaders, and must be tackled in far more comprehensive ways if we are to heal the damage of the past and head in more positive directions in the future.

One of the best ways to appreciate the unsustainability of current urban development patterns is to observe them firsthand. For example, my wife and I recently traveled up Highway 101 in California, a freeway running much of the length of the state. When we were more than 30 miles away from the main Bay Area cities we began to see the outposts of sprawl. "For Sale" signs appeared on either side of the road, and many fields were no longer being farmed – an indication that speculators were holding them waiting for prices to rise. As we traveled onwards through what was some of the richest farming country in the world until recently, the road soon became surrounded by a varying mixture of ranchettes, housing subdivisions, warehouses, office parks, strip commercial developments, big box stores, golf courses, and automobile dealerships.

Crossing a row of hills we descended toward San Jose and neighboring towns – the area known as "Silicon Valley" and envied the world over as a prototype of successful economic development. Whereas the great cities of the past created attractive squares, boulevards, and residential districts, Silicon Valley is organized around tacky commercial strips, crowded freeways, and generic office parks. There is very little public space in this landscape. Housing is astronomically expensive, roads are congested, and air and water pollution are both significant problems for the region. Workers routinely must commute 60 miles or more to their jobs. Rather than being the world's success symbol, by many tokens the development of this area has been a failure.

As we reached the southern border of Oakland it became apparent that the region's development boom had not touched the city's lower-income, African-American neighborhoods. The blocks we passed by were characterized by empty lots, dilapidated houses, and boarded-up factories. This land, urbanized 50 to 100 years ago, has now been abandoned by many employers and wealthier residents, leaving these impoverished areas to cope with a depleted tax base, toxic contamination, declining schools, and a lack of jobs.

We finally reached home after traveling in heavy traffic for two hours through an urbanized landscape that showed few signs of the beauty, culture, livability, and ecological richness that attracted many people to the Bay Area in the first place. Similar patterns of unsustainable urban development are occurring the world over, though they take somewhat different forms from place to place.

In contrast, a growing number of communities offer at least partial examples of development that is socially and ecologically healthy. The downtown revitalization of Portland, Oregon, the transit systems of Paris, Toronto, and Curitiba, Brazil, the pedestrian districts of Copenhagen and most other European cities, the democratic budgeting process of Puerto Alegre, Brazil, the ecological wastewater treatment marsh of Arcata, California, and many other examples offer hope for the future. Many such cases will be described later in this book. These examples suggest that different ways of developing our society and our landscape are possible, and that the action of dedicated individuals can make a difference.

To better understand the overall situation, let us look at some key dimensions of current urban development more closely.

Land use and growth management

One main set of problems concerns land use. Although most pronounced in automobile-dependent communities of the United States, Canada, and Australia, suburban sprawl is advancing almost everywhere else as well. The problem is not just that this style of development tends to be low density – some sprawl may in fact occur at moderate to high densities – but that it possesses many other characteristics that work against the evolution of livable, walkable communities. Sprawl development is often fragmented (the landscape becomes a mosaic of inwardly oriented projects that don't relate to one another or are separated by oversized roads), discontiguous (developments leapfrog out into the countryside), homogeneous (each project contains only housing, offices, or stores), poorly connected (street networks are characterized by cul-de-sacs, loop roads, or other poorly connected patterns), and ecologically destructive (new development fails to take into account natural landscape features and helps generate pollution and excessive resource use). All of these characteristics undermine livability and sustainability. Though certain forms of suburban sprawl have been known throughout history, by far the largest amount has been built since the Second World War. Factors leading to this rapid suburban and exurban expansion include a continual rise in automobile ownership, government road-building, subsidies for suburban homeownership, race and class prejudice, and ideals of country living based on the English tradition of the picturesque country estate.[1] As researchers such as Harvey L. Molotch, John R. Logan, Mark Gottdiener, and Marc Weiss have shown, sprawl has also been facilitated by "growth coalitions" of development interests and politicians which have orchestrated suburban development.[2] In the early 1920s Sinclair Lewis' fictional Babbitt epitomized this boosterish growth mentality, seeking suburban expansion with very little sense of the true social and environmental costs.[3] A century later, attitudes are much the same in many communities throughout the world.

Sprawl consumes enormous amounts of land. The US Department of Agriculture estimates that some two million acres of open space are lost to urbanization annually in the United States.[4] It also degrades or destroys wildlife habitat and natural ecosystems,[5] leads to dependency on the automobile, and produces enormous equity problems, among other effects. Suburban expansion tends to take place at the expense of older urban areas, which are then plagued by abandoned factories and empty storefronts. Older industrial cities such as Detroit and St Louis have lost up to half the population they had in their heyday. In the 1980s and 1990s suburban populations grew ten times faster than central-city populations in the largest US metropolitan areas,[6] as many businesses and residents left older urban areas for the 'burbs. Similar though less extreme versions of this phenomenon are occurring in Britain, Europe and much of the rest of the world.

Figure 1.1
Vanishing open space. The Tassajara Valley southeast of Oakland, California is targeted by developers for thousands of homes.

Figure 1.2 The growth of a metropolitan area. The rapid growth of metropolitan Toronto. Successive layers represent areas built in 1797–1870, 1870–1920, 1920–50, 1950–70, 1970–2000.

Figure 1.3 The American dream? A 2000-square-foot house with a three-car garage and sport utility vehicle (SUV) has become the way of life in many newer US suburbs.

Figure 1.4 Lake in the desert. Private lakes such as this one in Las Vegas have been created by depleting groundwater.

Figure 1.5 The malling of the countryside. Regional malls such as this English example are appearing in Europe as well as North America, undercutting local shops and economies.

Although rapid suburban expansion is sometimes seen as necessary to handle regional population increase, most of the physical growth of urban areas stems from inefficient land use rather than population needs. According to the US Environmental Protection Agency (EPA), the amount of land consumed by urbanization in 34 large US metropolitan areas grew 2.65 times faster than the rate of population growth during four decades in the late twentieth century.[7] Sprawl occurred even in some Northeastern and Midwest metropolitan areas that *lost* population. The problem, in other words, is not so much population growth as the ways we use land. (See Figures 1.1. to 1.5, above.)

Transportation and automobile dependence

Growing automobile use is one of the foremost crises of urban development in many countries. "Vehicle miles traveled" per capita in the United States is increasing at around 3.8 percent a year, meaning that the average person drives twice as much as he or she did 25 years ago.[8] In the United Kingdom, passenger car travel has risen ten times since 1952.[9] Traffic congestion – the number of hours people spend sitting in traffic – is increasing rapidly as well.

The share of travel undertaken by automobile in preference to other modes is growing steadily in most countries. Ninety-one percent of Americans drive to work rather than walking, biking, or taking public transportation, according to the 2000 census. The percentage of US residents who travel anywhere by foot is negligible, in part because our communities are not designed for pedestrians. These same trends are present elsewhere as well. The negative effects of growing automobile use include declining quality of life in neighborhoods impacted by traffic, excessive use of nonrenewable resources, air pollution, global warming, public health problems ranging from obesity to childhood asthma, and tens of thousands of deaths each year due to traffic accidents and pollution. (See Figures 1.6 and 1.7.)

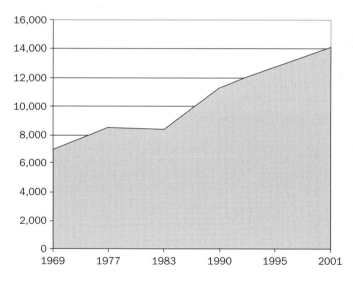

Figure 1.6 Growth in miles traveled annually per capita (US). Sources: US Bureau of the Census, *Statistical Abstract of the United States*, Washington, DC: US Government Printing Office, 2002; US Department of Transportation, *2001 National Household Travel Survey*, Washington, DC, 2001.

Figure 1.7 The vehicle-dominated world. Many people's main experience of urban areas is now the landscape seen through car windshields.

Energy and resource use

Due in large part to the nature of recent development and lifestyles, our resource use remains at unsustainable levels. Depletion of petroleum reserves is one area of concern – global oil production is expected to peak and then start declining early in this century. But other resources such as water are also being depleted worldwide, not just in desert locales such as Las Vegas and Southern California, but in metropolitan areas such as Atlanta and São Paolo as well. The concept of recycling water remains little explored even in arid regions. Permeable paving and natural drainage swales – techniques which allow rainfall to recharge the local watertable rather than running off rapidly often causing erosion or floods – are rare.

As many environmentalists have pointed out, most resource use in our society still takes place in linear fashion, with raw resources being acquired, used once, and then discarded as waste. In 2000, 85 percent of household waste in Britain was still trucked to landfills, rather than being recycled.[10] Overall recycling rates in even the most progressive US states are still barely at 50 percent. We are a very long way from achieving "closed-loop" resource cycles, in which materials recirculate extensively, waste is eliminated, and new inputs come from renewable sources.

Pollution and environment

One consequence of our current resource use patterns is the generation of a great deal of pollution and waste, much of it toxic. Between 50,000 and 100,000 synthetic chemicals are in commercial production,[11] and "persistent organic pollutants" such as dioxins and

PCBs are now found in most parts of the world. Some $31 billion dollars worth of pesticides are sold worldwide every year, many for use in urban areas by homeowners or building maintenance workers. One study of trees at 90 sites in both tropical and temperate countries found *no* locations that were free from contamination by DDT, chlordane, and dieldrin.[12]

Urban areas are now littered with "brownfield" sites – contaminated land formerly used by industrial, commercial, or military-related enterprises. The firms that created such problems have often moved on or gone out of business, and in many cases new owners or public agencies are left to coordinate and pay for cleanup. A lot of these sites remain for years as vacant land, dissuading many people from locating near them.

Every few years new forms of pollution emerge into the public spotlight. Rachel Carson's book *Silent Spring* in 1962 awoke millions to the dangers of pesticides and other toxic chemicals in the environment. Damage to the Earth's ozone layer caused by chloro-fluorocarbons became a cause for worldwide concern in the 1980s. Global warming had been considered by a few visionary scientists at least as early as the 1970s, but did not emerge into the mass media until the 1990s. In the early 2000s much attention was focused on the damage pollution and harmful fishing techniques have caused to the earth's oceans. And so forth. New crises continue to emerge. The basic problem is that our industrial economy does not take the full effects of its resource use and waste production into account, a prerequisite for a more sustainable society. Many of these wastes and pollutants also have serious implications for public health, and have been found to increase rates of illnesses ranging from childhood asthma to cancer.

Inequality and poverty

Development patterns worldwide have gone hand-in-hand with huge and growing inequities between economic groups, racial or ethnic communities, and neighboring cities and towns. Increasingly affluent and secluded or gated suburban neighborhoods exist a few miles from impoverished central cities or squatter settlements.[13] The concentration of poverty in central urban areas in the US has worsened in recent decades,[14] leading to a growing chorus of commentators calling for regional strategies to redistribute tax base and improve access to jobs and decent education within lower-income communities. Most new jobs are being created in the suburbs, while many poor people live in central cities, creating an enormous problem of access to work. The growing dearth of affordable housing in metropolitan areas is also a huge impediment for less-well-off individuals. The building industry concentrates on producing housing for the middle and upper ends of the market, and avoids production of the decent rental or ownership housing needed by many less-well-off citizens. Local politicians and citizens often resist affordable housing as well, from the misguided belief that it will hurt property values in well-off areas. In the developing world, equity gaps between rich and poor are even more dramatic. In Third World nations 50 percent or more of the population may be living illegally in informal housing constructed of cardboard, plywood, or other scavenged materials. Many also eke out livings in the informal economy, while globalization produces enormous wealth and an upper class working in modern skyscrapers.

Sense of community

Declining sense of community is another concern that is far less tangible but no less serious. Most urban and suburban development during the past 50 years has been relatively generic, with little sense of place, history, or cultural distinctiveness. Every place looks like everywhere else, a "geography of nowhere," as author James Howard Kunstler has put it.[15] Many of us now live in suburbs without any real downtown or walkable public spaces other than the mall. Parks, greenways, public squares, and sidewalks are often missing. The same generic chain stores and restaurants appear everywhere. Branches of multinational corporations displace locally owned businesses. Driving everywhere, suburban residents spend much of their time alone in their cars. Women, children, teenagers, the disabled, and the elderly are often isolated in their homes, especially if they are unable to drive.[16] Partly as a result of all these factors, political scientists such as Robert Putnam have documented a long-term decline in the extent to which citizens participate in community groups and social institutions.[17] This trend has troubling implications for the health of democratic government. Putnam believes that this decline of "social capital" is at least partly related to the physical nature of our cities and towns.

* * *

One could go on at length about the problems of current development patterns. But most of us already know that something is seriously wrong. A long line of literary works has also covered many of these issues, including the writings of Lewis Mumford in the early and mid-twentieth century, Jane Jacobs' *Death and Life of Great American Cities* (1961), Rachel Carson's *Silent Spring* (1962), Ian McHarg's *Design with Nature* (1969), Dolores Hayden's *Redesigning the American Dream* (1984), James Howard Kunstler's *The Geography of Nowhere* and *Home from Nowhere* (1993, 1996), Robert Putnam's *Bowling Alone* (2000), and recent works by New Urbanist designers such as Peter Calthorpe, Andres Duany, and Elizabeth Plater-Zyberk.

In many ways what is being enacted is a "tragedy of the commons," to use the term that biologist Garrett Hardin made famous in the 1960s. Developers, corporations, and ordinary citizens have maximized private benefit at the expense of shared public goods – including the quality of our cities and towns, not to mention the environment. Individualism has triumphed over collective well-being. In a conservative era dominated by free market philosophies, there has been little political interest in recognizing that someone, usually the public sector, must stand up for the common good. Developers build on every available piece of land at the urban fringe, landowners feel it is their right to seek maximum economic return from "their" property, owners of factories and automobiles each contribute their pollution to the metropolitan airshed, and developers often produce buildings with little thought to the streetscape and public environment that is being created. Government, which exists in large part to protect and advocate for public interests, is routinely attacked by those who would like to privatize everything. Individualistic attitudes have been institutionalized through law, government policies, corporate practice, advertising, the media, and many other structural elements of our society. These structural forces also reinforce individualistic, consumption-oriented values, types of behavior, and modes of thought within individuals.

Put together, these forces create a world that is dangerously unstable. At any time it may produce social, political, or environmental crises. Dramatic tragedies aside, they cause an enormous amount of needless suffering and environmental devastation. Addressing this situation is one of the central tasks of the twenty-first century. The field of planning will be crucial to this challenge, and politicians, professional planners, architects, environmentalists, developers, and citizens of all sorts will have roles to play in meeting it.

The role of planning

"Planning" refers to a wide range of systematic activities designed to ensure that desired goals are achieved in the future. These goals could include environmental protection, urban development, particular forms of economic activity, social justice, and many other ideals. As a formal profession, planning is most often concerned with managing land development at the urban and regional scales, but the field has broadened enormously since its origins, and now can be said to encompass the act of planning for desired future conditions at all scales of endeavor, within both public and private sectors.

The roots of planning as an organized profession go back to the late nineteenth century, when the extremely rapid growth of cities following the industrial revolution led to a profusion of urban problems such as inadequate sanitation, water supply, and transportation, and housing. In cities throughout Europe and North America, hundreds of thousands of industrial workers and their families were crammed into tenement housing without adequate light, water, or sewer facilities. Planning historian Sir Peter Hall has labeled this late-nineteenth-century industrial city the "city of dreadful night."[18] Many professions responded to the urban crises of this period. Engineers devised large-scale sewer and water systems. Architects and public health workers proposed housing regulation to ensure access to light and air. Landscape architects formed the backbone of the mid-nineteenth-century Parks Movement, and joined with architects to promote turn-of-the-century City Beautiful ideas. Meanwhile, early planning visionaries such as Ebenezer Howard, Patrick Geddes, and Catalán engineer Ildefons Cerdà applied broad and holistic styles of thought to urban problems, coming up with concepts such as Howard's "garden city."

After early development of zoning in German cities such as Frankfurt in the 1890s, city planning as a profession became formally established in Britain and North America during the 1910s and 1920s when municipalities first created planning commissions, planning staffs, and zoning laws. Development projects began to come under greater public scrutiny. Rapid construction of roads and bridges took place to accommodate urban growth and a new transportation technology: the automobile. Harvard established the first US academic city-planning course in 1909, and many other universities followed suit. Anxious to establish the legitimacy of their field, the new planning professionals soon left behind the idealistic visions of Howard and Geddes in favor of more pragmatic approaches. Quantitative measurement of urban data became preferred to designerly visions or normative advocacy. Unfortunately, city planners also often found themselves with little power within municipal political systems and grew used to a disempowered role as statisticians, bureaucrats, and overseers of the zoning code.

During the twentieth century, new planning specialties arose in response to particular problems of the growing metropolis, such as transportation planning, community development planning, and environmental planning (see Figure 1.8 at the end of this chapter). A quantitative, scientifically oriented form of regional economic analysis also became widespread in the 1950s and 1960s. Seduced by positivistic science and the worldview known as modernism, planners assumed the stance of objective experts using scientific method, modern technology, and economic analysis to determine paths of urban development. This approach is typified by Walter Isard's 1975 statement that

A regional scientist is not an *activist* planner. . . . The typical regional scientist wants to surround himself with research assistants and a computer for a long time in order to collect all the relevant information about the problem, analyze it carefully, try out some hypotheses, and finally reach some conclusions and perhaps recommendations. His findings are then passed on to key decisionmakers.[19]

Much of this approach still lingers. One cause of today's urban problems is that planners in the past often abandoned holistic understanding of urban environments and active involvement in favor of detached quantitative analysis. Marxist critics of the field during the 1960s and 1970s also argued that planners had ignored the realities of political and economic power within society, and were simply facilitating capitalist development rather than dealing with pressing needs for equity and improvement in social conditions. Environmentalists, historic preservationists, and neighborhood activists fought against some of the worst excesses of modernist planning, such as urban renewal in which entire neighborhoods of older housing were bulldozed to create sterile new apartment blocks. As modernist planning became increasingly discredited in the late twentieth century, the stage was set for something different.

Planning remains a rapidly growing profession, both in its traditional urban context and in many non-traditional contexts as well. Virtually every city or town these days has a planning department, usually composed in the United States of a "current planning" division that oversees development permitting and the zoning code, and an "advanced planning" division responsible for developing longer-range plans. But planners also work within many other types of government agencies – regional agencies, state and national governments, international development agencies, transportation authorities, park districts, and public utilities among them – as well as for private corporations, development companies, consulting firms, community development corporations, and various types of nonprofit nongovernmental organizations (NGOs). Virtually any sort of systematic, forward-looking activity can be labeled "planning," and people with or without city planning degrees fill an enormous variety of planning-related positions. Planning overlaps to some degree with policy analysis, though it tends to be more action-oriented, and with other specialties such as urban design, engineering, law, and political organizing. Planners may wear many hats at different times, and may hold jobs with a wide range of titles other than planner. But it is the activity of planning that matters, whatever the label, and how that forward-thinking activity can help promote sustainable development in whatever the context at hand.

Overall, twentieth-century urban planners were successful at meeting some goals but not others. Cities in the industrialized world now have decent sanitation, water, roads, and zoning and building regulation designed to ensure public safety and protect property values. Planning was also fairly successful in organizing society to accommodate new technologies such as the automobile – although without sufficient reflection on the long-term impacts. As a result we now face a different set of problems than existed a century or more ago. So the need now is to develop a twenty-first-century planning paradigm that better meets long-term human and ecological concerns. A number of writers, planners, architects, landscape architects, developers, politicians, and local activists have begun exploring steps to create more sustainable cities. The profession's central organization in the US, the American Planning Association, is also seeking ways to support sustainability planning.[20] Some argue that sustainability planning represents a new agenda for planning.[21] However, the process of redefinition is just beginning, and faces huge hurdles because of politics, economics, and public attitudes. Much work remains to be done to figure out what sustainability planning is and how it can best be nurtured.

In redefining the field we need to recognize that planning has limits – not every aspect of urban or regional development can or should be planned out. History has supplied many examples of grand visions by architects or social theorists that proved too rigid or inappropriate in practice. The misplaced faith of twentieth-century planners in scientific knowledge and research is also a cautionary tale. Some theories of urban development – such as the organic theory of urban form developed by Christopher Alexander in books such as *A Pattern Language* (1977) – hold that cities are best developed incrementally, learning from experience in the process, rather than according to some overarching master plan. Towns and villages often developed in this slow, incremental way historically, before the advent of large-scale city-building in industrialized societies, and many of these places are among the most delightful urban communities currently. Certainly an increased degree of community involvement is necessary within planning as well, compared to many twentieth century planning processes.

However, even if more incremental and community-based development is desired it helps to have a framework within which such development can fit, ensuring that it meets certain basic criteria and perhaps prohibiting inappropriate actions. The trick will be to plan with a large degree of humility regarding planners' ability to predict the future, and with a great openness to learning from experience and collaborating with the wide range of constituencies affected by any initiative.

Levels of government and scales of planning

Insofar as the public thinks about it at all, planning in North America is often associated with the process by which local governments approve development and build roads or other infrastructure. But an enormous variety of planning activities occur at many levels of government and within the private sector as well. Maintaining a sense of how these different scales of planning interrelate and of how action at some levels can reinforce efforts at other scales is essential. Each level of planning will be discussed in greater detail in subsequent chapters, but a brief introduction is appropriate here.

At the very broadest level, a growing number of global agreements and programs constitute an international scale of planning that is relatively recent, depending in part on the existence of international institutions which are themselves recent. Chief among these institutional actors are the United Nations agencies, the World Bank, the International Monitary Fund, many bilateral development agencies, and nonprofit organizations such as the International Council on Local Environmental Initiatives (ICLEI), which helps local governments around the world develop and implement sustainability related programs. Most large international development agencies were established in the period after the Second World War, when the industrialized nations sought to assist in global recon- struction, ensure the success of their own model of development, or promote their own political hegemony, depending on one's point of view. The nonprofit sector activities are even more recent, with many groups established after the 1960s. ICLEI, for example, was established in 1990. With consciousness of common global problems such as climate change spreading, broad frameworks of goals and policies at the international level are likely to be expanded in the years ahead. For those issues that are less than global, a related area of opportunity lies in regional agreements between several countries with common borders or issues. An example is the series of agreements between Canada and the United States to protect environmental quality in the Great Lakes.

Nation-states represent the next lower scale at which planning typically takes place. Although in countries such as the United States and Canada federal governments have relatively little power or interest in directly regulating land use and physical planning (the spatial development of cities and their associated infrastructure), national govern- ments in other places have taken much more active roles. The British government prepares a very extensive set of Planning Policy Guidance documents for local and regional agencies, and has taken an active role in structuring and overseeing local government institutions. National governments in Sweden, France, the Netherlands, Japan, and other countries have also been heavily involved in urban and regional development policy. Moreover, in nations such as the US which have little explicit planning policy, federal laws and programs related to transportation, taxation, housing, civil rights, and environmental protection profoundly affect urban and regional development without explicitly acknowledging that they are doing so.

Planning by states or provinces in those countries large enough to contain them is a further level of large-scale planning. In Canada, the provinces have great power to review land use decisions, allocate infrastructure, and revise local institutions. The province of British Columbia has played a strong role in park and land use planning, for example, while the province of Ontario completely restructured local governments in the Toronto area in 1998. In Australia, the six state governments also have a high level of control over urban planning issues with the power to override or restructure local governments. In the United States, a number of states have adopted statewide growth management planning frameworks, as well as housing programs, environmental review requirements, pollution regulation, and waste reduction policies. The planning and construction of large-scale transportation projects is also usually handled at the state level.

Regional planning has often been sought by those wanting to address problems such as air and water pollution, watershed planning, transportation, affordable housing development, and equity. These problems typically cross the jurisdictional boundaries of local city and county governments. Such planning might occur at the level of a metropolitan region (consisting of one or more central cities along with dozens of outlying suburbs and counties), or on the scale of watersheds, airsheds, commutesheds, bioregions, economic regions, or cultural regions. Unfortunately, regional planning is the scale least represented by government institutions. There are few strong regional agencies in the United States particularly, and local, state, and national governments typically feel threatened by the possibility of regional agencies usurping some of their power.

Local government – including cities, towns, counties, and a wide variety of special agencies set up for specific purposes such as managing school districts or park systems – is the level at which the most detailed urban planning is usually done. Most municipal governments have a full-time planning staff whose responsibility is to process development applications and to prepare longer-range plans and policies. Many consultants, nonprofit organizations, neighborhood associations, and business groups are active participants in planning at the local level as well. In addition to city-wide general or comprehensive plans, local governments in turn prepare plans for smaller scales of development, including large areas within cities, smaller neighborhoods, and, occasionally, specific sites of special importance. Municipalities may also prepare "functional" plans to deal with specific topics not covered in their comprehensive plans.

Neighborhood and site planning is typically done by developers, urban designers, architects, and landscape architects, operating within the context of municipal zoning codes, plans, and design guidelines (if these exist). The character of particular places, their relation to local ecosystems, and the nature of community facilities are typically determined at this level. Neighbors and community groups frequently play a significant role in affecting plans at this scale. City officials will directly participate as well when streets, parks, or other public spaces are involved.

In theory goals and policies set at each of these levels establish the context within which planning is carried out at smaller scales. The reality, of course, is not so clearcut. In the absence of specific incentives or penalties, broad global, national, state, or regional goals may be overlooked at lower levels. Decision-makers at any given level are also often not used to keeping broader perspectives in mind, and may have little personal stake in doing so, especially if they stand to benefit politically or financially by particular courses of action that consider only the local interest. Some states require that different scales of plans adopted by local and regional governments be consistent with one another, but many do not. Integrating different scales of planning, then, is a task of raising awareness at every level of the overall context into which specific actions fit, and structuring incentives, mandates, and educational processes that help encourage such thinking.

Recent trends in urban planning and design

Although political situations often seem stacked against sustainability initiatives, recent developments in the planning field offer hope that constructive approaches to current problems can be found. In the past couple of decades a number of new planning and urban design movements have sprung up to challenge twentieth-century development directions, incorporating a new awareness of social and ecological issues. Taken together, these efforts help lay the foundation for sustainability planning.

Within architecture and urban design the movement known as the "New Urbanism" appeared in the early 1990s and has become a strong force for reevaluating the physical layout of communities. Originating with a number of architects promoting neotraditional community planning – based in large part on the traditional American small town – the movement held its first annual congress in 1993 and published a charter in 1996. Founders include Peter Calthorpe, the husband-and-wife team of Andres Duany and Elizabeth Plater-Zyberk, Daniel Solomon, Stefanos Polyzoides, Elizabeth Moule, and Douglas Kelbaugh. In Britain, similar design principles have been championed by the Prince of Wales' Foundation. On the continent, architects Leon and Rob Krier have designed similar new communities in a denser, more traditional European form.

In large part a reaction to the placeless, unwalkable landscape of suburban sprawl, the New Urbanism calls for improved design at building, neighborhood, city, and regional levels to create more walkable, livable communities.[22] New Urbanists for example call for narrowing streets, adding sidewalks, placing porches on the front of houses, tucking garages behind them out of the way, creating street grids or other connecting street patterns instead of cul-de-sacs, and organizing neighborhoods around mixed-use centers and attractive public spaces. Most of these design elements are similar to those used in communities before the age of the automobile. The movement is open to criticism on a number of fronts – in particular for being focused on better-designed suburban development, often for upper income groups, rather than the creation of truly "urban" places, and for not incorporating green building design and landscaping – but nonetheless represents a much-needed attempt to establish a new model of neighborhood design that has many sustainability advantages.

Just a few years later, in the mid-1990s, a movement for "Smart Growth" gathered steam. With its origin in state efforts to control metropolitan growth, Smart Growth focused especially on mechanisms to promote more compact, economically efficient urban development, in particular through preserving open space, limiting the outward expansion of cities, promoting infill development, and redesigning communities in ways similar to those proposed by the New Urbanism. Many of the same individuals have been active in both movements. One of the main motivations in this case was to reduce infrastructure costs for local governments. Although states such as Oregon, Vermont, New Jersey, and Florida had been pursuing growth management programs since the 1960s in some cases, Maryland took particular leadership in the new round of Smart Growth policy through its 1997 legislation. This policy framework tied state infrastructure funding to local establishment of Priority Growth Zones, channeled new state facilities into existing town centers, and made additional funding available for open space acquisition. Other states such as Minnesota and Washington followed suit.

A third, somewhat more diffuse movement called for "Livable Communities." A series of annual conferences on this theme has been held in Carmel, California, since the 1970s, and organizations such as Livable Oregon have promoted this theme within state and local planning. Walkability, diverse and mixed-use development, and the establishment of a wide range of urban amenities to make cities more livable have been main themes of this movement.[23] Livability advocates build on the previous writings of twentieth-century humanist urban critics such as Lewis Mumford, William H. Whyte, Jane Jacobs, and Bernard Rudofsky.

A related, likewise diffuse assortment of professionals, in large part coming from the field of public health, has sought to promote "healthy cities." This term appears to have been coined at a Canadian conference in 1985, and refers to a wide range of issues related to pollution, toxic chemicals, safety, homelessness, education, community, and urban quality of life. Organizations such as the World Health Organization (WHO) and the International Healthy Cities Foundation have helped lead this movement internationally, and healthy city programs have been started in more than 1000 cities worldwide.[24] Recent recognition of public health problems stemming from obesity and lack of exercise are likely to dovetail with this movement.

Beginning in the 1980s, advocates for "Environmental Justice" called attention to the ways in which society saddled minority and low-income neighborhoods with the negative side-effects of urban development – especially exposure to toxic chemicals and industrial pollution.[25] Environmental Justice leaders challenged mainstream environmental groups in the US to broaden their agenda and staff diversity, most notably in a series of letters to the "Gang of 10" heading the major national organizations. In the 1990s and 2000s this movement looked in addition at the inequities involved in other fields such as transportation planning.[26]

Also in the 1980s and 1990s the field of "Landscape Ecology" emerged, based on the work of Richard Forman, Frederick Steiner, and others. This discipline developed a new way of looking at landscapes in urbanizing or rural areas, emphasizing "patches" of wildlife habitat, "corridors," "edges," and so forth. The result is a set of terms and concepts that makes possible more systematic approaches to preserving and restoring natural landscape elements within cities and towns.[27] Landscape architects and planners such as Michael Hough, Anne Whiston Spirn, John Tillman Lyle, and Rutherford H. Platt also explored the relation between cities and natural landscapes during this period, arguing that the two are profoundly interrelated.[28] Meanwhile, more specific environmental planning movements for creek and wetland restoration, use of native species in land-scaping, ecological sewage treatment, and green building have contributed to more environmentally oriented development practices.[29]

Lastly, many efforts to increase public participation in urban development decisions have arisen over the past four decades. "Participatory planning" has become a prime objective in local government, indeed has been required in many cases, leading to an expanded emphasis on public meetings, workshops, design charettes, and consensus-based goal-setting. Realizing that much of what planners do is mediate between various interests, planning theorists have established the perspective known as "communicative planning," which emphasizes processes of communication, participation, and consensus-building.[30] Related movements for "Community Design" and "Community Based Planning" have called for planners and urban designers to collaborate with local residents in addressing neighborhood problems and designing new buildings, public spaces, and community facilities. In conjunction with these efforts for public participation, some observers have called for a more activist role on the part of planners in advocating for underserved constituencies within local government beginning with movements for equity plannng and advocacy planning established in the 1960s.[31]

In short, in recent years many related planning movements have been converging toward a similar end: more environmentally, socially, and economically sustainable communities. Although the concept of sustainability itself was first applied to other fields such as forestry, agriculture, energy use, fisheries management, and international development, by the late 1980s and 1990s the profession began to take it up, and "sustainable city" programs emerged in a few pioneering locales. Some resulted from grassroots activism, some were based on municipal initiative, some benefited from the support of national governments, and some were facilitated by multilateral entities such as the European Community and United Nations agencies. Main areas of activity were the west coast of North America, where Seattle, Portland, Olympia, San Francisco, San Jose, and Santa Monica established programs, Canadian cities such as Toronto and Vancouver, and Britain and continental Europe, where "Local Agenda 21" programs inspired by the Rio Earth Summit proliferated. Most of these municipal initiatives were small and preliminary, often focusing on energy and resource use. But the concept of sustainability began cropping up in many other fields as well, including transportation, urban design, and architecture. As the 1990s progressed urban planning academics started to explore the subject, conferences and books adopted it as a theme, and universities began to offer related courses. Academic and popular literature on urban sustainability increased exponentially. Clearly interest in the topic has grown enormously, but just as clearly the obstacles to actually bringing about sustainable urban development are huge and there is a need to better define the subject, figure out strategies for implementation, and develop long-term programs to bring it about.

This book aims to address those needs by exploring how planning initiatives at different levels can fit together to produce more sustainable cities and towns. The ideas presented here are not just for professional planners, but for citizens, local activists, students, elected officials, developers, architects, engineers, and others concerned with how cities and towns can become more sustainable, livable places. Whatever our roles, each of us can help bring about a more sustainable future, sharing strategies, working together, and supporting one another in this process. (See Figure 1.8.)

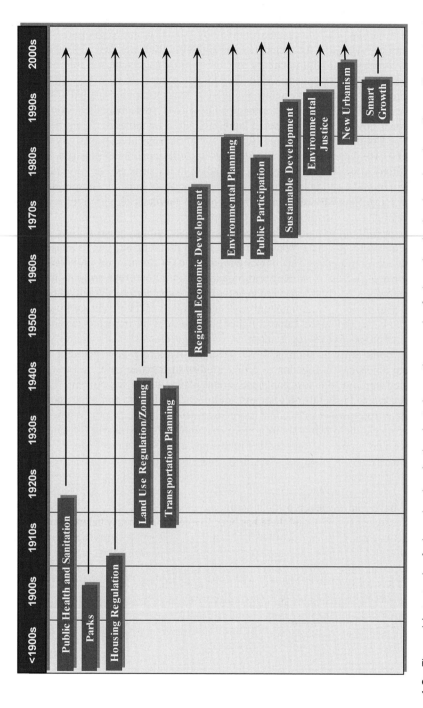

Figure 1.8 The evolving agenda of urban planning. As the chart shows, the agenda of urban planning has broadened greatly over the past century.

2 Sustainable development

The concept of "sustainability" in its modern sense emerged in the early 1970s in response to a dramatic growth in understanding that modern development practices were leading to worldwide environmental and social crises. The term "sustainable development" quickly became a catchword for alternative development approaches that could be envisioned as continuing far into the future.

The verb "sustain" has been used in the English language since 1290 or before and comes from the Latin roots "sub" + "tenere," meaning "to uphold" or "to keep." The *Oxford English Dictionary* traces the adjective "sustenable" to around 1400 and the modern form "sustainable" to 1611. However, the word appears to have been used mainly in legal contexts until recently, as in "the Defendant has taken several technical objections to the order, none of which . . . are sustainable" (1884). The phrase "sustainable development" appears to have been first used in 1972 by Donella Meadows and other authors of *The Limits to Growth*, and by Edward Goldsmith and the other British authors of *Blueprint for Survival* in that same year. Once in use, the term became one of those inevitable expressions that so neatly encapsulates what many people are thinking that it quickly becomes ubiquitous.[1] Yet the conceptual roots of the term "sustainability" go far deeper, and have to do with the evolution of human attitudes toward the environment within Western culture.

Roots of the concept

Environmental issues have almost always been concerns of human societies. For millennia people have had to develop their communities and livelihoods within the context of pre-existing ecosystems, and there has been frequently an uneasy balance between human and non-human worlds. Parts of the Mediterranean were extensively deforested during antiquity, and environmental collapse may have contributed to the decline of the Mayan cultures. Many societies have had elaborate rituals and institutions devoted to maintaining what we might call "sustainable" resource use, in particular norms and taboos around the use of "common pool resources" such as fisheries, forests, water sources, and grazing land.[2] Ecosytems or individual species have often been prominently represented within many spiritual traditions, especially those of indigenous peoples living close to the land.

As early civilizations developed they profoundly changed pre-existing natural environments. Indigenous cultures cleared forests, managed landscapes by setting fires, domesticated or transplanted species, and hunted or drove into extinction many non-human forms of life. Few landscapes anywhere were undisturbed by human contact. Writers beginning with Aristotle put forth terms such as "second nature" to refer to elements of the natural world that had been influenced by interaction with humans. So in some ways current sustainability debates are the modern version of age-old concerns about how to maintain human societies within the context of natural ecosystems.

However, although many early civilizations bumped up against ecological limits, the coming of the Industrial Revolution in the late eighteenth and nineteenth centuries made the impacts of human actions far more dramatic. Skies in parts of Britain, continental Europe, and North America were blackened with coal smoke. Forests were cut down to produce lumber or charcoal for iron smelting, and rivers and streams were fouled with sewage and industrial wastes. Clearcutting and inappropriate agricultural practices such as plowing across contours often led to erosion and flooding. One response to the negative effects of industrialization was a strong strain of romantic or transcendentalist philosophy in which nature was asserted as an antidote to industrial civilization. Writers such as Henry David Thoreau and John Muir and Romantic poets such as William Wordsworth, Percy Shelley, John Keats, and Alfred Lord Tennyson extolled nature as a spiritually rejuvenating alternative to industrial society. As Muir put it, natural things were "the terrestrial manifestations of God." This spiritual perspective underlaid the "preservationist" strand of the early environmental movement.

Another response to the pressures of industrialization was more pragmatic: to study environmental impacts and find alternative strategies that avoided them. In 1864 George Perkins Marsh published the first systematic consideration of how humankind was altering the natural landscape in his book *Man and Nature*, based on detailed observation of environmental changes in southeastern France and New England. Marsh focused in particular on deforestation, which he saw as leading to increased runoff, floods, landslides, other ecological disasters. Warning that humans were upsetting the balance of nature, he prophesied that long-term ecological decline would lead to decline in human populations.[3] A few decades later, German foresters developed "sustained yield" techniques of forest management.[4] Applied particularly to the Black Forest in southwestern Germany, these concepts influenced Americans who trained at continental forestry schools in the late nineteenth century, including Gifford Pinchot, who later became President Theodore Roosevelt's chief forester. Pinchot and others imported European concepts of sustained yield resource management back into the United States, where they influenced the growing conservationist movement. In contrast to the preservationist viewpoint, which ascribed intrinsic value to nature, this sustained yield approach was relatively utilitarian, concerned with preserving natural resources for future human use. Such anthropocentric attitudes, in which ecosystem elements are viewed as valuable mainly in terms of their potential human use, are still prevalent today.

A mixture of preservationist and conservationist sentiments is contained in Aldo Leopold's mid-twentieth-century notion of a "land ethic" – a human responsibility to care for particular lands and ecosystems, discussed most fully in *A Sand County Almanac* (1948). Although trained in utilitarian forest management perspectives, Leopold came to believe that humans had an ethical responsibility to steward and safeguard natural ecosystems, and that these had intrinsic value apart from human use. This more eco-centric perspective helped lay the groundwork for the rise of deep ecology in the 1970s and 1980s, an even more radical philosophy which sought to put the well-being of the global environment first, with human priorities revised to reflect their role as just one small element of the global system. These varying perspectives on the relation between humans and ecological systems helped lay the foundation for late twentieth-century sustainability debates.

Public concern about the relation between industrial development, urban expansion, and the environment grew steadily after the Second World War. War production had stimulated a huge expansion of petrochemical industries that in the post-war period created many new pollution, toxic materials, and resource use problems. Works such as William Vogt's best-selling *Road to Survival* (1948) and Fairfield Osborn's *Our Plundered Planet* (1948) helped tie the rise of ecological problems to this growth in industrial development. Other writers called attention to the alienation and conformity of 1950s industrial society in bestselling works such as David Riesman's *The Lonely Crowd* (1953) and William H. Whyte's *The Organization Man* (1956). Meanwhile, in many books between the 1920s and the 1970s the great urban planning critic Lewis Mumford linked large-scale urbanization, technology, and warfare, warning of the dangers of the "technopolis," in which anti-humanistic technology was the primary value.[5] In books such as *The Culture of Cities* (1938), *The City in History* (1961), and *The Urban Prospect* (1968), Mumford advanced instead an ideal of the city as an organic community, designed on a human scale, oriented towards human needs, fueled by a life-enhancing economy, surrounded by undeveloped lands, and with streets filled with people instead of automobiles. This vision can be seen as similar to recent sustainable city ideals.

Modern environmentalism – in which advocates became far better organized, adopted an increasingly broad agenda, and brought about a wave of environmental legislation – is generally dated to the late 1960s and early 1970s.[6] During this time social critics, futurists, feminists, and environmentalists critiqued existing notions of development and proposed alternative paradigms emphasizing the spiritual, the natural, and the human over values of profit and economic progress. Particularly significant were Rachel Carson's book *Silent Spring* (1962), which first called attention to the dangers of pesticides and other toxic chemicals in the environment, Kenneth Boulding's *The Meaning of the Twentieth Century* (1964), Barry Commoner's *The Closing Circle* (1971), and Theodore Roszak's *Where the Wasteland Ends* (1972).[7] Barbara Ward and René Dubos' book *Only One Earth* (1972) was also influential. The report of an unofficial commission set up by Maurice Strong, Secretary-General of the 1972 United Nations Conference on the Human Environment, this widely distributed volume warned about threats to global survival and included an explicit description of the greenhouse effect[8] and warnings about "unsustainable" growth in automobile usage.[9]

Current events also helped change public consciousness. Views of the Earth from space, first taken by astronauts in the 1960s, helped people conceptualize the planet as a whole for the first time. The debacle of the Vietnam War helped throw into question the prevailing pattern of US economic and political control over Third World countries, and exposed the underside of "the military-industrial complex" that Dwight Eisenhower had warned against in his farewell address. Earth Day splashed environmental problems onto front pages and magazine covers in 1970. The 1972 UN Conference on Environment and Development, held in Stockholm, for the first time brought together public officials and NGOs from around the world and gave them a forum to share ideas and strategies. The 1973 energy crisis hit the pocketbooks of millions of people, many of whom suddenly realized that their fossil fuel use could not continue to expand forever.

At a more philosophical level, in the late 1960s humanistic psychologists such as Abraham Maslow and Carl Rogers pointed out ways in which human potential is shaped by the surrounding social and cultural environment, and ways in which human nature can perhaps be shaped in healthier directions in the future. Their work helped counter pessimistic views of human nature as warlike and competitive, which had been reinforced by violent events earlier in the twentieth century. The implication of this optimistic humanism, endorsed as well by spiritual philosophers such as Pierre Teilhard de Chardin[10] and later by feminist "stage theories" of psychological development,[11] is that people and perhaps entire societies can evolve towards more conscious, compassionate, and sustainable modes of existence, given the right conditions.

By the 1970s the ground had been laid for new perspectives on global development.[12] In *Limits to Growth* (1972) Meadows and other MIT researchers modeled trends in global population, resource consumption, and pollution, and found that regardless of the range of assumptions they entered the model showed the human system crashing in the mid-twenty-first century. But they argued that "it is possible to alter these growth trends and to establish a condition of ecological and economic stability that is sustainable far into the future."[13] The sooner such efforts began, Meadows and her coauthors believed, the greater the likelihood of their succeeding. This conclusion was opposed by conservative economists such as Julian Simon, who argued that economic mechanisms would naturally take care of resource problems by reducing consumption or substituting other resources for those depleted.[14] In some cases, this market-based process clearly did happen. However, 20 years later Meadows and her colleagues revisited their model and found its basic predictions still accurate. Indeed, they warned that the human population had reached a situation of "overshoot" in terms of resource limits, and would need to take strong action to correct unsustainable trends.[15]

Meanwhile, in their own 1972 book *Blueprint for Survival* Goldsmith and other editors of the British journal *The Ecologist* drew on the work of the *Limits to Growth* group as well as nineteenth-century British economist John Stuart Mill in calling for the creation of a stable global society.[16] More synthetic and polemical than the Meadows group, Goldsmith et al. began with a sweeping critique of industrial society, stating that

> The principal defect of the industrial way of life with its ethos of expansion is that it is not sustainable. . . . Radical change is both necessary and inevitable because the present increases in human numbers and per capita consumption, by disrupting ecosystems and depleting resources, are undermining the very foundations of survival.[17]

In ways that presaged much later sustainability literature, they then systematically reviewed various strategies for resource management, agriculture, and social and political reform. As they quite eloquently put it,

> Our task is to create a society that is sustainable and will give the fullest possible satisfaction to its members. Such a society by definition would depend not on expansion but on stability. This does not mean that it would be stagnant; indeed, it could well afford more variety than does the state of uniformity that at present is being imposed by the pursuit of technological efficiency. We believe that the stable society . . . is much more likely than the present one to bring the peace and fulfillment that hitherto have been regarded, sadly, as utopian.[18]

Once introduced, the concept of sustainable development diffused rapidly not just through the networks of environmental activists but also among economists, ethicists, and spiritual leaders concerned about the course of global development. A 1974 conference of the World Council of Churches issued a call for a "sustainable society," and the earliest book with the word "sustainability" in the title appeared in 1976, a volume entitled *The Sustainable Society: Ethics and Economic Growth* by Lutheran theologian Robert L. Stivers. Herman Daly's writings about a "steady state economy," discussed further in the next chapter, were also influential at this time. The sustainability literature got one of its strongest pushes from Lester Brown and others at the Worldwatch Institute, a Washington, DC-based organization which in the late 1970s began publishing an extensive series of papers and books related to global sustainability, including the *Worldwatch Papers* and annual *State of the World* reports.[19] The tide of literature on sustainability swelled in the 1980s with the International Union for the Conservation of Nature (IUCN)'s influential *World Conservation Strategy* (1980), the President's Council on Environmental Quality's *Global 2000 Report* (1981), and above all the 1987 report of the World Commission on Environment and Development, chaired by Norwegian Prime Minister Gro Harlem Brundtland. These documents documented the growth of global environmental problems and critiqued notions of "development," although generally accepting the desirability of continued economic growth. Even fiction writers contributed to the reevaluation of development trends. Ursula Le Guin's science fiction novels explored alternative societies wrestling with problems of overconsumption, inequality, and environmental destruction. Ernest Callenbach's *Ecotopia* books laid out a vision of a harmonious and sustainable society created when local activists in Northern California secede from the United States.[20]

With the release of the Brundtland Commission report *Our Common Future* in 1987 and the United Nations Rio de Janeiro "Earth Summit" conference in 1991, calls for sustainable development entered the mainstream internationally. The influence of the IUCN and Brundtland reports in particular flowed from the broad participation of main-stream governmental officials within these bodies, which gave their findings an air of authority going beyond the "alarmist" reports of the *Limits to Growth* researchers, *Global 2000*, or the Worldwatch Institute. The Brundtland Commission in particular received input from literally thousands of individuals and organizations from around the world. Initiated at the request of the United Nations Secretary-General, it followed in the footsteps of two other highly respected UN-sponsored commissions, the Brandt Commission on North–South Issues and the Palme Commission on Security and Disarmament Issues. A more authoritative body to explore the topic would have been hard to find. Following Brundtland and the Rio Earth Summit, national reports such as the *Sustainable America* report of the President's Council on Sustainable Development (PCSD) in 1996 attempted to establish sustainable development directions for particular countries. The 1996 United Nations Habitat II "City Summit" in Istanbul also took slow but significant steps towards establishing global consensus on how the sustainability agenda can be applied to urban planning. The tide of academic and professional literature related to sustainability grew steadily during the 1990s and early 2000s, and although some initially expected the subject to be a passing fad, has shown no sign of diminishing in the early years of the new century.

Definitions and perspectives

Despite several decades of discussion, no perfect definition of sustainable development has emerged. The most widely used is that of the Brundtland Commission: "develop-ment that meets the needs of the present without compromising the ability of future generations to meet their own needs."[21] However, this formulation is open to criticism for being anthropocentric and for raising the difficult-to-define concept of needs. (Does every household really need two cars? A VCR? A 2000-square-foot house on a 5000-square-foot lot? What happens if every household worldwide has these things?) Many groups have also criticized the Brundtland Commission's approach for being too accommodating to the interests of the industrialized nations and for not questioning the desirability of continued economic growth.[22]

Other definitions include that given by the World Conservation Union in 1991: "improv-ing the quality of human life while living within the carrying capacity of supporting ecosystems." This version raises the problematic notion of "carrying capacity," which is useful to think about for educational purposes but extremely hard to pin down in practice. It is one thing to say that the carrying capacity of a given watershed is a certain number of deer; deer populations can be counted and analyzed over time, and are relatively rooted to a particular place. It is far more difficult to say that a given region or the planet as a whole can support a certain number of human beings, when humans readily transport themselves and the resources they use over vast distances, and can substitute some resources for others if these become scarce.

Still other writers prefer to define sustainability in terms of preserving existing stocks of "ecological capital" and "social capital." This approach builds on the economic wisdom of living on the interest of an investment – in this case the earth's stock of natural resources – rather than the principal. For example, British economist David Pearce argues that sustainable development "is based on the requirement that the natural capital stock should not decrease over time."[23] Although conceptually appealing, this approach likewise has an anthropocentric flavor, and involves difficult questions of measurement and whether resource substitution should be allowed.

Most sustainability advocates throw up their hands when faced with the definitional question and fall back on the Brundtland formulation. My own preference is to use a relatively simple, process-oriented definition emphasizing long-term welfare: "Sustainable development is development that improves the long-term health of human and ecological systems." This definition avoids fruitless debates over "carrying capacity," "needs," or sustainable end states, while emphasizing the process of continually moving towards healthier human and natural communities. In theory at least the directions of this process can be agreed on through participatory processes in which all relevant stakeholders are represented, and progress can be measured by means of various performance indicators. (See Box 2.1.)

Box 2.1 Some definitions of sustainable development

Theme: Meeting the needs of future generations
"Sustainable development is development that meets the needs of the present without compromising the ability of future generations to meet their own needs."

Brundtland Commission (1987)

Theme: Carrying capacity of ecosystems
Sustainable development means "improving the quality of human life while living within the carrying capacity of supporting ecosystems."

World Conservation Union (1991)

Theme: Maintain natural capital
"Sustainability requires at least a constant stock of natural capital, construed as the set of all environmental assets."

David Pearce (1988)

Theme: Maintenance and improvement of systems
"Sustainability . . . implies that the overall level of diversity and overall productivity of components and relations in systems are maintained or enhanced."

Richard Norgaard (1988)

Theme: Not making things worse
Sustainable development is "any form of positive change which does not erode the ecological, social, or political systems upon which society is dependent."

William Rees (1988)

Theme: Sustaining human livelihood
Sustainability is "the ability of a system to sustain the livelihood of the people who depend on that system for an indefinite period."

Otto Soemarwoto

Theme: Protecting and restoring the environment
"Sustainability equals conservation plus stewardship plus restoration."

Sim Van der Ryn (1994)

Theme: Oppose exponential growth
"Sustainability is the fundamental root metaphor that can oppose the notion of continued exponential material growth."

Ernest Callenbach (1992)

Theme: Grabbag approach
"Sustainable development seeks . . . to respond to five broad requirements: (1) integration of conservation and development, (2) satisfaction of basic human needs, (3) achievement of equity and social justice, (4) provision of social self-determination and cultural diversity, and (5) maintenance of ecological integrity."

International Union for the Conservation of Nature (1986)

Although writers on sustainability share the same basic concerns about the directions of global development, there are also several recurrent debates between them. One main rift is between those who maintain a faith in technology, scientific rationality, and economic growth and those who don't. The former approach often fits well with the mainstream conservation movement within industrialized nations and with large international development agencies and research institutes that are used to engaging in detailed scientific, economic and policy, analysis. The aim becomes to achieve ecological goals by quantifying environmental impacts, analyzing economic policy options, fine-tuning regulation of private industry, and adjusting market incentives. In contrast, others believe that sustainable development is fundamentally incompatible with current capitalist economic structures, attitudes, and lifestyles. For example, Australian sociologist Ted Trainer argues that "a sustainable society must be based on non-affluent living standards, on highly self-sufficient and small-scale local economics, and on zero economic growth."[24] This camp has definitely been in the minority in official circles, but finds considerable support at the grassroots level.

A second main division is between those who focus on ecological crises and those who emphasize social needs and equity. "Deep ecologists" and mainstream environmentalists in the industrialized nations tend to fall into the first camp, while social ecologists and Third World grassroots activists take the latter perspective. Activists in the so-called "developing world" often see First World concern about the global environment as a way to deny them the advantages that industrialized countries already enjoy, and criticize sustainability advocates in North America and Europe for not focusing sufficiently on the problem of First World over-consumption. Some also criticize the Brundtland Commission's work for embracing conventional concepts of economic growth without paying attention to over-consumption and Third World exploitation. However, many others recognize "the intimate connection between the ecological crisis and the broader issues of social and economic justice," as *Ecologist* co-editor Nicholas Hildyard puts it,[25] and have sought to conceptualized "sustainable development" in a way that takes both environmental and equity needs into account.

A third area of contention concerns the extent to which indigenous peoples should be used as models of sustainability. On the one hand, many deep ecologists and social activists agree with Helena Norberg-Hodge and Peter Goering that "traditional societies are the only tested models of truly sustainable development."[26] Writers such as Jerry Mander point to the wisdom of native cultures that have learned to live relatively harmoniously with the land, and argue that such cultures illustrate a quality of spirit that is a necessary antidote to Western materialism.[27] On the other hand, others dismiss this viewpoint as romanticism, and argue that indigenous peoples frequently behaved in unsustainable ways themselves. The Plains Indians, for example, reportedly stampeded large herds of buffalo off cliffs, and Paleolithic hunters may have caused the mass extinction of many species. There is probably something to be said for both points of view, though on the whole traditional peoples seem to have lived with a reverence for land and nature that industrial society would do well to learn from.

A final area of potential confusion concerns gradual changes within ecological science itself, in particular the move away from the notion that ecosystems naturally reach a point of balance or harmony, towards a more process-oriented view that acknowledges the somewhat chaotic, unpredictable, constantly changing nature of natural systems. The former viewpoint, developed following the traditional ecological theories of Eugene P. Odum and others, might imply a search for steady-state conditions of human development. The latter perspective would allow for more continual change as long as it headed in directions that nurtured human and ecological well-being.

As the preceding history suggests, advocates of sustainable development have brought a number of different perspectives to the table. A good starting place is to look at four main groupings of writers: environmentalists, economists, equity advocates, and spiritually and ethically oriented writers. Environmentalists tend to be motivated by the threat of ecological crises; they range from environmental managers working within large corporations (adopting a more-or-less utilitarian attitude toward the environment) to deep ecologists and Earth First! sympathizers (adopting more ecocentric attitudes). Economists use the language and tools of economics, a quasi-science that emphasizes monetary valuation of things and the goal of efficiency. The tendency of economic writers is to bring environmental and social issues into an economic framework of analysis, for example by viewing sustainable development as a process of maintaining natural capital, or by seeking market-based mechanisms for cleaning up environmental pollution. Equity advocates often focus on inequality, exploitation, and First World over-consumption, and develop detailed analyses of how concentrations of political and economic power lead to exploitation. Such individuals and groups often mobilize politically against economic globalization and to regain local control over economic activity. Spiritual writers and ethicists dwell on the need for a transformation of values and mind-sets as a precondition to sustainable development. By reconnecting with the earth, each other, and our own relation to the universe, this viewpoint suggests, humans will become better able to coexist with one another and the planet. Ecofeminist critiques of development follow a similar path, arguing that specifically male values, mind-sets and institutions are much of the problem.

Such a categorization is simply a useful way of organizing the sustainability literature, and parallels the "Three Es" – environment, economy, and equity – that are often seen as the goals of sustainable development. It should be stressed that many writers combine more than one approach. Box 2.2 provides a general overview of how some authors may be viewed in relation to these groupings, with lists of names arranged in rough order of chronology.

Modernist, postmodernist, and ecological worldviews

Any pivotal concept like "sustainable development" must be seen against a backdrop of the slow, massive shifts in outlook that shape history at particular times. In this case, the sustainable development movement can be seen as part of a larger reaction against the modernist worldview that dominated global development during the twentieth century and that continues its influence today, although often under a postmodern guise. (Whether postmodernism should be viewed separately from modernism is an ongoing debate, as will be discussed further in a later chapter.) In contrast, sustainability can be seen as a key goal of an ecological worldview that has been slowly gaining adherents for many decades, and that represents a potential alternative to both these others.

The modernist worldview has taken on different manifestations at different times in the visual arts, in literature, in architecture, in science, and in philosophy. However, it is based on a number of core elements: (1) a desire to leave traditional forms behind and to create a new, "modern" world often oriented around technology; (2) a faith in science, rationality, and an objective viewpoint; (3) a search for universals often connected with science; (4) methodological approaches that break problems down into their constituent parts and that tend to view the world atomistically and mechanically; and (5) a frequent discomfort with normative statements and value-based discourse.[28]

Between the 1920s and 1970s modernist architects cast aside traditional or classical forms and experimented with sleek new designs that often used new materials such as glass, steel, and concrete. The modernist movement in architecture was represented by the Congrès des Arts Modernes (CIAM) and the 1938 Charter of Athens, authored in large part the most famous modernist architect, Le Corbusier. Although many modernists endorsed a humanistic political philosophy with laudable social goals, their design aesthetic emphasized forms of development – in particular the slab-like "towers in a park" scheme emulated by low-budget US public housing and urban renewal – that were later seen as anti-human. The style and works of many modernists also exhibited an arrogance that led quite understandably to a backlash.[29]

In the urban planning field, modernists moved away from the ecological holism of Geddes and Mumford to embrace the social sciences and highly quantitative forms of analysis. The ideal of the planner as a detached, objective expert took over. At the same time, planners adopted an unquestioning faith in material progress and economic

Box 2.2 Perspectives on sustainable development

Environmentalists
Environmental concerns paramount; ranges from "environmental management" to "deep ecology"

Predecessors
Thomas Malthus
Henry David Thoreau
19th c. German forestry

Conservationists
Gifford Pinchot

Preservationists
John Muir

20th c. Natural resource scientists
Aldo Leopold
Rachael Carson
Barbara Ward
René Dubos

Global environmentalism
Donella Meadows
Lester Brown/Worldwatch Institute
World Resources Institute
Brundtland Report
Earth Summit/Agenda 21
President's Council on Sustainable Development

Deep ecologists
Arne Naess
Bill Devall/George Sessions

Bioregionalism
Kirkpatrick Sale

Environmental management
ISO 14000

Economists
Economics as the focus and language of choice; emphasis on incorporating environmental concerns into an economic framework

Predecessors
John Stuart Mill
Kenneth Boulding
E.F. Schumacher

Steady state economics
Herman Daly

Environmental economics
David Pearce
Michael Redclift

Ecological economics
Robert Repetto
Robert Costanza
Kerry Turner
Johan Holmberg
Richard Norgaard

Restorative economics
Paul Hawken

Local self-reliance
David Morris

Ecological footprint analysis
William Rees

Economic democracy
Martin Carnoy
Derek Shearer

Socially responsible investment
The CERES principles

Equity advocates
Structural inequality, exploitation, and First World overconsumption as primary concerns; emphasis on resisting economic globalization, reclaiming the commons and local control over development

Predecessors
Marxist, Socialist, Anarchist critiques of capitalism

Social ecologists
Murray Bookchin

Dependency theory
Andre Gunder Frank

Development critics
Edward Goldsmith
Nicholas Hildyard/*The Ecologist* magazine
Frances Moore Lappe
Helena Norberg-Hodge
Arturo Escobar
Anti-WTO Activists

Third World activists
Vandana Shiva
Martin Khor

Environmental justice
Robert Bullard
Carl Anthony
Benjamin Chaves

Spiritual writers and ethicists
Focus on a transformation of values and mindsets; reconnection with the Earth and each other; search for an alternate paradigm to 20th-century modernity

Predecessors
Teilhard de Chardin
Gregory Bateson
Paul Goodman
Ivan Illich

New paradigm writers
Ervin Laszlo
Fritjof Capra

Environmental ethicists
Baird Callicott
Timothy Beatley

Ecopsychology
Theodore Roszak

Green politics/ecofeminists
Charlene Spretnak
Petra Kelly
Carolyn Merchant

Spiritual writers
Gary Snyder
Thomas Berry
Matthew Fox
Thich Nhat Hanh
Dalai Lama

development. Towards the end of the twentieth century these goals came into question owing to the bleakness, ecological degradation, inequities, and questionable livability of the resulting urban environments.

Within international development, modernist attitudes meshed well with the rise of post-Second World War development practices relying on large-scale infrastructure and technology. The "Green Revolution" – though which Western countries convinced developing nations to substitute fertilizers, pesticides, and hybrid seeds for indigenous agricultural practices – is a classic example. Biotechnology may represent a more recent version of this approach, which relies on science, technology, and large inputs of non-renewable resources and capital to increase agricultural yields. Throughout the Third World, countries also rushed to emulate modernist First World urban development by building automobile infrastructure, huge industrial plants, and North American-style suburbs, often with disastrous results.

To some extent the postmodernist viewpoint represents a rethinking of the values and assumptions of modernism. The ideal of universal development principles or design ideas has been shattered – these have been shown to bring about frequently disastrous results. Instead the postmodern perspective acknowledges the value of many different cultures and viewpoints. "Anything goes" might be the mantra. Within architecture, postmodernism is characterized by a mixing of styles and forms within a single building. Buildings often become playful, borrowing from here and there, as in Philip Johnson's famous AT&T building in New York, which emulates a piece of Chippendale furniture. The results are often a welcome relief to bland, faceless modernist design. But the motivations underlying postmodern design are far more than just playfulness. Whereas modernism followed Le Corbusier's dictum "Form follows function," Nan Ellin points out that postmodernism might be said to follow a number of new principles with less commendable motives: "form follows fiction" (Disney World, Las Vegas), "form follows fear" (gated communities, sanitized semi-public spaces such as malls), "form follows finesse" (projects designed by overly narcissistic architects trying to carve out niches for themselves), and "form follows finance" (urban landscapes most fundamentally shaped by flows of capital).[30]

As geographer Michael Dear notes in his study of Southern California, urban regions have been fragmented into a postmodern melange of edge cities, gated communities, and social groups more connected to global electronic networks than to particular places.[31] The relatively simple model of a central city and suburbs that prevailed until recently is fading as a wide variety of different spaces and cultures emerge within the postmodern urban environment. However, urban geographer David Harvey has argued that postmodernism may not be a radically new state – the underlying logic of capitalist production has not changed in his view, simply some of its surface manifestations.[32] Shiny new suburban office towers, regional malls, and gated communities are just new window-dressing for the same dynamics of economic power that fueled modernism.

The main problem with postmodernism as a philosophical framework lies in estab-lishing grounds for ethical and moral judgments – that is, for action of any sort that might seek goals such as a sustainable society. For many, the result of the postmodernist outlook is a nihilistic relativism that denies the existence of any shared values or grounds for social change. If anything goes, is there any point in trying to build cities one way as opposed to another? Are there grounds for adopting certain planning policies, economic development strategies, or design guidelines as opposed to others?

The ecological worldview, in contrast, acknowledges cultural diversity but seeks to ground the development of society in fundamental values that we all share by virtue of being human and sharing a small planet. This perspective emphasizes interdependence, based in part on scientific understandings of the radical interconnectedness of the "web of life."[33] It views the world in terms of overlapping complex systems and organic unity, rather than as an atomistic collection of people and material things, as in positivistic science and neoclassical economics. It emphasizes flexible, evolving systems that can learn and adapt. Unlike postmodernism, the ecological perspective holds the possibility of justifying ethical belief and action, in that these are necessary to sustain social and ecological systems. This ecological worldview – and the challenge of sustainable development in particular – can be seen as a grand narrative replacing the modernist ideals of technological and material progress.[34]

Differences between the modernist, postmodernist, and ecological viewpoints are summarized in Box 2.3.

Box 2.3 Modernist, postmodernist, and ecological worldviews

	Modernist worldview	Postmodernist worldview	Ecological worldview
Values	Universal values based on modern science	Pluralistic values based on cultural and cognitive traditions	Acknowledges pluralism but also a shared core value set based on common problems
Cognitive approach	Atomistic (break problems down into constituent parts; view world as collection of individual elements)	Acknowledges pluralistic ways of viewing the world	Emphasizes interrelationships, networks, systems
Core influences	Newtonian physics; neoclassical economics	Twentieth-century physics (relativity, uncertainty principle)	Ecological science; chaos theory; systems theory
Political implications	Reinforces centralized political authority	Undermines centralized political authority	Emphasizes flexible and evolving relationships between different political institutions
Preferred planning modes	Rational, comprehensive planning	Decentralized local planning to meet pluralistic community needs; communication to gain consensus on directions	Emphasizes communication and education to help evolve public understanding; advocacy planning to achieve shared goals; evolving incentives and mandates between different levels of government

Modernism advances a very strong value set, one that places priority on scientific and technological tools and methods. Within planning and urban development, modernist outlooks have underlain the expert-driven, technocratic planning common during the mid-twentieth century, within which planners determined urban problems through abstract quantitative analysis and saw themselves as impartial analysts and researchers. Neo-classical economics, with its even stronger value set oriented around economic efficiency, growth, and material progress, went hand-in-hand with this mind-set. The modernist approach forbade planners from acting in any normative fashion, even while it advanced such strong values of its own.

Postmodernism works against value-based planning for a different reason – all viewpoints are seen as equally valid. Since truths are seen to be relative to culture and the existence of any universal beliefs is questioned, no rationale remains for choosing a certain path of development over others. But still, as Harvey points out, the values of capitalist economics underlie the postmodern perspective. Radical pluralism itself can also be seen as a value. These give postmodernism a strong though unacknowledged normative bias.

The ecological viewpoint respects different cultural perspectives, and it values maintaining this diversity. However, it also calls for common values and rules that are fundamental to survival on a small planet. Thus without being backed by modernist science (although supported by more recent scientific findings showing a radically interrelated universe), universals can be reached. Many of these points of global agreement have been expressed since the 1940s in United Nations conventions and declarations, in particular the UN's Universal Declaration of Human Rights in 1948, and have been expressed recently through the Agenda 21 agreement emerging from the 1992 Earth Summit and ongoing efforts to develop an Earth Charter.

Sustainability, then, can be seen as one of the core values and goals of an emerging ecological worldview that weaves together recent developments in physics, ecology, and psychology along with core elements of many of the world's great spiritual traditions (which support the importance of ethical action within an interdependent world). This cognitive outlook sees the world in terms of interdependence and co-evolving complex systems, and supports values, ethics, and actions that likewise emphasize interdependence. Environmental, economic, equity, and spiritual or ethical perspectives on sustainability can all fit with this worldview. "Sustainability" itself is a code word for other values – principally the sustaining and nurturing of life on the planet – that become a starting point for action in urban planning as in other fields. Acknowledging this normative foundation implies a conscious direction to future action that is very much needed. (See Figure 2.1.)

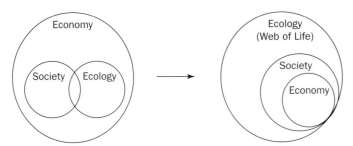

Figure 2.1 The transition from an economic to an ecological perspective. In an ecological world-view, sustainable development implies a perspective that sees economic values as only a subset of broader social and ecological values.

The role of values and institutions

Values are priorities that people adopt – consciously or otherwise – based on their world-views and assumptions about reality. These priorities then motivate behavior that follows from these more general cognitive outlooks. If they were being logically consistent some-one subscribing to a worldview based on free-market economics might value competition, entrepreneurship, and individual freedom. Someone subscribing to an ecocentric view of the world might value the integrity of natural ecosystems, closeness to nature, and a low-impact lifestyle. Someone with a strongly feminist perspective might value equality, civil rights, and social welfare policies aimed at caring for women, children, and the elderly.

Values either can be explicitly developed as the basis for action, or they can be adopted unconsciously as a set of de facto guidelines for how an individual or society leads life. Societies have often set out a few basic values as representing their core beliefs. For example "life, liberty, and pursuit of happiness" is perhaps the most basic statement of American values, while the Canadian constitution refers instead to "peace, order, and good government." However, in practice daily social or individual life may be based on a hodgepodge of different and frequently conflicting values.

Talk of values has unfortunately often been co-opted by "family values" conservatives in the United States and by similar reactionary groups elsewhere. These groups often use "values" language in hypocritical ways – valuing "family" may mean advocating for a traditional, patriarchal model of the family with the wife staying at home, rather than endorsing good child care, education, health care, and parental leave policies to support today's working parents; valuing "life" may simply mean opposing abortion, rather than actively nurturing human welfare. (One common joke is that for "pro-life" protestors life begins at conception and ends at birth.)

However, defining sets of progressive values can be extremely useful in bringing about social change – and in planning activities of all sorts – in that stating values helps groups clarify their goals and then move on to look at what politics and programs might achieve them. For example, the Global Green Party, building on the work begun by the German Green Party in the early 1980s, adopted a list of six core political values at a 2001 conference in Canberra that includes Ecological Wisdom, Social Justice, Participatory Democracy, Non-violence, Sustainability, and Respect for Diversity. For its part the Green Party USA has adopted a somewhat broader set of ten key values that adds "Decentralization," "Community-based Economics," "Feminism," and "Personal and Global Responsibility."

The "Three Es" of sustainable development can be considered to represent a rather condensed value set as a basis for change. Some sustainability advocates have sought to expand this list to include concepts such as "empowerment," "education," and the like. Others might come up with much larger sets of values such as those of the Green Party USA. In the end the exact formulation is not as important as the fact that sustainability-oriented politics is based on *something* – some set of core beliefs and priorities that can then be planned around, and that reflects global needs for healthy societies and eco-systems. Developing such values explicitly helps eliminate the deep gulf that often occurs between stated and de facto values in a society, producing a politics of hypocrisy in which constructive change is difficult. Arguably the United States is in such a situation, professing to value democracy while tolerating a weak and corrupted version, claiming to value peace while waging wars, and touting the virtues of free markets while in fact subsidizing large quasi-monopolistic corporations.

Within societies values are in turn propagated and shaped by a wide variety of institutions – social, political, cultural, and economic structures and traditions that to a large extent determine the ways we see the world and live our lives. These institutions include systems of laws, courts, and government; corporations, advertising, and the media; and a large number of informal rules and codes of behavior. A substantial literature, led in large part by the "structuration theory" of British sociologist Anthony Giddens, has grown to examine how such institutions structure values and behavior within society.

Changing institutions, then, is a way to change values, and vice versa. Working to reform institutions – for example, election processes, government agencies, planning codes and procedures, and tax structures – can be seen as central to establishing a context in which more sustainable development can come about. The role of institutions will therefore be a recurrent theme in later chapters. Institutions, values, and worldviews all form part of the context in which people develop their individual approaches to sustainable development.

3 Theory of sustainability planning

The concept of sustainable development relies on a number of underlying themes that help determine its agenda. As we have seen these viewpoints arise from the environmental movement and the "small planet" consciousness of the late 1960s and early 1970s, and also from the much longer-term reaction to the modernist perspectives of the mid-twentieth century and the Cartesian mind-set that gained sway during the Enlightenment (and that has been reinforced by the ideologies and institutional forces of capitalism). Based on the ecological worldview discussed in the previous chapter, interest in sustainability has emerged as individuals and organizations seek to address the shortcomings of the modernist worldview and its models of development.

The more specific characteristics that make sustainability planning different from business-as-usual in the profession include a long-term approach to decision-making, a holistic outlook integrating various disciplines, interests, and analytic approaches, a questioning of traditional models of growth and acceptance that limits to these exist, a new appreciation of the importance of place, and proactive involvement in healing societies and ecosystems. These stances can help reorient planning debates in constructive ways to address current development challenges. They encourage people to conceptualize problems in different ways than in past mainstream practice, "providing a different kind of dialogic space in which particular conceptions of the good might be fostered" as Susan Owens and Richard Cowell (2002) have put it.[1]

Elements of the sustainability planning approach

A long-term perspective

Quite obviously, sustainability planning seeks to bring about a society that will not only exist but thrive far into the future. This expanded time horizon is implicit in the word "sustain." It is also why the 1972 publication of *Limits to Growth* was so earth-shaking, in that for the first time researchers used scientific methods to attempt to predict the future of human civilization 100 years or more in advance. Their finding that global society might well crash halfway through the twenty-first century – as well as subsequent information about resource depletion, overpopulation, global warming, and species loss – has caused millions of people worldwide to question prevailing development directions.

Operating mainly within local government, which tends to operate from a relatively short-term and small-scale perspective, planners often have had difficulty adopting a long-term viewpoint. Typically, planning documents address a five-to-twenty-year time horizon, though occasionally plans covering up to 50 years are appearing these days. Local politicians often consider only a one-to-four-year timeframe – until the next election. Cost–benefit analyses for large infrastructure projects consider at most a 20-year timeframe. Indeed, as will be discussed further in Chapter 4, economic analysis is fundamentally incapable of considering costs and benefits more than about 30 years into the future, owing to the existence of discount rates which assume that the economic consequences of particular actions are worth more in the present than similar effects in the future. All of these factors reinforce a relatively short-term viewpoint within local government planning.

Much longer perspectives are needed that take into account human and ecological well-being 50, 100, or 200 years from now – during future generations, in other words. To do this, planners may need to more specifically assess how near-term actions can lead long-term goals. They may need to figure out new ways to illustrate for the public how particular buildings, transportation systems, patterns of land use, or economic development programs help or hinder sustainability far into the future. In the area of land use, for example, local officials may need to address questions such as who will clean up industrial land after it is used, how future options for parks, open spaces, or recreational corridors can be preserved within new development, and how mechanisms can be put in place to encourage the eventual recycling of strip development, malls, and shopping centers into more balanced and intensive communities. In each case, near-term modifications to project design or public policy may be able to improve the future flexibility and sustainability of development. Tools such as sustainability indicators and ecological footprints, discussed later in this book, are also ways to encourage attention to the long-term impacts of present development.

In order to understand the history and evolution of urban environments and how current problems arose, time horizons need to be expanded not just into the future, but into the past as well. It is crucial to understand how urban places have coevolved with environmental, economic, social, political, or technological factors, and how specific planning actions have influenced this evolution. For example, land subdivision and development permits lock in patterns of land use that will persist for hundreds of years. Road systems are very difficult to change once established; their long-term implications need to be considered up front. The cul-de-sacs and disconnected fragments of recent suburban sprawl cannot be easily rearranged. Zoning systems and land use designations have frequently frozen in place development concepts that then acquire fierce political constituencies if anyone attempts to change them. Understanding how such mechanisms have worked in the past is the prelude to better planning in the future.

The relatively new discipline of environmental history examines this sort of coevolution between places, cultures, environments, and modes of production, although not necessarily with a focus on planning. In his book *Nature's Metropolis* (1991), William Cronon, for example, presents a history of Chicago that shows how the location of the city, its growth, and its current character stem from factors such as the location of natural resources in the upper Midwest, decisions by railroads and other industries to locate or build infrastructure in certain places, and the environmental characteristics of the often swampy prairie south of Lake Michigan.[2] Similar analyses can be developed for any city or town.

A long-term perspective also means being able to look at small, incremental changes in the present and to see how they can interrelate and reinforce one another to build a more sustainable society in the future. Lack of this understanding leads to a fragmented and disjointed urban environment in which individual actions do not reinforce one another, but cultivating it can help planners organize otherwise isolated projects and initiatives so that they add up to a satisfactory whole. For example, if several new buildings are added piecemeal to a moribund downtown district they may not do much to help revitalize the area. In a run-down urban environment, each developer may design an inwardly oriented building that barricades itself from the street and does little to create attractive public spaces. But if developers and planners adopt a strategic vision in which individual building projects work jointly with many other physical planning and economic development initiatives to create a vibrant, pedestrian-oriented new downtown community, then an entirely different picture emerges. Planners and developers may identify other sites for additional infill or cleanup. They may redesign the streetscape to make it more pedestrian friendly, with wider sidewalks, more street trees, café seating, and bulbouts on the street corners to slow traffic and improve pedestrian safety. They may identify a site for a new mini-park or public square, or where a creek can be restored as an urban centerpiece. They may adopt a set of design guidelines so that all new downtown buildings contain attractive retail spaces along the sidewalk, with parking behind the structures rather than in front. They may take steps to preserve and renovate historic buildings. They may adopt an economic development strategy to encourage certain types of locally owned small businesses. Along with these initiatives, planners may take parallel steps to implement an urban growth boundary outside of town and to prohibit big-box retail that would kill smaller downtown businesses. Each of these steps is small and incremental in itself, and citizens may not initially see them as significant or related. But together they form part of a long-term strategy for something very exciting.

A holistic outlook

As this example suggests, a second main characteristic of sustainability planning is a holistic outlook that sees the relationships between things, embodying an ecological understanding of the world. In practice this means two things for planning. First, it means linking the different planning specialties that have historically been compartmentalized, such as planning for housing, transportation, land use, and environmental quality, as well as integrating goals of planning such as the Three Es. Such integration counteracts the separations between disciplines that frequently occurred during the modernist era, in part owing to the emphasis on technical specialization. Second, a holistic perspective means linking different scales of planning, searching for ways that planning efforts at international, national, regional, local, neighborhood, and site scales can reinforce one another. This likewise has often not been considered by planners and governmental officials, who understandably have frequently been preoccupied with their own immediate level of government.

In the early twentieth century many of the first planning theorists and practitioners adopted such a broad and holistic viewpoint. Figures such as Ebenezer Howard, Patrick Geddes, Lewis Mumford, Daniel Patrick Burnham, Charles Eliot, and Frederick Law Olmsted addressed many aspects of the urban context, including natural resources, parks, transportation systems, building and community design, regional planning, and social dynamics. Howard's classic 1898 *Garden Cities of Tomorrow*, for example, was not just a physical prescription for decentralizing the overcrowded nineteenth-century industrial city into a constellation of new towns, but a social and economic vision featuring collective property ownership and development of industries that could support local residents. Daniel Burnham's *Plan of Chicago* (1909) included proposals for grand avenues and public spaces, a regional park system, transportation corridors, and recreational facilities along the waterfront. For 50 years Mumford was perhaps the ultimate synthesizer of diverse issues, writing passionately about architecture, social issues, urban development, regional planning, economic systems, the natural landscape, and the impact of technology. His two monumental histories of urbanization, *The Culture of Cities* (1938) and *The City in History* (1961), were for years the bibles of urban planning students.

However, as the modernist era gained momentum, planning became increasingly fragmented into technocratic specialties. Transportation planning, housing, land use, urban design, environmental planning, and economic development became separate disciplines within both academia and the professional world. Consulting firms often specialized in only a few of these areas. Separate professional associations were set up for many of these specialties. Urban designers became almost a separate species from policy analysts and economists, each with very different ways of seeing the world. The result was the loss of a perspective on how different aspects of urban environments fit together into a whole.

The task now is to weave different perspectives and specialties back together and to reinforce a sense of how all urban development actions relate to one another. What is needed are planners who can see the whole and integrate the different components of urban development. Urban design must be integrated with public policy and economics, transportation planning with land use and housing, quantitative methods with qualitative analysis. Other disciplines such as architecture, landscape architecture, sociology, and environmental science must be brought in as well.

It is particularly essential to weave together planning goals such as "environment, economy, and equity" – the "Three Es" of sustainable development, which will be explored at greater depth in Chapter 4. Remedial focus may be needed on two of the three "Es" that are generally underrepresented within our current economic and political decision-making structures: environment and equity. For these to be considered equally with economic objectives, planners will have to play a very active role, calling these issues to the attention of decision-makers, involving advocacy groups who can speak to these issues within public debates, and nurturing urban social movements that can counterbalance entrenched economic interests. A minority of the profession has sought to do this historically, for example, under the "equity planning" banner, but much more is needed.

The Three Es can be seen as symbolizing the horizontal integration of issues at a single scale of planning. But also needed is a vertical integration of these objectives across different scales of planning – national, state, regional, local, neighborhood, and building and site scales. Actions at each level must be seen in terms of their impacts on other levels. In practice, much the reverse has usually happened. In the United States, for example, the federal government has maintained the fiction that land use is entirely a local concern, and has been blind to many of the ways that its actions affect local development. (For example, federal policy makers have refused to consider the role that federal funding of highways and home mortgage loan insurance has played in stimulating suburbanization.) Most states have likewise refused to establish planning or growth management programs, leaving such considerations up to cities and counties. Meanwhile, local governments have vigorously resisted considering the effects of their planning and development policies on regional, state, or global levels, often being preoccupied instead with securing local economic development and tax base.

Understanding the interrelationships between different scales of planning takes practice, training, and the experience of different contexts. Many of our governmental and legal institutions are set up in ways that discourage such coordination of scales. For example, having many taxes collected by local government gives local leaders every incentive to increase municipal tax base at the expense of broader interests, competing with neighboring municipalities for limited tax revenues by zoning for sales tax-producing regional malls and office parks rather than affordable housing, which typically consumes more in government services than it produces in tax revenues. This phenomenon is known as "fiscalization of land use." Likewise, having so much authority currently vested in national governments – rather than international treaties and institutions such as the United Nations – discourages global thinking and strong collective action to deal with problems like global warming.

At this point in history institutions are much stronger at some scales of planning than others. International institutions are weak and still relatively recent. It is only since the establishment of the United Nations in 1948, other institutions such as the World Court, the UN Charter on Human Rights and other treaties, and the major UN conferences of the 1970s and 1990s that a global level of democratic coordination has begun to emerge. Regional planning institutions are also weak in many countries. In the United States, for example, planning at a metropolitan level is usually overseen by relatively weak Councils of Governments (COGs) – voluntary associations of local government that have no statutory authority over most aspects of land use and urban development. As a result, most local governments ignore these regional agencies and their guidelines.

On a personal level it is understandable that many people focus on their homes, neighborhoods, and daily lives without a sense of how these affect regional or global sustainability. The impacts of personal actions on the daily environment are immediate; those affecting larger scales are more remote and may occur far in the future. The culture does not often encourage connecting the global with the local, or other forms of critical thinking. Increasing this holistic understanding at the personal level is perhaps the most fundamental task before us.

A variety of initiatives can help integrate different scales of planning (many of these will be discussed more later as well). Action by higher levels of government to set general goals for planning – with local and regional governments performing the implementation – is one way to begin to structure a framework for sustainable development. Various United Nations conferences and organizations have begun to establish such principles at a global level. Many national governments have also adopted a variety of land use, environmental, and civil rights policies, while some state governments have adopted detailed land-use planning goals. Higher-level governments can also offer incentives and mandates to lower government to encourage them to implement such goals. Technical support and coordination between different levels of government are extremely useful as well. Educational events and conferences can increase understanding between planners and other activists working at various scales. These and other mechanisms can boost intergovernmental understanding and coordination, better integrating the different scales of planning. This sort of flexible, evolving, mutually supportive framework between different levels of government often goes under the name of "governance," as opposed to more rigid models of government institutions in the past.

Acceptance of limits

The notion of "limits to growth" has long been at the core of sustainable development debates, and indeed was the title of the 1972 book that first used the term. In a society based in large part on a belief in "progress" as quantitative growth in production and consumption, this concept is truly revolutionary. The possibilities of applying this theme to planning are numerous. First and perhaps most fundamental is the challenge of planning for economic development that produces qualitative improvement in human and environmental conditions, rather than continual quantitative growth in consumption and material use. This task requires prioritizing forms of economic activity that produce goods and services that people really need, such as affordable housing, education, health care, healthy and nutritious food, cultural and recreational opportunities, and the like. It also means avoiding quick and easy forms of economic development that may produce imme- diate tax revenues but little long-term sustainability benefit – retail strips, auto malls, petrochemical industries, and large manufacturing complexes aimed at producing vast quantities of low-end consumer products.

Living within limits also has clear implications for land use planning, in that it encourages a compact city form and the prevention of sprawl. Planning tools such as Urban Growth Boundaries (UGBs) may be particularly useful in this regard because they literally set limits to urban expansion that can be seen and understood by the general public. The notion of limits applies as well to water use, especially within arid or semi- arid areas, and may mean planning for extensive water conservation and recycling. In fact, such steps are already bringing about a transformation of landscaping practices in parts of the southwestern US, where concepts such as xeriscaping – landscaping using drought- tolerant, often native vegetation – are taking hold. The fields of energy and resource use provide other challenges to live within limits, especially in terms of reducing and eventually eliminating fossil fuel consumption.

Lastly and perhaps most challenging of all will be the task of planning for a stable global population, so that the species can live within planetary limits. Planners have not traditionally addressed population issues, which are often highly controversial. But population growth profoundly affects local growth management efforts and quality of life, as well as broader sustainability topics such as global resource use. Planning for com- munities with excellent health, education, and family planning services can help reduce the population growth rate, especially if these resources are available to women. At national and international levels, planning for forms of development that improve the condition of the world's least affluent countries is extremely important, both to reduce internal population growth rates and to reduce the immigration pressures that challenge already-developed nations.

A focus on place

Though it may seem less significant at first, a focus on "place" has been a key principle of many sustainable city advocates as well as other recent movements such as the New Urbanism. Such an emphasis means nurturing the health and distinctiveness of specific, geographical locations. Focusing on place is particularly necessary right now because development throughout much of the last century has done exactly the opposite – creating an aspatial, global realm of homogeneous, interchangeable communities with little connection to local landscapes, ecosystems, history, culture, or community. This process is facilitated by economic globalization, which tends to produce standardized products and generic urban environments with little authentic connection to past cultures or ways of life. One result of this non-place-oriented approach has been that it has facilitated the exploitation of natural ecosystems and local human communities – since the exploiters tend to feel little connection with the people and landscapes being exploited – and to undermine the ethic of stewardship called for by Leopold and others.

The anti-spatial trend of the modernist era was epitomized by urban planning theorist Mel Webber's concept of "community without propinquity," advanced in the 1960s as the era of automobiles and unlimited personal mobility seemed to promise a new tech- nological utopia without the need for local community.[3] In Webber's vision, technology such as the motor vehicle, telephone, and television would allow people to maintain social ties without regard to distance. To be sure there is some value in this ideal, and these technologies have had many advantages. Yet the costs of ignoring local places and communities are clearly very high as well.

Many other elements of current urban development tend to diminish attachment to place. Housing is increasingly mass-produced by large-scale homebuilding corporations, which often produce a generic product without any relation to local landscapes, climate, history, or culture. Office parks and malls likewise follow generic models. Long-distance commuting for work or shopping tends to further detach people from the places where they live. Unique, locally owned businesses are displaced by regional malls full of chain stores. And so on. Within the field of planning itself, abstract economic or social science methods of analysis have displaced the more place-oriented strategies of urban design. The latter were influential within early-twentieth-century planning, but were seen as unscientific by many mid- and late-twentieth-century planning experts. Even today planners far too often sit poring over their computers, analyzing problems abstractly rather than going out and looking at real places, talking to real people, and gaining appreciation of the problems and complexities of actual communities.

A reaction against aspatial planning began in the 1980s and 1990s. In 1979 planning scholars John Friedmann and Clyde Weaver stated their belief that the next wave of regional planning would have to emphasize "territory" as opposed to "function."[4] Sociologist Manuel Castells analyzed the struggle of local groups to maintain and assert their own identities – frequently place-oriented – in the face of a global network society.[5] Academic geographers argued for a new emphasis on how "space" is produced by political, economic, social, or cultural forces, partly on the assumption that ignoring this dynamic allows corporations and powerful elite forces to control our physical world unrecognized.[6] New Urbanist architects and planners sought to build communities with a strong, though usually very traditional, sense of place. Timothy Beatley and other authors suggested that reasserting the importance of "place" within planning can assist local stewardship efforts.[7] Such a reassertion can be seen as key to the effort to resist economic globalization, and to support local identities and communities of resistance instead.

Closely related to "sense of place" is the concept of "livable communities," which likewise has become much discussed during the past decade because of a collective realization that many of our communities do not meet this criterion very well. They often lack neighborhood centers, downtowns, parks, local stores, recreational facilities, or even sidewalks. Increasing traffic makes streets unsafe and unpleasant, while many neighborhoods are so isolated and disconnected from town centers that residents must get in a car to travel anywhere. These and many other factors have decreased the livability of cities and towns so steadily that we often take this situation for granted.

Since we live in environments that have often been very damaged, in ecological, social, and cultural terms, much of the work of sustainable development involves healing specific places. This approach may mean helping to restore urban ecosystems, for example creeks, wetlands, or wildlife habitat. It may mean looking for ways to redress inequities or to respond to the racism and deprivation that have led to the impoverishment of particular neighborhoods. It may mean working to restore a locally-oriented, socially responsible economy in particular cities or towns. Or it may mean taking advantage of thousands of other opportunities to build a better future in particular places.

Active involvement in problem-solving

A final key element of the sustainability planning perspective concerns the role of professionals, politicians, and ordinary citizens, and the need for all of these groups to participate more actively and constructively in problem-solving. Sustainable development requires enormous changes in the ways that things are currently done, which will only come about through the dedicated work of many people.

Although many of us must fill professional roles as technical experts, mediators, and facilitators who must be dispassionate and removed from advocacy, sustainable development requires a proactive, normative, and personally involved stance whenever possible. This balance is not easy. Many planners and other professionals are profoundly uncomfortable with the notion that they should be seen as advocates of any particular viewpoint. Yet it is a myth of positivist science that such a thing as a totally detached, "objective" role exists. By adopting a passive, technocratic attitude planners may inadvertently facilitate unsustainable forms of development. Planners, politicians, or citizen activists may often need to go out of their way to express the needs of underrepresented groups, stand up for the environment and future generations, and try to compensate for unequal power dynamics within governmental decision-making. With every given situation, planners can actively look for ways to create a more sustainable situation either immediately or 5, 10, 20 or more years down the line. This is not to say that they should force their beliefs on others. But they do need to actively call the attention of the public, politicians, and community interests to the long-term perspective, to the need to balance economic, environmental, and equity objectives, to the interconnections between issues, to the interests of underrepresented groups, and to the implications of decisions at each level of government for other scales of planning. And they need to figure out how to actually get things done, in particular how to navigate through often gridlocked institutional bureaucracies in order to bring about positive change. This alone often requires a highly active, entrepreneurial approach.

Particular action is usually needed in giving voice to underrepresented points of view – such as lower-income communities, future generations, other species, people on other parts of the globe, and natural ecosystems. Planners and government officials may want to invite spokespersons or advocacy groups to meetings, inform them about planning processes, bring underrepresented perspectives to the attention of others within planning debates, and frame issues in such a way that all constituencies are heard. Planners can also help nurture grassroots movements for equity, environmental protection, or other sustainability-related goals through technical assistance, donations and support grants, strategic advice, and service on organizational boards or staffs.

The Code of Ethics and Professional Conduct adopted by the American Institute of Certified Planners (AICP) in 1978 and revised in 1991 in fact requires planners to take an active role that is generally consistent with sustainability planning. These requirements vis-à-vis the public are listed in Box 3.1.[8] Under this Code of Ethics, if planners cannot do these things within a particular position, then it may be unethical for them to continue. In choosing employment, planners should avoid situations in which they are likely to be placed in such ethical conflicts, and if stuck in such a position they must be willing to consider moving to another job. Increasingly, many people with planning degrees work for nonprofit housing developers, community development corporations, nonprofit advocacy groups, regional agencies, and private foundations – groups in which they can take on advocacy roles that may not be available within local government. These employment sources may prove better opportunities to work actively on behalf of sustainable communities and to reconcile personal ethics with the need to make a living.

Box 3.1 Planners responsibility to the public

(From the American Institute of Certified Planners (AICP) Code of Ethics and Professional Conduct)

1 A planner must have special concern for the long range consequences of present actions.
2 A planner must pay special attention to the interrelatedness of decisions.
3 A planner must strive to provide full, clear and accurate information on planning issues to citizens and governmental decision-makers.
4 A planner must strive to give citizens the opportunity to have a meaningful impact on the development of plans and programs. Participation should be broad enough to include people who lack formal organization or influence.
5 A planner must strive to expand choice and opportunity for all persons, recognizing a special responsibility to plan for the needs of disadvantaged groups and persons, and must urge the alteration of policies, institutions and decisions which oppose such needs.
6 A planner must strive to protect the integrity of the natural environment.
7 A planner must strive for excellence of environmental design and endeavor to conserve the heritage of the built environment.

An important part of activist planning is to structure processes so that positive long-term change can come about. This requires proactive vision and strategic thinking. Good communications skills, a sense of humor, abilities to work with different constituencies, and a savvy, entrepreneurial approach are also extremely useful. These skills can be learned and honed to allow planners to be effective in working for change in difficult situations. Such efforts may mean cooperating with a wide variety of other professionals – architects, landscape architects, engineers, policy analysts, lawyers, and politicians. They may also require playing multiple roles, for example, serving as an expert resource, a facilitator, a networker, an organizer, and an advocate at different times. Planners may find themselves working in different capacities at different times in their careers. But throughout, it is important to develop the ability to identify and seize opportunities to bring about constructive change.

* * *

A number of writers have proposed additional sustainability planning themes. In an article in the *Journal of the American Planning Association* (2000) Philip R. Berke and Maria Manta Conroy propose six principles: harmony with nature, livable built environments, place-based economy, equity, polluters pay, and responsible regionalism. In their book *The Ecology of Place*, Timothy Beatley and Kristy Manning (1997) suggest additional themes for what they view as "the new planning paradigm," including fundamental ecological limits to development, reduced consumption of nonrenewable resources, a restorative and regenerative approach to development, quality of life, community, equity, and full cost accounting. Innumerable other specific principles and strategies for sustainability planning have been proposed or pursued.

Clearly many different strategies are important and complement the more general approaches emphasized in this section. However, I have tried to distinguish here themes that most strongly differentiate sustainability planning from planning as usual. Among the main reasons why conventional planning has led to unsustainable development is that it often fails to take a long-range perspective, has not sufficiently integrated different disciplines and goals, hasn't considered the interrelation of actions at different scales, hasn't focused on place, and hasn't emphasized constructive action, in particular to inject underrepresented viewpoints into policy debates. In all of these ways sustainability planning is different from planning-as-usual. The challenge is to figure out ways of applying these sustainability themes within difficult, real-life situations. But with care and dedication, ways to do this can usually be found.

Past perspectives on planning

To be applied most effectively, sustainability planning must build on and situate itself in relation to past planning theories. In other words it must make use of the past body of wisdom regarding how human communities can and do develop. Accordingly, this section outlines major existing branches of planning theory, and asks how sustainability planning relates to these.

Rational comprehensive planning

Probably the dominant planning strategy during much of the field's history has been the so-called rational comprehensive planning model. In the United States particularly this approach can be seen, as flowing from a longstanding identification with the philosophy of pragmatism, seen, according to Hilda Blanco, as an "application of rational process and . . . the best knowledge available to address social problems."[9] Under the rational model, planners analyze situations, define goals, identify obstacles that prevent these from being accomplished, develop alternative solutions, compare these, decide on a preferred approach, implement this, and then evaluate its success.[10] The focus of such a straightforward, linear planning process may be a city, a county, a region, a park district or utility, a nation, or any other type of organization. Whatever the jurisdiction in question, planners follow predictable steps to develop a final plan, which may contain both written policies and graphics or maps. Planners' actions within this theoretical model may also include collecting and analyzing background data, making projections, evaluating whether existing programs are working, and recommending modifications to these if necessary. The rational planning process draws heavily on social science methods, especially in the initial data collection and analysis of problems, and can be said to be "comprehensive" if it considers a broad spectrum of interrelated issues and policies.

The roots of this model extend back at least to Patrick Geddes' philosophy of "survey, analysis, plan" in the early years of the twentieth century. (Geddes emphasized survey methods that involved planners getting out into the field and observing environments firsthand as well as utilizing more abstract quantitative data.) Rational, comprehensive planning came to the fore beginning in the 1920s as the urban planning field became professionalized and planners sought systematic and credible methods of determining urban planning policy. In the United States, Secretary of Commerce Herbert Hoover prompted the federal government to pass the Standard State Zoning Enabling Act of 1923 and the Standard City Planning Enabling Act of 1927, establishing legal authority for local planning; in Britain a series of Town Planning Acts beginning in 1909 created similar powers. Many US state governments began requiring that cities create municipal Master Plans (also called General Plans or Comprehensive Plans), and the rational comprehensive model was generally used to do this.[11] Following from their Master Plan and often serving as elements of it, cities also frequently develop Area Plans (also called Specific Plans or, at a very detailed level, Sector Plans or Precise Plans) for particular neighborhoods or locations within the city, and Functional Plans to develop detailed policy in particular issue areas. Various versions of the rational comprehensive model are also now used to develop watershed plans, habitat conservation plans, ISO 14000 plans for improved environmental performance of industrial plants, national spatial plans, and Local Agenda 21 plans to implement principles stemming from the 1992 UN Earth Summit. These specific types of planning that relate very directly to sustainability objectives will be discussed at greater length in later chapters.

The rational comprehensive method has many strengths. It provides a clear, straightforward method of formulating policy and programs, and is useful at many different levels of planning. It also meshes well with the use of indicators to measure sustainability problems and the effectiveness of policies. Rational planning appears logical to many members of the public, tends to be respected by local political leaders, and can be designed to offer opportunities for public involvement.

However, this model has been criticized on a number of grounds. For one thing, it is often seen as overly expert-driven, based on a mind-set in which detached, "objective" planning analysts determine policy rather than letting public concerns drive the planning process. It has also been criticized for relying on quantitative analysis of data rather than taking into account less tangible, qualitative elements of the urban environment. Because the process can be abstract and expert-driven in this way, it may fail to develop the public and political buy-in necessary for policies to work. Planners may become seduced by the expert or facilitative roles required in the rational comprehensive model, and may simply focus on fulfilling these responsibilities and allowing the model to work rather than taking on more active entrepreneurial or advocacy roles to ensure that plans are implemented, goals are met, and underrepresented interests are considered in the process.

The rational model has also often overlooked the realities of political power, meaning that much hard work has gone for naught when political or economic elites simply follow their own agendas instead. Strategies that incorporated advocacy and hard-nosed political organizing might have been necessary to bring about change instead. Lastly, rational comprehensive planning has often failed to take into account important social and environmental issues that were not part of the intellectual mainstream of the time. In particular, planners operating according to the values and methods of modernist science and economics often have not adequately incorporated environmental or equity objectives into their supposedly rational plans. As a result, these plans have often contributed to unsustainable development practices.

In the late 1950s and 1960s several modified versions of the rational model appeared, responding primarily to the misplaced certainty embodied within its more-or-less scientific approach. Charles Lindblom proposed a "disjointed incrementalism" that viewed planning as a process of "muddling through" day-to-day decisions while avoiding overarching large-scale judgments that could not be supported.[12] Amitai Etzioni proposed an method of "mixed scanning" under which planners were supposed to incorporate elements of both broad perspectives and local angles.[13] Under this approach, planners surveyed the broader scene first to get a sense of perspective and then focused in on particular issues and strategies. This method in particular might help lead to the more holistic approach of sustainability planning.

Neo-Marxist planning theory

In the late 1960s a strong theoretical critique of the rational planning model arose from a number of sources. Writers such as geographers David Harvey and Doreen Massey, sociologists such as Henri Lefebvre and Manuel Castells, and, somewhat later, academic planning theorists such as John Friedmann, Norman and Susan Fainstein, Robert Beauregard, and John Forester called attention to the fact that previous planning theorists had paid insufficient attention to power dynamics within the city.[14] The result, as Harvey argued forcefully, was that equity issues were not addressed within planning, indeed, that planning often allied itself with powerful economic or political forces and increased *in*equity.

Whether or not specifically based in Marxist theory, which had developed its own specific language and set of constructs, such theorists began developing detailed studies and critiques of power dynamics within urban development. Researchers such as John Logan and Harvey Molotoch analyzed the role of "growth coalitions" of developers, real estate interests, and local politicians in promoting boosterish local development.[15] Clarence Stone studied the role of elites in directing the rapid expansion of Atlanta.[16] John Mollenkopf produced a detailed study of how development interests dominated local politics on Long Island.[17] John Friedmann argued for a focus on "social mobilization" in order for planning to succeed.[18] And Brian Stoker and others developed "regime theory" to look at how ruling elites ran cities.[19]

Overall, the range of neo-Marxist theories provided a rich and nuanced understanding of urban dynamics. But it was less successful at moving beyond critique to show how better forms of planning might come about. Its most useful theoretical contributions in this regard came through the study of urban social movements (discussed later) and what evolved into the "communicative action" school of planning theory.

Participatory and communicative planning

One main response to the shortcomings of the rational comprehensive planning method was the rise of participatory planning beginning in the late 1960s and 1970s. In the United States, new federal programs such as Model Cities required extensive public involvement; this requirement was also written into environmental review legislation beginning with the National Environmental Policy Act (NEPA) in 1972. In the United Kingdom, Parliament amended town and country planning legislation in 1968 to incorporate a statutory right to public consultation. Researchers such as John Forester began to realize that much of what planners do every day is to meet with various constituencies, network, share information, and facilitate communication. A new theoretical perspective emerged which is often referred to as "communicative" planning, in that it emphasizes both public participation (sometimes within a consensus-based format) and ongoing processes of communication between planners, citizens, developers, government officials, and other parties as the main mechanism through which things get done and people learn. To be useful, as Patsy Healey argues, communicative theory must also be based on awareness of how knowledge and patterns of communication are socially constructed;[20] this perspective thus overlaps with the institutionalist and social learning perspectives discussed later.

Many different sorts of public participation are possible, ranging from relatively tokenistic advisory boards or blue-ribbon commissions to actual community direction of the planning process. In a classic 1969 article social worker Sherry Arnstein proposed a "ladder of citizen participation," beginning with relatively nonparticipatory situations in which planners sought to manipulate the public or viewed meetings and educational materials in a paternalistic way as "therapy" for the public, through relatively tokenistic stages of "informing," "consultation," and "placation," to situations in which the public gained real power through "partnership," "delegated power," and "citizen control."[21] Later planning theorists developed more nuanced theories of collaborative planning, based on the work of German philosopher Jurgen Habermas, in which many different constituencies learned from one another over time through communicative action.[22] Habermas believed strongly that for this process to work it must be free from domination and distortion (as when different parties twist information for particular purposes), and otherwise meet the criteria of a "discourse ethics."[23]

At a minimum public workshops and hearings have now been included in almost every planning endeavor. Frequently planners go well beyond this to conduct surveys, hold focus groups, facilitate consensus processes, or conduct urban design "charettes" in which residents map out the preferred form of new development in their neighborhood. Participation is possible at every scale of planning, though often through different methods. At the site scale, one extreme is represented by the cohousing movement in which future residents design and plan their community themselves with the help of technical consultants. At the neighborhood scale, various design and planning processes run by cities or nonprofit organizations involve the public through workshops, charettes, and/or project boards composed of neighborhood residents and constituencies. At the city scale, the same forums may also be held and many different boards and commissions set up to exert citizen control over the planning process. Electoral politics also of course comes into play. At regional and national scales direct participation is more difficult, but public perspectives are often incorporated by having representatives of diverse constituencies meet repeatedly over months or years to develop consensus, by conducting surveys and workshops, and by relying on local elected officials to express the views of their constituents.

If done well, participatory planning can help develop policies that are responsive to public needs, in particular the needs of constituencies that may not be represented within the political establishment or planning staff, such as lower-income communities and communities of color. Both the goals of the planning process and the recommended policies and programs may be improved through such input. The process of communication may increase understanding of the issues on all sides and may build mutual respect for the diversity of viewpoints. It may also strengthen political buy-in and develop a political coalition that can help implement recommendations. All of these factors make participatory planning extremely attractive for sustainable development purposes, especially in improving equity and responding to the needs of our least-privileged communities.

However, participatory or communicative planning theory is also open to criticism. There is no guarantee that consensus-based planning or efforts directed by local communities will produce good results – they may simply reinforce short-sighted, parochial viewpoints. Inserting broader, longer-term perspectives into such debates is crucial. There are also great difficulties in defining what the "community" is or who represents it.[24] Is it the residents who may live in a place, others who may use it occasionally, or still others at greater distances who may be affected by particular local actions? Is it just human residents, or other species as well? Is it just current residents or does it include future generations? Consensus processes often produce lowest-common-denominator solutions and vague goals with few specific means to implement them. Avoiding these outcomes may take skilled facilitation by planners. In the ideal situation all stakeholders potentially affected by a decision would be directly represented within the decision-making process. If we consider that these stakeholders might include future generations, people on the other side of the world, other species, and entire ecosystems, this is clearly impossible. Planners may need to inject such viewpoints into the debate themselves, while also maintaining a facilitating role, or else may need to identify advocates who can adequately represent these viewpoints within the consensus process. They may need to actively anticipate all potential viewpoints, and include them into their decision-making whether or not these have been well expressed by members of the public. Needless to say, balancing the interests of different stakeholders in this way is not easy.

Participatory or census-based planning without buy-in from outside political interests and institutions can be a fruitless exercise. Individuals and groups may spend years on collaboration, only to see their recommendations ignored. For example, Amy Helling documents how regional planners and hundreds of citizens in the Atlanta area spent five years and $4.4 million in the early 1990s on a collaborative visioning project to develop a Vision 2020 framework. The resulting report was then ignored by the existing power structure and other regional agencies, which had not bought into the process.[25] Luckily in that case the state government and US Environmental Protection Agency (EPA) eventually stepped in to force better coordination of regional transportation, land use, and air quality planning. But the earlier idea that citizen collaboration could bring about change without buy-in from regional elites proved to be deeply flawed.

Participatory processes also demand enormous time, money, and commitment from both nongovernmental organizations and planners. Groups must be able to send staff to countless meetings over several years, and individuals need to possess the patience to work through complicated issues in slow discussion. Many organizations simply do not have the resources or experience to engage in such processes, or they may believe that there are more efficient ways to work for change. Participants frequently become burned out, and only those with the biggest gains to make or axes to grind persist. Power imbalances between participants can also skew results. For such reasons participation must be carefully designed to produce real representation and real results in a timely and cost-effective manner. Ironically, that does not necessarily mean having hundreds of public workshops or elaborate multi-year consensus processes. Instead, it may imply having a limited number of well-structured events or ongoing committees that are appropriate to the planning task at hand. Along with these, as Diane Warburton points out in a study of community, participation, and sustainable development, there is a large need for education, social learning, and consciousness-raising, similar to the process that Brazilian educator Paolo Freire once called "conscientisation."[26] Only by developing deeper public awareness of issues and commitment to cooperative and constructive work can participatory efforts be successful.

Advocacy planning

The concept of advocacy planning arose in the 1960s partially to compensate for past decades of non-participatory, top-down action by local governments. Pioneered particularly by Paul Davidoff, a lawyer, planner, and professor at Hunter College in New York, supporters of advocacy planning recommended that planners work with particular constituencies such as low-income communities to make sure their viewpoints were effectively represented. Advocacy planners might even be employed by these constituencies, serving in effect as lawyers or technical experts working on their behalf.[27]

Such advocacy is often clearly needed if change is to occur on environmental and social issues. Advocacy planning acknowledges that the core dynamic of social change in many cases is a power struggle in which some groups – developers, corporations, well-heeled neighborhoods, and so on – have far greater access to resources, expertise, and political clout than others. Entire urban or suburban landscapes may be shaped by unseen concentrations of political or economic power.[28] Ignoring these powerful forces and hoping that processes of consensus building by themselves can produce change may be naïve, when what is needed is grassroots organizing, advocacy, and coalition-building to create alternative power centers in society.

On the other hand, planners have to walk a fine line between advocacy and maintaining their credibility as technical experts and process facilitators. They may need to make clear when they are taking on one role as opposed to another. When serving as an information source, their research needs to be thorough and respectable, not slanted so as to manipulate data for the needs of a particular constituency. It is also often difficult to take on advocacy roles while employed within a city or regional government, or while maintaining academic or professional ties. Professional agencies tend to look down on advocacy. Nevertheless, planners in advocacy positions can play an enormously useful role by involving underrepresented constituencies in decision-making processes, injecting a full range of viewpoints into debates, and in the long run changing the nature of urban planning debates.

Advocacy planning dovetails with the rise of NGOs as a significant force in urban planning and development around the world. Since the mid-twentieth century a wide variety of nonprofit groups have been founded to undertake many sustainability-related tasks, ranging from advocating for environmental protection to undertaking local economic development and affordable housing construction.[29] Although most of these groups are still relatively small and weak, they have achieved many successes, and have in many cases have become a valuable counterweight to other sectors of society such as business, labor, and government. Indeed, the rise of "civil society" worldwide is one of the most hopeful phenomena of recent decades. But for this movement to more profoundly affect the development of human communities, NGOs will need to increase their professional expertise in fields such as urban planning. They will need to advocate their viewpoints convincingly to both government and the private sector, undertake large amounts of housing construction themselves, and counter private-sector developers' plans with more ecological or equitable alternative scenarios. Professional planners, historically hired mainly by government or consulting firms, thus may find themselves working increasingly in advocacy roles for activist NGOs or in research roles for nongovernmental institutes.

Theories of urban social movements

Behind specific advocacy groups are often urban social movements (USMs) – broad upwellings of public concern about particular issues that may be represented by various constellations of NGOs and political leaders. Examples might include historic preservation, neighborhood preservation, civil rights, and environmental movements, as well as campaigns to protect or improve particular places or achieve local policy change. Such grassroots uprisings serve as a counterweight to the political power represented by elite social and business groups,[30] and are considered by Friedmann to represent a social mobilization tradition within planning. Indeed, one of the foremost scholars of such movements, Manuel Castells, argues that USMs are crucial to establish new, locally-oriented senses of identity that can counter the power of global corporate culture and elite networks in the twenty-first century.[31]

Urban social movements have tackled a wide variety of planning issues, such as calming traffic, claiming civil rights for gay residents, protesting against sterile, modernist urban renewal, and fighting for ecological restoration of creeks, rivers, and parkland. But these movements face a wide variety of challenges in building their power, maintaining it, and channeling their efforts in the ways most likely to bring about constructive change. In particular, unless they are institutionalized through the establishment of strong NGOs and political leaders, USMs may quickly dissipate when grassroots leadership and public interest move on. Not all USMs are constructive and inclusive efforts – NIMBY ("Not in My Backyard") groups opposing affordable housing or other much-needed local services are an example of negative local action – but many represent the sincere efforts of residents to bring about positive change in the local environments over which they have some control. Active support and advice from planners and other professionals can help these advocacy-oriented groundswells of civic attention bear fruit. Even non-advocacy-oriented planners can acknowledge their importance by making sure they are invited to participate in local plan-making and consensus oriented planning processes.

Institutionalism

Several waves of theory over the past 50 years, coming mainly from sociology and political science, have stressed the role of institutions in structuring the context in which action takes place. These institutions include not just government agencies and large organizations, but the whole panoply of laws, customs, cultural norms, and social traditions that affect how we live, think, and act. This perspective emphasizes how debates and mind-sets are shaped by the prevailing structures of society. The implication is that to plan effectively, we need to look at these institutions and think about how changes to them can reinforce constructive action in the future. The work of Giddens has been pivitol in this regard,[32] as was the pioneering work *The Social Construction of Reality* published in 1966 by sociologists Peter L. Berger and Thomas Luckmann.[33] Institutional theory is consonant with the approach labeled "social learning" by Friedmann,[34] in that it emphasizes an evolutionary approach to change, with Healey's communicative planning approach, which she sees as dependent on institutionalist analysis and action aimed at "building up the *institutional capacity* of a place,"[35] (emphasis in the original) and with other theories of critical thought in which the emphasis is on how various forces organize knowledge, how generative metaphors structure cognition, and so forth.

Recent writings about "social capital" can be seen as fitting into the institutionalist perspective. In *Bowling Alone* (2000), for example, political scientist Robert Putnam analyzes how television, the mass media, suburbanization, pressures of time and money, and generational change have led to a decline in civic engagement and social capital in the United States.[36] These factors in large part represent or stem from the institutions of society. The solution, Putnam believes, lies first in naming our problem, and then in working "to create new structures and policies (public and private) to facilitate renewed civic engagement."[37] In his view this means focusing on a long list of initiatives, particularly on the education of young people, promoting community service programs, making workplaces more "family-friendly and community-congenial," rewarding socially responsible economic activity, promoting more integrated and pedestrian-friendly communities with better public spaces, supporting spiritual communities of meaning, fostering forms of electronic entertainment that reinforce community engagement, promoting group activities, and reforming the political system to make it more participatory and democratic.[38] Though it doesn't offer a single easy strategy to bring about change, such a package of mutually reinforcing initiatives can potentially change the institutions and patterns of social learning that from this point of view determine the overall direction of society.

* * *

Despite these broad theoretical approaches, day-to-day urban planning perhaps most often resembles Lindblom's "disjointed incrementalism" or "muddling through."[39] The reasons for this are understandable – chaotic local politics, unpredictable events, the importance of seizing transitory opportunities, and bureaucratic obstacles or inertia. Still, the risk is always that in a "muddling through" process planners or other decision-makers lose sight of broader goals and strategies. They may be sidetracked into technocratic or bureaucratic roles that have little to do with actively addressing urban problems. Worse yet, they may facilitate types of development that later will be seen as unsustainable.[40] Situating planning within a broader theoretical framework can thus help give greater coherence to day-to-day activities, fit them into an overall strategy, and clarify long-term goals.

Situating sustainability planning within planning theory

While following the themes mentioned earlier, sustainability planning may need to draw on many different planning theories and strategies. To be most effective it may need to weave together a range of theoretical perspectives, for example at different times paying particular attention to rational planning methods, communicative processes, underlying structural forces of political and economic power, social movements, and the role of institutions. In keeping with its holistic approach, sustainability planning therefore may be best conceptualized as a "meta-theory," situating its particular perspectives and agenda on top of the best possible foundation of existing social and political theory.

The elements of this foundation will surely evolve over time. In recent decades, for example, new theoretical approaches such as environmental history, institutionalism, and communicative action have emerged, all of which shed light on key questions of how human society evolves and how it might relate more sustainably to the Earth's environment. Movements such as deep ecology, ecofeminism, and environmental justice have also appeared, and their theoretical perspectives deepen our understanding of potential sustainable development directions. Although individual theorists and activists must often specialize in one intellectual approach or another, different theoretical perspectives should not be seen in isolation, since each contributes valuable knowledge to the overall picture. The implications of each for the overall challenge of creating a more sustainable society should be emphasized wherever appropriate, and the connections between theoretical frameworks pointed out, to avoid unnecessary fragmentation of knowledge that will undercut the ultimate objectives.

This holistic approach stems naturally from the ecological forms of cognition underlying the sustainability perspective, which emphasize connections between things, a dynamic view of intellectual as well as natural systems, and a view of systems as a whole. As discussed previously, this outlook on the world is fundamentally different from that of mid-twentieth-century positivistic science, often referred to as the Cartesian worldview, which still greatly influences many disciplines, and which has led partisans to fiercely endorse one theoretical perspective or another, rather than finding broader understanding.

Planning and power

All that being said, some theoretical mechanisms may be particularly important in explaining why unsustainable development has come about, and may be important as well for understanding how more sustainable directions might emerge. In particular the realities of power – along with the institutions and ideologies that support power – must be understood if sustainable development is to come about in the long run, or if specific planning efforts are to be successful in the short run.

Plans are only effective if they are implemented, inspire action, or otherwise help bring about changes in the world. Whether this happens depends in large part on whether those with power buy into planning processes, learn new ways of looking at the world, or come to see their own interests as dependent on change. Politicians, developers, large businesses, unions, neighborhood associations, and many other groups wield power within societies, and planning at any level takes place within the context of these power relationships. Planning typically takes place through the auspices of government, but as has been well documented certain groups often dominate government at various levels. Local developers or growth coalitions often dominate municipal government, for example. Citizen coalitions, environmentalists, historic preservationists, civic reformers, and other groups may represent opposing power bases. These constituencies themselves may have varying goals and be in conflict with one another. Public agencies and institutions, including planning staffs, also hold considerable power, in that they are able to structure processes, set agendas, and either expedite or stifle particular initiatives within bureaucracies. Planning efforts that ignore the reality of such power dynamics are likely to fail.

No matter what other theoretical approaches are employed, an understanding of the nature and dynamics of power is essential for successful planning at any level. Sustainability planners will often need to work with various groups inside and outside government to ensure that sufficient political backing exists to implement change. Many of the larger power dynamics within our current society are extremely difficult to change, such as the relative influence of monied interests within government. Over many decades these interests have shaped public opinion, the news media, and the culture itself in certain directions, for example in support of materialist cultures and a certain national identity. It may require enormous time and effort to help this context evolve toward a situation that is more democratic, community-oriented, and open to addressing current environmental and social problems. However, in the meanwhile planners can seek opportunities to build constituencies for smaller-scale near-term improvements, while working for longer-term structural change and social learning.

Planners' roles

With this background of sustainability planning theory in mind, how do planners and other professionals work towards sustainable development individually? Much of the answer lies in recognizing the different roles that may be useful in different contexts. It has long been recognized that at whatever level of government, planners may play a number of roles at different times.[41] Often they serve as process organizers – initiating planning efforts, keeping them on schedule, and ensuring that final results are achieved. Within this role they may also serve as facilitators of meetings and workshops. Planners may work to negotiate between various powerful interests, brokering compromise. They may serve as technical experts supplying information to all parties. (This traditional, apparently nonpartisan role is what the public often expects to see.) They may take on a role as political organizers, either to level the playing field so that underrepresented constituencies are heard within the process, or to assemble the necessary political constituencies so that plans will actually be implemented. Planners are often educators, ensuring that information is spread to whoever needs it and helping various participants understand issues and contexts. And last but not least, planners can adopt a motivational role as visionaries, cheerleaders for civic initiatives, or dramatic speakers inspiring action.

All these roles are important at different times. Planners and related professionals may wear many different hats in the course of a single day, and the exact role at any given time may depend on the context and the other constituencies involved. It may be necessary to keep roles distinct and carefully separated, or it may be appropriate to play several roles at once. Hitting the right balance in all cases tends to require substantial experience, diplomacy, initiative, communicative abilities, humor, and understanding of situations.

Sustainability planning implies a different balance between some of these roles than undertaken by the planning profession historically. The technical expert role is likely to be much less than previously, and even when operating as an expert it is important to always be asking to what use technical information can be put. Likewise, although managing public planning processes will still be a central role, it is incumbent on the planner to be more than just a detached manager, but to pay close attention to the results of the process, and to facilitate public understanding and political action.

Meanwhile, more active planning roles assume increasing importance – those of facilitating consensus, supplying vision, educating the public, and serving at times as an advocate and organizer. A savvy, entrepreneurial stance may be required as planners and other professionals navigate within institutional bureaucracies and weave their way across political minefields, trying at all times to advance debates and achieve the most constructive, real-world results possible. At times this will be a risky and difficult endeavor. But the results are likely to be worthwhile in terms of producing small- or large-scale movement towards sustainable communities.

4 Planning and the Three Es

As perhaps the pre-eminent symbol of their holistic approach, discussions of sustainable development often focus on how to simultaneously meet goals in the areas of environment, economy, and equity, usually referred to as the "Three Es." In the past these objectives have often been separated within urban planning, governmental decision-making, and development discussions of all types. During the 1980s and 1990s, for example, public debates pitted "environment against economy" within controversies such as logging in the Pacific Northwest and air pollution controls for Midwest power plants to reduce acid rain in the Northeast. More recently, growth management efforts in metropolitan areas have often been fought by construction unions, developers, and freeway builders on the grounds that these environmentally oriented policies will destroy jobs. Equity initiatives, such as "living wage" ordinances, are also opposed on economic grounds, and even environment and equity have come into conflict at times, such as when recycling facilities with their noise and traffic are located in low-income neighborhoods.

Reconciling these seemingly conflictual goals – and developing new decision-making approaches that reconcile the needs of all three perspectives – is much of what sustainability planning is about. Some might argue that a construct such as the Three Es is unnecessarily simplistic, or might want to add additional "Es" such as Ethics, Education, and Empowerment. Certainly what is important in the end is the underlying ability to weave many different conceptualizations of core values together. But the Three Es represent an excellent starting point in this effort, a relatively simple, straightforward set of criteria that are intuitively understandable to most people. Many civic organizations ranging from the Regional Planning Association of New York and New Jersey to the Bay Area Alliance for Sustainable Development have used the Three Es framework. This model also serves the purpose of elevating Equity and Environmental goals to the level of Economic objectives within day-to-day decision-making, which alone would be truly revolutionary if achieved, and an enormous step towards more sustainable development. So we will focus on this approach here as a useful construct for achieving the broader holistic vision of sustainable development.

Sustainability and the environment

The first of the Three Es, Environment, is what many sustainability advocates have historically focused on. As we have seen, sustainable development has strong roots in the mid to late twentieth-century environmental movement with its increasingly broad definition of the "environment" and its focus on integrating global and local environmental issues. A number of writers in fact see "sustainable communities" as representing a third wave of the modern environmental movement, beginning in the 1990s.[1] (In this view, the environmental legislation and activism of the 1960s and 1970s represented the first wave; 1980s attempts at more flexible, market-oriented, and negotiated approaches to environmental clean-up constituted the second.)

However, environmental goals have varied enormously over time and between different groups of advocates. It is important to explore different debates, viewpoints, and themes within this movement to understand its implications for sustainable development.

One main theme within environmentalism has been the dramatic broadening and redefinition of the term "environment" in recent decades. The early conservation groups founded in the late nineteenth century, such as the Sierra Club in 1892 and Audubon Society chapters in 1896, focused primarily on wilderness and wildlife issues. These topics remained the movement's main priorities through the first half of the twentieth century, even for progressive thinkers such as Leopold. But in the 1960s the agenda and political power of the environmental movement expanded to the point where a number of authors begin their histories of modern environmentalism with this decade.[2] Air and water quality, pesticides and other toxic chemicals, energy use, nuclear power, nuclear arms control, environmental justice, urban growth, international development, and global climate change became concerns of environmental organizations between the 1960s and 2000. Many people began to see every element of the world around them as part of their "environment," including air quality within their homes, the nature and origins of their food, the amount of traffic on local streets, and the quality of public spaces in their towns and cities. This broadening of perspective has been rooted in the gradual emergence of an ecological worldview that stresses the interconnectedness of all things. Such a holistic perspective is of course also one of the foundations of sustainability planning.

A second, related change in environmentalism – affecting its goals and character – is the partial shift from the anthropocentric and utilitarian attitudes that have underlain industrial society and capitalism towards more ecocentric approaches. The philosophy of Thoreau and Muir, with its emphasis on the intrinsic value of nature, found much public support in the late twentieth century and began to gain ground on the utilitarian conservationist outlook of Pinchot and many others, which was dominant within mainstream environmentalism for much of the century. Deep ecology, spiritual ecology, and recent other strains of environmentalism support this fundamental rethinking of the relation between humans and the natural world. Debates such as over logging of ancient forests, for example, reflect the emerging of a point of view that these ecosystems have value in their own right and should not be put to human use at all, even if they could be managed in a "sustained-yield" fashion.

As previously suggested, the notion of "limits" is another main environmental theme underlying sustainability debates. From the time of Thomas Malthus in the early nineteenth century and probably long before, thoughtful observers have wondered if and when human civilization would reach the limits of a small planet. The *Limits to Growth* debate of the early 1970s, which as we have seen helped give rise to the term "sustainable development", raised serious questions about planetary limits. In contrast is the viewpoint known as "technological optimism," represented by Simon, which maintains faith that human ingenuity, technology, or the hidden hand of economics will avert catastrophe.[3] To a certain extent Simon's viewpoint has been proven correct – resource depletion has occurred at a slower rate than expected by some environmentalists. Simon even won a bet with environmentalist Paul Ehrlich over how much prices of five metals would rise between 1980 and 1990 (the aggregate price fell by almost one half). Yet in a longer time frame "limits to growth" arguments clearly have much merit. They are especially valid when applied to urban landscapes – land is a limited commodity for which there is no substitute, and the loss of open space and agricultural land can be visually confirmed every day.

On a practical level, integrating environmental goals into planning and activism implies developing as much knowledge as possible about local ecosystems and their history, as well as about environmental law and regulation, environmental planning tools such as Environmental Impact Reports, and best practices of ecological planning and restoration. Such understanding makes possible better decisions about how to balance the range of possible environmental goals with economic and equity objectives. Much of this knowledge, if possible, should be based on detailed, first-hand observation of particular places, since this is the best way to fully appreciate the character of ecosystems and to understand how humans have interacted with them.

In terms of urban development, environmentally oriented principles related to sustainability include compact urban form (which saves open space, reduces driving, and often produces more livable, walkable communities), transit-oriented development (which likewise reduces driving and fossil fuel use), close-loop resource cycles (ensuring that water, metals, wood, paper, and other materials are reused or recycled), environmental justice (integrating environmental and equity concerns), pollution prevention (steps to prevent pollution in the first place rather than clean up later), the "polluter pays" principle, and the restoration of creeks, shorelines, habitat, wildlife corridors, and other ecosystem components within cities and towns. These and other strategies are ways to move towards a radically greener society, one which can coexist with the Earth's limited resources and often fragile ecosystems in the long run.

Sustainability and economics

Environmental goals often seem in stark contrast to those of economics, and at a broader level the ecological worldview seems at odds with the perspectives of many economists, especially those subscribing to free market philosophies. Understanding ways in which contemporary economics might better fit with sustainable development is a challenging task.

Many pros and cons of market-based capitalist economics from a sustainability point of view are well-known. Such an economic system can be very good at regulating supply and demand, allocating resources, and providing incentives for entrepreneurship and innovation. It is also clearly good at generating wealth and a high level of material comfort for many. These benefits are very significant, and mean that we wouldn't necessarily want to scrap market economics in an ideal world even if we could.

However, our current capitalist economics – both in theory and practice – has many flaws from a sustainability perspective. There is the problem of valuation: it is extremely difficult to put a price on social and environmental goods (such as human health, equity, and environmental quality) so as to factor them into economic decision-making. There is the problem of public goods: it is difficult to make meaningful economic decisions regarding things such as clean air, safe streets, or attractive public spaces that everyone uses but nobody pays for. There is the problem of externalities, those enormous social and environmental impacts of production and consumption that are generally not incorporated into economic decision-making.[4] The price of gasoline, for example, does not reflect the externalities of driving, which include air pollution, water pollution, traffic congestion, degradation of urban quality of life, deaths due to traffic accidents, and the costs of maintaining access to petroleum around the world.

Then there is the problem of discounting the future: the existence of interest rates and inflation means that future costs and benefits are less valuable than those in the present, and that it is very difficult to incorporate the long-term effects of actions into economic equations. Most cost–benefit models are literally incapable of considering impacts more than 30 years into the future. Thus, current economic theory is structurally handicapped in adopting the long-term perspective required by sustainability planning.

Other challenges abound. There is the problem that supposedly "free" markets are distorted by subsidies and regulations. There is the problem that demand and human "needs" are manipulated by corporations and advertising, usually to increase the level of material consumption. There is the problem that capitalism tends toward concentration of wealth and monopoly of power, both of which undermine equity. There is the problem that global trade distances consumers from the true costs of their economic decisions, making it difficult for them to understand these costs, while displacing social and environmental harm onto far-distant people and ecosystems. Capitalist economics assumes continuous expansion in consumption of material goods and resources, a phenomenon that conflicts with the environmental notion of "limits." Last but not least, there is the problem that economic power tends to subvert democratic institutions and shape cultures to meet its own ends, meaning that economic objectives constantly threaten to overwhelm environmental and equity goals.[5] (See Box 4.1.)

All of these deficiencies of capitalist economics – and of economic analysis generally – undercut sustainability. Yet many economic tools are useful or necessary in the process of moving towards a more sustainable society. A number of alternative strategies have been proposed over the years to restructure capitalist economics to meet environmental and equity as well as economic goals.

One of the most radical proposed reforms is known as steady-state economics. Noted social philosopher John Stuart Mill first raised this concept in the mid-nineteenth century, but the main advocate since the 1970s has been US economist Herman Daly, who has worked at the World Bank as well as teaching at Louisiana State University and the University of Maryland. Under steady-state economics, human population and consumption are held at constant levels with a minimum throughput of resources, while qualitative, technological, and moral evolution occurs instead of quantitative increases in material production. The endless growth in material consumption, in other words, is brought to a halt. Daly has proposed progressive resource depletion quotas as a mechanism to nudge the economy towards this steady level of consumption. The government would set a maximum quantity of nonrenewable resources that could be consumed each year, and then adjust the level downwards annually. From that point market mechanisms would do the rest. Prices would rise correspondingly, promoting conservation and substitution of alternative materials. Other economists such as the late Kenneth Boulding have floated the more radical idea – perhaps partly tongue-in-cheek – of applying the same sort of quota system to managing human population. Each couple would be issued birth permits, and these would be bought and sold like any other commodity. Although such a system may seem far-fetched, it represents a logical extension of economics to the problem of overpopulation.

The notion of a steady-state economy was most widely written about in the 1970s, when Daly published two volumes on the subject. Needless to say, the concept has not caught on in recent years; our current economy has a very entrenched addiction to material growth, and the basic economic indicators repeated on the evening news directly reflect growth in material production rather than overall quality of life. Yet the steady-state economy remains an important theoretical alternative to growth-oriented capitalist economics, one that may reemerge in somewhat different guise in response to future resource or environmental crises.

A much more pragmatic approach to reconciling environmental and economic goals has been the discipline of environmental economics, which first appeared in the 1970s.

Box 4.1 Evaluation of market-based economics from a sustainability perspective

Virtue of efficiency	Good at setting prices, regulating supply and demand, allocating resources
Virtue of motivation	Good at providing incentive for entrepreneurship, innovation, and creativity
Virtue of production	Good at producing a large number of material goods and generating high level of material comfort for many
Virtue of flexibility	Good at substituting resources and technologies and adjusting prices to counter resource scarcity (assuming markets are not overly monopolistic or constrained)
Virtue of analysis	Provides an important set of tools to analyze economic costs, benefits, and returns from particular projects
Problem of valuation	Difficult to place an economic value on social and environmental goods
Problem of public goods	Little incentive to incorporate common-pool resources into economic decision-making
Problem of externalities	Many costs of action not included in economic decision-making
Problem of discounting the future	The existence of interest and discount rates means that future costs and benefits are less valuable than near-term ones, making it difficult to incorporate the long-term effects of actions into decision-making
Problem of manufactured demand	Perceived human "needs" are manipulated by corporations
Problem of equity	Capitalism tends toward concentration of wealth and economic power, producing inequality
Problem of the distancing effect of trade	As production moves farther away from consumers, they do not perceive the true costs and externalities of economic actions
Problem of growth	Most economics assumes continuous growth in production and consumption
Problem of market distortions	Subsidies, regulations, and the manipulation of demand by producers mean that there is never such a thing as a "free" market
Problem of democracy	Economic power tends to subvert democratic institutions for its own benefit, usually undercutting non-economic objectives such as equity and environmental protection

Concerned with how best to use economic mechanisms to reduce pollution, resource use, and other environmental impacts of production, this field seeks to revise a range of mainstream economic tools such as cost–benefit analysis to better include the environment.[6] Often the focus is on attaching economic valuations to elements of the environment so as to include them within economic equations. However, this approach doesn't fundamentally challenge any of the basic assumptions of capitalist economics, and can be seen as a somewhat technocratic and reformist response to the challenge of sustainable development.

In contrast, ecological economics is a more fundamental reform movement within economics that also uses the tools and language of neoclassical economics but seeks to locate the human economy within a much larger context of ecological interactions.[7] According to Robert Costanza, Herman E. Daly, and Joy A. Bartholomew (1991), "Ecological economics sees the human economy as part of a larger whole. Its domain is the entire web of interactions between economic and ecological sectors" (p. 3). This valiant effort to reconcile economic and environmental worldviews meets some of the deficiencies of conventional economics, but still succumbs to others. It generally does not challenge growth or technology, and still has difficulty trying to fit intangible or qualitative social and economic goods into a quantified economic framework.

Paul Hawken and Amory Lovins have argued for a restorative economics – a "natural capitalism" – that uses the enormous power of markets to bring about environmental restoration rather than exploitation. Mechanisms to assist in this process might include higher prices for nonrenewable resources and waste disposal and green taxes to internalize the environmental externalities of productions.[8] Their approach might be seen as a more advocacy-oriented version of ecological economics, in which specific policy mechanisms are used to integrate economics into a broader framework including all three Es.

Since the 1960s efforts at local self-reliance have at times posed an alternative to conventional, export-oriented global capitalism. Within international development, nations such as India sought import substitution policies in the 1960s especially. These efforts, which sought to promote locally produced products and restrict imports from abroad, were disparaged by advocates of export-oriented capitalism, and often did not work well because of the difficulty of producing a wide spectrum of goods locally in the face of international competition and political pressure. Within North America, proponents of local self-reliance have likewise advocated local economic development strategies that focus on promoting small, locally owned businesses rather than courting multinationals.[9] Such movements have often coalesced around efforts to keep Wal-Mart and other "big box" retailers out of certain towns (the entire state of Vermont has also sought to stave off Wal-Mart). Alternative currency networks such as Ithaca Hours have also attempted to promote local self-reliance through the dramatic strategy of introducing a new currency that can only be used locally, with notes representing one hour of labor instead of dollars. Anyone receiving such a note as payment can then redeem it at other local businesses for other products or services. Such local networks represent the modern version of ancient barter systems.

Also since the 1960s a sporadic movement for economic democracy, championed by American consumer advocates such as Ralph Nader and Mark Green, has sought to exert democratic control over corporations within the US. Such advocates have argued that states originally gave corporations very limited and strict charters, and that the idea of corporations as an independent power base within society with full legal rights and few responsibilities to the public is anti-democratic.[10] They frequently quote Thomas Jefferson, who wrote in 1816: "I hope we shall crush in its birth the aristocracy of our monied corporations which dare already to challenge our government to a trial of strength, and bid defiance to the laws of our country."

Large-scale efforts at new corporate chartering procedures have so far made little progress. But in an era of growing corporate excess perhaps such initiatives will come. A related effort is the large and growing movement for socially responsible investment, which seeks to use shareholder power and consumer choice to influence corporate activity. Begun in earnest in the late 1970s and early 1980s with the first socially responsible investment funds, this movement has grown so much that in the early 2000s it claimed to represent a quarter of all investment (much of this through large pension funds which have cautiously adopted social investment "screens" on their portfolios). Shareholder campaigns against corporations that pollute or use sweatshop labor, boycotts, and anti-globalization organizing have also helped put exploitative companies on the defensive, and have forced greater sensitivity within businesses to environmental and equity concerns. (See Box 4.2.)

Box 4.2 Alternative economic approaches to promote sustainability

Steady-state economics (Herman Daly, John Stuart Mill)	Responds to the problem of growth by seeking to hold population and consumption constant with minimum throughput of resources, and instead seeking qualitative, technological, and moral growth
Environmental economics (David Pearce)	Concerned with reforming economics to better incorporate externalities, future effects, etc.
Ecological economics (Richard Norgaard, Robert Repetto)	More radical position sees human economy as part of larger web of ecological interactions
Restorative economics (Paul Hawken, Amory Lovins)	Aims to harness economic energy for sustainable development, for example through green taxes
Local self-reliance/ import substitution (David Morris)	Emphasizes local ownership, production, consumption, resources
Socially responsible investment	Activist movement to influence corporate behavior through collective purchasing and pressure

---

Other principles such as "full-cost accounting" and the "polluter pays" principle have been developed in an attempt to reform economics. Both of these refer to situations in which the social and environmental costs of public and private decisions are factored into decision-making, particularly the costs of pollution and eventual clean-up of facilities. Decision-makers would then have a strong or economic incentive to adopt sustainable development and resource use strategies. A related term, the "precautionary principle," warns corporate and governmental decision-makers that if they cannot fully under-stand the effects of their actions, they are best advised to take the least harmful and most sustainable approach.

Such concepts and endeavors hold potential for developing forms of economics that are more compatible with sustainable development. But the balance is still uneasy. Fundamental changes in economic values and processes will be necessary to accommodate environmental and equity goals. As Michael Redclift has put it, "Sustainable development, if is to be an alternative to unsustainable development, should imply a break with the linear model of growth and accumulation that ultimately serves to undermine the planet's life support systems."[11] Leading values behind the free market capitalism that dominate current global development include efficiency, individual choice, growth in consumption, materialism, and privitization of common resources. Although they may have advantages in terms of motivating an efficient growth-oriented economy, these values frequently displace those of environmental protection and social equity, and will need to be put in a much better balance with these other Es.

Sustainability and equity

Equity is the third and by far the least well-developed of the Three Es. To be sure, it has long been a focus of many community activists, labor unions, and social justice organizers. However, these constituencies often have relatively little power, and equity concerns frequently take a back seat in planning and political discussions. Often there is literally nobody in the room who will speak up for disenfranchised segments of the population. Equity goals are often poorly understood and articulated by decision-makers, unlike concerns for the environment or economic development. There is little organized constituency for equity at most levels of government, and many powerful forces work for *in*equity, that is, far greater concentration of wealth and power. Yet virtually all policy-makers interested in sustainability have been forced to acknowledge the importance of addressing equity concerns.

In a global context, a diverse groups of writers including Doreen Massey, David Harvey, Edward Goldsmith, Vandana Shiva, Martin Khor, L.S. Stavrianos, and Arturo Escobar have called attention to growing inequities in economic power and distribution of resources, and the ways that these inequities are played out spatially for different human communities around the world.[12] Some, beginning with Andre Gunder Frank and others in the 1960s, have charged that processes of economic globalization create Third World dependency on the First World, and lead to situations in which the benefits of development are exported to the North or given to elites in the South rather than benefiting the poorest and most needy.[13] Equity advocates within the developing world also focus on First World over-consumption and argue that the industrialized nations of the North have no right to advise nations of the South on how to develop if they can't rein in their own consumption. They often view First World nations as exporting the risks and externalities of economic production, exploiting low-wage labor internationally, and seeking control of global resources. Attempts by multinationals based in the North to control the genetic and biological resources of the South have met with special resistance in recent years.

Within First World nations, a somewhat different assortment of inequities has become urgent. These include wealth and tax base disparities between rich and poor communities (especially wealthy suburbs and impoverished central cities), concentrations of poverty, inequitable distribution of affordable housing and transportation infrastructure, inequitable representation within decision-making, and environmental justice questions regarding differential exposure to environmental hazards.

To take up the first of these concerns, growing imbalances of resources between rich and poor communities have become worrisome to many. These disparities occur in large part because of the fact that our metropolitan areas are fragmented into many smaller cities that receive wildly varying amounts of money from local tax revenues. Typically suburban jurisdictions have seen their coffers benefit from mall, office park, and upper-income housing development, while central cities have seen their tax bases decline as businesses and affluent residents leave and once-thriving commercial streets become lined with empty storefronts. Meanwhile, their resource needs rise in order to provide social services for less-affluent communities, to repair aging infrastructure, and at times to clean up brownfield lands left over from past industrial development. Essentially the financial benefits of our current sprawl patterns of development accrue to some local governments, while others are left bearing the costs.

These regional disparities are worsening as suburbs expand. (Interestingly, before the rapid suburbanization of the last 60 years central cities were often better off than suburbs, since they had a rich concentration of businesses and affluent residential districts.) A main result of growing regional inequities has been the creation of highly isolated concentrations of poverty in central cities. Myron Orfield, a Minnesota state legislator who has studied this problem nationally, finds a disturbing growth of "extreme poverty neighborhoods" in which more than 40 percent of residents are below the federal poverty line and "transitional poverty neighborhoods" where 20 to 40 percent of residents are in this category.[14] William Julius Wilson has also written extensively on the plight of the "truly disadvantaged" in central city neighborhoods where decent jobs no longer exist.[15] This sort of concentrated poverty leads to social isolation and a wide variety of problems that reinforce one another, structurally entrenching a situation of inequality.

A related area of inequity in US urban areas has to do with the provision and distribution of affordable housing. Affordable units have been in woefully short supply in many metropolitan areas in recent decades, since for-profit housing developers prefer to build middle- and upper-income housing rather than low-income units. This general deficiency is aggravated by the fact that many local governments, following local prejudice and political sentiment, actively resist accommodating lower-income populations (which are often members of minority racial or ethnic groups). In an old practice called "exclusionary zoning," many cities and towns have zoned their land for large-lot or single-family development, thus ensuring that only relatively pricey housing is created. Meanwhile, such cities resist zoning for multifamily units such as apartments or condominiums, which often provide cheaper rental housing for minorities. Courts have ruled such practices illegal, but many cities still resist zoning to create housing for a diverse population. American housing policy has also strongly favored homeownership over rental housing (which better meets the needs of the truly poor), tends to segregate affordable housing units in a limited range of locations, and often fails to provide the necessary community services, social services, amenities, or transportation to make affordable units become part of functional neighborhoods for residents of all income categories.

Inequities are also perpetuated these days through NIMBYism. Even though many studies have shown that well-designed, scattered-site affordable housing projects do not decrease surrounding property values, existing residents often fight them based on that fear as well as a general dislike of others different from themselves. Such attitudes often accompany, or are camouflaged by, opposition to any building type that represents higher density than single-family homes. Many planners and elected leaders are easily swayed by NIMBYs and fail to stick up for affordable housing or other appropriate forms of development. Cities also fail to allocate funds to subsidize housing for the poorest of the poor, and often seek to meet state or regional affordable housing requirements through senior housing (on the theory that low-income senior citizens are acceptable to local neighborhoods).

Unequal distribution of transportation funding is a further source of inequities within urban regions. Federal transportation money is funneled through state and regional agencies that have often favored freeways or commuter rail systems that serve suburbia over forms of public transit that serve central cities. Increasingly advocacy groups have challenged such policies on civil rights grounds. For example, in the mid-1990s the Los Angeles Bus Riders Union sued that region's Metropolitan Transit Commission over its policy of funding enormously expensive subways and commuter rail instead of cheaper bus service serving low-income residents. In a court settlement the agency agreed to increase bus service and reduce fares. Transit riders in New York have waged a similar campaign. In Washington, DC, community activists complained bitterly that Metro's Green Line subway serving predominantly black Northeast DC and Anacostia was the last major line to be built. In the San Francsico Bay Area, social justice advocates staged demonstrations in the late 1990s at the Metropolitan Transportation Commission over its policies of fully funding freeway construction but not bus transit serving central city populations. The agency relented and shifted $375 million within its Regional Transportation Plan to meet that purpose.

US federal spending of many sorts throughout the twentieth century was highly skewed in favor of suburban sprawl – favoring relatively well-off groups within society – rather than the preservation and restoration of urban centers. The interstate freeway system represented a massive subsidy for suburbanization, opening up millions of acres of land around cities to sprawl development. The federal tax deduction for home mortgage interest likewise favors suburban homeowners, who typically have larger mortgages and larger incomes from which to deduct the interest. This provision gives nothing to renters, who are primarily middle- and lower-income individuals and often live in older urban areas. Although this subsidy for homeownership has had some advantages in terms of allowing middle- and working-class families to own their own homes, its overall equity implications appear negative, and its impacts in terms of land use have been disastrous. Federal spending on military production and large-scale waterworks likewise fueled the growth of suburban, sunbelt areas such as Los Angeles, San Diego, Atlanta, and Phoenix, while older, more urban industrial areas in the Northeast and Midwest suffered from disinvestment. The 1980s deregulation and subsequent 1990s federal bailout of the savings and loan industry subsidized a massive, unnecessary construction of suburban office buildings and malls. And so on. The inequities implicit in such federal policies – and their effects on urbanization generally – have rarely been acknowledged by decision-makers.

Within most nations inequities have been built into public decision-making processes. Historically many lower-income or minority groups have not been involved in these processes, have not had the skills or knowledge necessary to participate, or may have had more pressing concerns such as earning a living and surviving. Information may be presented in technical language that ordinary citizens have trouble understanding, especially immigrants who speak different languages at home. Public meetings or hearings on development are frequently held during daytime hours, when many working individuals cannot attend. Many advocates or lower-income residents can also simply not afford to spend countless unpaid hours trying to affect public decisions. In contrast, developers are in effect paid for their own extensive involvement (through making profits off subsequent development). Upper-income communities frequently have the time, experience, political contacts, ability to litigate, and access to decision-makers necessary to affect policy. They also may have a sense of empowerment that lower-income communities or color do not – they know they can affect the political process, and may have prior experience of doing so.

A final main area of inequity within US urban development is encompassed by the term environmental justice. This movement initially called attention to ways that minority groups or communities or color are disproportionately exposed to toxic chemicals, pollution, and unwanted land uses such as dumps and incinerators. It has since expanded to include other urban planning subjects such as the lack of parks and recreational facilities within lower-income communities, disparities in transportation service, and the need to restore inadequate infrastructure.

The environmental justice movement has its roots in the 1960s, when citizens' groups first publicized problems such as inner-city children eating lead paint chips and Native American communities in the Southwest suffering from radioactive uranium mine tailings. The movement spread in the 1980s as awareness grew of how communities of color were exposed to a wide range of hazards and many activists realized that affluent white constituencies dominated the mainstream environmental movement and the staffs of national environmental organizations. One of the first organized actions came in 1982 when community groups rallied opposition to a proposed PCB landfill near African-American neighborhoods in Afton, North Carolina.[16] More than 500 people were arrested, including Dr Benjamin Chavez, the former director of the National Association for the Advancement of Colored People, and Dr Joseph Lowery of the Southern Christian Leadership Conference. A national conference in New Orleans on toxic substances and minorities the next year was the first major effort to systematically link discrimination, environment, and social justice. A 1983 General Accounting Office study, done at the request of Rep. Walter Fauntroy (D-DC), found that three out of every four landfills in the southeastern United States were located near predominantly minority communities.[17] In 1987, a study of 415 hazardous waste sites by the United Church of Christ Commission for Racial Justice found risk to minorities nearly double that to whites, further cementing the link between environmental risk and minority communities.[18]

In 1990 a group of non-Anglo activists sent a letter to the Group of Ten CEOs of national environmental organizations alleging "racism" and "whiteness" of the environmental movement. A second letter soon followed from the Southwest Organizing Project, signed by more than 100 activists and community-based groups alleging that people of color were "the chief victims of pollution." At the time it was found that there were no African-Americans or Asian-Americans and only one Hispanic among 250 Sierra Club staff, and only five persons of color among 140 staff members of the Natural Resources Defense Council (NRDC). These organizations soon took steps to remedy this situation, and two years later the National Wildlife Federation claimed that 23 percent of its staff were members of minority groups.

In 1991 the First National People of Color Environmental Justice Conference brought leaders of the growing movement together in Washington. A second such conference was held in 2002. The Clinton Administration responded to environmental justice advocates by releasing Executive Order 12898 on Environmental Justice in 1994, setting out "Federal actions to address environmental justice in minority populations and low-income populations." This Executive Order required each federal agency to make environmental justice part of its mission and to identify strategies for achieving it. An additional Executive Order in 2000 improved access to federally funded or assisted programs for persons with limited English proficiency.

Other steps to improve equitable access to governmental decision-making processes in the US have been taken a number of times since the 1960s. Title VI of the federal Civil Rights Act of 1964 laid the groundwork by stating that "No person in the USA shall be excluded from participation, denied benefits, or subjected to discrimination in any federally-funded program, policy, or activity on the basis of race, color, or national origin." The National Environmental Policy Act of 1969 required public participation in determining environmental impacts and alternatives to proposed action. The Intermodal Surface Transportation Efficiency Act of 1991 (ISTEA), the Transportation Efficiency Act for the 21st Century of 1998 (TEA-21), and other legislation mandated effective public participation within specific local or regional planning processes. Meanwhile, many local governments have enshrined public participation within their General Plans or Master Plans, and have gone to considerable lengths to carry out neighborhood planning, notify citizens of proposed planning decisions, and incorporate feedback. But all of these programs represent only a start towards the difficult goal of involving underrepresented communities in urban-planning-related decision-making.

Efforts to improve interjurisdictional equity in terms of tax resources have proven even more difficult. The main device proposed is regional tax sharing, but to date this has only been implemented in one US metropolitan area, Minneapolis-St Paul, where since 1974 40 percent of the increase in sales tax revenue has been put into a regional pool distributed by population. A similar initiative was debated for the Sacramento region in 2002. Elsewhere tax sharing has had little political support. Still, it is a device increasingly discussed as a means to level out resource disparities.

Courts have intervened in recent decades to mandate that many state governments to some extent equalize school funding between jurisdictions. These legal decisions have resulted in state "equalization" programs in California, New Jersey, and elsewhere to provide some base level of funding for each pupil. Also, the distribution of "community development block grants" (known as CDBG funds) from the federal government to cities to some extent promotes equity, in that these funds are allocated on the basis of population and are a remnant of larger federal "revenue-sharing" programs first instituted under President Richard Nixon in the early 1970s. In general any strategy that collects taxes at a higher level of government – such as state and federal government – and then redistributes revenues to cities and counties on the basis of population represents a way to overcome entrenched inequities in resources between local jurisdictions.

Programs to ensure regional fair-share housing provision likewise form a mechanism to improve equity. In California, the state Department of Housing and Community Development requires cities to update their General Plan Housing Elements every five years to accommodate "fair-share" amounts of housing in different income categories, as determined by formulas developed by regional agencies. Courts in states such as New Jersey and Connecticut have likewise required cities and towns to accept affordable housing. In practice, however, localities remain very resistant to doing this and often fail to comply with such mandates. Unfortunately, relatively few penalties exist to compel compliance. Another strategy is for higher-level governments to offer incentives for local fair-share housing compliance. In the late 1990s, for example, the state of California offered unrestricted grant funds to cities that exceeded a certain percentage of their past affordable housing construction. Meanwhile, some local governments have adopted "inclusionary zoning" requiring all large development projects to include a certain number of affordable units, generally 10 to 20 percent. Such mechanisms have the potential for increasing the equitable distribution of housing options.

These are a few of the main ways that planning can help address the equity goals so vital to sustainable development. The big question is how equity objectives are to be advanced in the face of a society whose electorate and political and economic leadership tends to be uninterested in them. Developing equity initiatives may require that planners and other professionals adopt proactive or advocacy roles – speaking up for under-represented constituencies, reminding decision-makers of the interests of groups not represented at the table, working to help minority and lower-income groups become more familiar with decision-making processes and able to articulate their viewpoints, and so on. Much advocacy on behalf of equity has also traditionally been done by local organizing groups such as ACORN, by labor unions, and by local community development corporations (nonprofit organizations run by local residents and dedicated to neighborhood improvement). Calls for "equity planning" have been sounded within the planning profession for decades, most notably by Davidoff and Norm Krumholtz, former planning director of Cleveland. Many practicing planners are deeply sympathetic to equity concerns. Figuring out how to inject them into planning debates more systematically and successfully will be an essential element of the sustainability planning agenda.

* * *

Many of the more visionary planning pioneers during the last century sought in their own ways to reconcile goals of environment, economy, and equity. The influential British garden city theorist Ebenezer Howard, for example, is well known for his vision of a balance between city and countryside, but he also sought to integrate equity concerns into his Garden Cities by having collective land ownership and social organization. He went so far as to work out the economics of how residents would jointly purchase land and build housing. This dimension of his work was overlooked by many of his followers.[19] Mumford likewise paid attention to all three themes, as have more recent authors such as Kevin Lynch and Jane Jacobs.

But too often in practice these objectives have become separated. Economic development specialists have assumed that any form of new business development would help the community – without taking environmental impacts into account or considering the nature and wage levels of new jobs. Local environmentalists have often bought into "slow growth" movements without realizing that without accompanying efforts to promote affordable housing these would exclude lower-income residents and generate inequities. And some equity advocates have brushed aside environmental considerations in the search for new development for their communities. The task ahead for sustainability planning is to figure out creative strategies for reconciling these perspectives, simultaneously meeting all three sets of objectives in the context of particular places.

5 Issues central to sustainability planning

Although almost every development-related topic relates in some way to sustainability, the planning profession is better positioned to address some subjects than others, and arguably some planning-related issues are more urgent than others. Following is a discussion of key issue areas for sustainability planning, with attention to some of the practical implications surrounding action on each subject. I will cover more specific mechanisms for addressing each subject at different scales of planning in later chapters.

Growth management and land use planning

Stabilizing the outward growth of cities and suburbs – and in the process preserving agricultural land, wilderness, important natural habitat, and species – is one of the most pressing challenges for sustainability planning. Urban regions obviously cannot expand forever in the way that they have for the last century and a half, and their growth causes many secondary problems related to motor vehicle use, pollution, congestion, quality of life, and the segregation of groups from one another along the lines of income and race. Suburban sprawl has now merged into rural or exurban sprawl in many places, as even formerly isolated areas of the countryside experience increased residential and commerical development. Consequently the "compact city" has been a goal of many planners for decades, especially in places such as North America and Australia where sprawl is rampant and communities have a low population density. Although a few theorists at the libertarian end of the political spectrum argue that sprawl is not bad, in fact desirable in terms of enhancing individual choice and mobility,[1] such arguments usually leave out the immense social and ecological costs of this forms of development, and the fact that sprawling cities often do not in fact give residents the choice to live in good, affordable housing in well-located and walkable neighborhoods.

Compact cities represent a radically different model from most twentieth-century urbanization. If pursued rigorously this approach would call for most new development to be handled within the existing urban envelope through "infill." This category encompasses several main forms of development: building on vacant lots within the urban area, redevelopment of underutilized lands where, say, small or deteriorating buildings exist, and rehabilitation or expansion of existing buildings. The opposite of infill is often known as "greenfield" development in that it frequently takes place on agricultural fields at the urban edge. At present only about 30 percent of new development is accommodated through infill in even the most infill-oriented US metropolitan areas.[2] The United Kingdom, with its more limited land area and more urban tradition, has set a goal of 60 percent infill (actual 1998 development was 57 percent infill;[3] and Friends of the Earth in the UK has explored increasing the infill target to 75 percent[4]). In the European and Asian context infill is a bit more problematic than in the United States, Canada, or Australia, since cities there have traditionally been far denser and less redevelopable land exists in the form of old shopping centers, business parks, parking lots, or industrial sites. Some British authors complain that government-designed infill projects lack green space and integration with the natural landscape[5]; Peter Hall has warned that compact city efforts may amount to "town cramming."[6] However, improving design and putting a strong emphasis on creating parks, gardens, and restored ecosystem features within urban areas may be able to address these concerns.

Economic and practical considerations for developers often work against infill. Urban land is generally more expensive than land outside the existing urban area and the task of designing and gaining approval for an infill project more difficult. Housing must be built at somewhat greater densities to be economically viable, and ways must be found to make those densities equally attractive, for example by providing a good range of outdoor open spaces. (Densities of 12 to 40 dwelling units per acre – at least double the suburban densities of many places – can be achieved through efficiently designed two-to-three story development, as will be discussed at greater length in Chapter 12.) Moreover, the dream of settling on previously unbuilt land lies deep within the psyche of many cultures, and the ideal of owning a country estate has great appeal within European and Latin American cultures as well as in North American. So compact development requires a rethinking of the ideal of country living, and development of a desire instead for compact, walkable communities that contain many amenities not present in suburban or rural locations. Living well within limits is thus a theme at the core of the compact city vision.

Since the mid-twentieth century many urban regions have engaged in the related challenge of "growth management," that is, slowing or stopping outward growth and organizing land use to better fit with transportation systems, ecological and recreational needs, and existing development. Growth management can be accomplished through a wide variety of planning mechanisms applied at different scales. Basic strategies include Urban Growth Boundaries (UGBs), urban service limits, agricultural zoning, acquisition of conservation easements, transfer of development rights frameworks, and outright purchase of open space for parkland. These strategies will be explored in further chapters. National, state, or regional land use planning frameworks can be extremely useful in terms of helping local governments manage growth. But much of the burden is on municipal governments, both to limit their outward sprawl, and to figure out strategies to restore and reuse previously urbanized lands. (See Figure 5.1.)

Improved land use planning is not simply a question of increasing the compactness or density of human communities. It is also a task of achieving a better balance of land uses – homes, workplaces, shopping, and recreational or community facilities – which twentieth-century zoning has spread in separate swathes across the landscape. And it is a question of designing communities better to use land more efficiently and to provide a greater range of amenities in a smaller space.

Sustainable land use planning must start with the question, "What is the most appropriate thing to do with any particular site?" The answer may well be "nothing." It may be best not to build on a given parcel at all, especially if it is located far from existing communities, or instead to restore elements of its degraded ecosystems to something approximating their natural state before human intervention. Or if the site is strategically located within an urban area it may be appropriate to build more intensely than was originally intended. The most appropriate building or type of use may also be different than originally considered. If a parcel is located within an area already rich with jobs, it may be most appropriate to build housing to create a better local balance, despite existing zoning for commercial uses (the compatability of building types must also be considered).

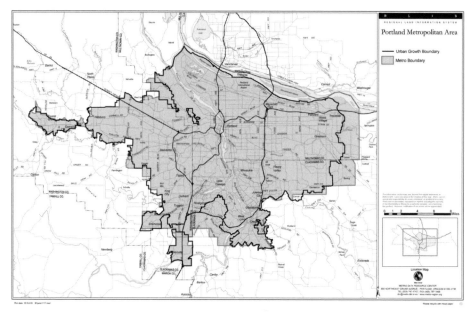

Figure 5.1 Portland, Oregon's Urban Growth Boundary, first adopted in 1979, prevents subdivision of land outside the designated urban area. Source: Metro Resource Data Center.

And vice versa. Some land uses may be generally inappropriate in most locations. Big-box retail development, for example, which has been pursued extensively in many industrial countries beginning in the late 1970s, may not be desirable anywhere within a more sustainable society, since this form of land use tends to generate enormous amounts of traffic, drives smaller, locally owned stores out of business, uses land inefficiently, and creates an over-scaled, pedestrian-unfriendly environment. Sprawling office parks may not be a particularly desirable model either, for similar reasons. Sustainable land use may focus instead on creating balanced, compact, mixed-use communities rather than these single-use monocultures that are oriented around certain models of economic efficiency rather than a broader set of human and environmental needs. Changing land use in these ways may require new concepts of property ownership, moving away from the notion that people should be able to do anything they want with a piece of land which is "theirs," towards land use that balances individual, social, and ecological interests. Such a transition represents a profound ethical change, to say the least. But it is an essential one if sustainable land use is to come about.

Better land use planning alone, however, will not be enough to solve problems caused by urban growth, as long as the number of residents in many areas continues to rise, not to mention the number of cars per resident, the size of houses, and the use of resources. Ultimately the growth of population and consumption will need to be addressed as well. Although "growth management" debates in planning circles are now restricted primarily to land use topics, these more basic questions need to be considered too. Doing so will involve rethinking fundamental social values regarding growth and the nature of progress.

Urban design

Better design of cities is a closely related challenge for sustainability planning. This includes not just the design of public spaces, streets, neighborhoods, and homes, but the configuration of park and greenway systems, regional growth patterns, transportation networks, water and sewer systems, and even industrial processes. Designing such systems requires thinking about how they relate to all other elements of a given community, combining physical planning (related to land use, infrastructure, and the design of places) with public policy frameworks (including tax frameworks and economic incentives) that can support such changes.

At its heart sustainable design is based on human and ecological values, rather than a value set dominated by economic efficiency and profit. This means creating urban places much like the ones Jane Jacobs described in 1961 in *The Death and Life of Great American Cities*: communities that are walkable, human scale, diverse, and oriented around a fine-grained and vibrant mix of housing, shops, and public facilities. These places do not have to be the highly urban neighborhoods that Jacobs described. They can also be village centers and neighborhoods in smaller towns, or rural groupings of buildings. To fulfill the environmental goals of sustainability, the physical design of such places should reflect local climates, ecosystems, materials, and flows of energy, water, and resources. Such design will better integrate human communities with the natural landscape, reduce automobile dependency, use resources more efficiently, and engender a sense of place identity.

Sustainability-oriented urban design might focus on restoring streams and greenways through cities. It might create neighborhood and village centers, and rebuild urban downtowns in cases where those have stagnated. It might make streets safe and pleasant for pedestrians and cyclists, through a variety of street design techniques. It might design compact communities that mix various types and prices of housing. Perhaps most difficult of all, it might transform the desolate commercial strips, malls, office parks, and industrial sites that currently plague our cities and towns into places that nurture human community as well as ecological health.

One main obstacle to these goals is economic. Humanistic and ecological design must be made financially feasible, and this will require a long-term evolution of the market-place for development. At first a number of relatively modest design improvements may be possible without adding greatly to costs, for example by providing sidewalks and connecting street networks within subdivisions, creating modest amounts of park space, preserving streams and wetlands and ranging new commercial buildings along the street to improve the pedestrian environment. In the longer term, very different types of buildings, landscaping, streets, and public spaces can be created. But for private developers to undertake many of these things may require steps to change economic incentives. True-cost pricing of energy, building materials, and automobile use might be one such step. Differential fees to penalize sprawl development or development that does not provide public spaces or amenities might be another. Zoning codes and subdivision regulations may need to be changed to require ecologically appropriate landscape design, pedestrian-friendly street design, and the provision of well-designed public spaces within new neighborhoods. Slowly, step by step, such changes can alter the reality of what is economically or pragmatically feasible. (See Figure 5.2.)

Figure 5.2 A city's "living room." On the site of a former three-car parking garage in Portland, Oregon, urban designers created a multi-purpose space (Pioneer Courthouse Square) that has been called the city's "living room."

Housing

The question of how growing populations should be housed has been present for millennia, and has led to a great many different innovations. In their time, the tenement house, the apartment tower, and the garden suburb were all innovative steps to improve housing quality. However, all have had their problems. Much of our population is still housed badly – in structures with little sense of character or quality, in units that are too expensive relative to income, or in neighborhoods with little sense of community, with few amenities nearby, and with a high degree of dependency on motor vehicles. A sizable number of individuals and families cannot afford housing at all and are homeless. Others make do with crowded or substandard conditions, even in the midst of the most affluent society the world has ever known. Even for those of us with adequate housing, our residential neighborhoods are often bleak, cookie-cutter developments with little sense of joy or individuality.

Decent housing is seen by many as a basic human right. There were efforts to insert this principle into the official statement of the Istanbul City Summit in 1996, strenuously opposed by the United States, which argued that adequate housing should be left to the market rather than guaranteed by the state. However, historically many nations including the US have provided subsidized housing for less affluent citizens. These units are known as "social housing" in much of the world. In Britain the county councils built an enormous amount of social housing during the first half of the twentieth century, and in Sweden, the Netherlands, Canada, and many other countries government played a similar role. Sometimes such housing turned out well; however, at other times such efforts suffered from deficient design or funding. In the US particularly, federal agencies adopted provisions assuring that federally funded housing would be built with standard designs and the cheapest possible materials. Very little attention was usually given to landscaping, which is crucially important to the livability of higher-density developments. The apparent intent was to stigmatize lower-income groups by ensuring that only low-quality housing was available to them.

In affluent modern societies there are certainly resources to provide far better homes for all residents. It is a question of priorities. But even with limited budgets better architectural and urban design can improve the livability and sustainability of housing, in particular design that uses land efficiently, varies the form, character, and color of individual units, employs high-quality non-toxic materials, and provides light, air, private or semi-private outdoor spaces, and opportunities for residents to personalize their surroundings.

Various levels of government can play a role in creating the context for better housing. National or state agencies can supply funding and set general guidelines for structural quality and public access to housing. Local governments can work with builders and communities to ensure that such housing gets sited well and actually built, and can adopt design guidelines as well as building codes that set appropriate standards. Both levels can provide incentives and funding to ensure that private sector builders actually construct housing affordable to less-well-off residents, or can set requirements that builders include this within their market-rate projects. Through all of these steps, communities can move towards housing that better meets human and ecological needs. (See Figures 5.3 and 5.4.)

Figure 5.3 Well designed affordable housing. Affordable housing, such as this development in San Jose, can be made attractive and human-scale.

Figure 5.4 Compactly designed single-family housing. Single-family detached housing is possible within compact neighborhoods such as this one in Mountain View, California.

Transportation

Transportation systems have been a powerful force in determining the form and character of cities since the mid-nineteenth century, when first horsecar and then streetcar lines began the decentralization of urban areas. Some level of transportation infrastructure is certainly necessary. But our urban environments are now overly dominated by motor vehicles – in terms of the land area given over to roads and parking, the design of suburbs and street networks, the conversion of streets from multi-use public spaces to automobile thoroughfares, the generation of noise and pollution, the severe limits we currently have on walking or bicycling, problems with public safety, and the general character of our built environment.

Addressing this imbalance between motor vehicles and other human needs does not mean getting rid of cars and trucks altogether. It means using them less, or at least stopping the continual increase in "vehicle miles traveled" in both absolute and per capita terms. This can be done through action in three main areas:

1 providing good alternative modes of travel, in particular stressing mobility by walking, bicycling, and public transit;
2 changing land use and urban design policies to support these alternative modes and to reduce the number and length of trips that people need to take everyday; and
3 reforming transportation pricing to incorporate the full social and environmental costs of driving into the prices of fuel, road use, parking, motor vehicles, and vehicle registration.

Starting with the first of these areas, alternative modes of travel include not just walking, biking, buses, light rail, subways, and commuter trains, but new transit technologies such as bus rapid transit (high-tech buses that travel on their own lanes and have the ability to pre-empt traffic signals) and on-call vans that can serve residents in our current dispersed suburban communities. Most of these alternatives must be coordinated by regional or subregional agencies and transit companies, while much of the funding is supplied by federal and state government. Non-polluting technologies can be applied in many cases, such as hydrogen fuel-cell engines for buses from which the only emission is water vapor. Bicycle and pedestrian planning is generally undertaken by local government, and these modes represent the cheapest and most environmentally friendly forms of transportation. The end result within more sustainable communities is likely to be a reversal of the usual hierarchy of transportation priorities. Instead of automobiles coming first, then transit, and last of all bicycling and walking, these human-powered forms of transport would be preferred whenever possible, especially for short-range trips, which will become a greater percentage of overall trips as land uses change to bring destinations closer together. (See Figures 5.5 to 5.9.)

Land use changes – the second main requirement for reduced automobile use – must be brought about primarily by local governments, which can ensure that development is relatively compact, mixed-use, connected in terms of its street system, and contiguous to existing urban areas. These characteristics help reduce the number and length of trips that people need to take. Achieving a jobs–housing balance within communities is a related goal. Land use planning strategies to decrease automobile use typically cluster development around transit corridors, transit stations, or neighborhoods in much the same way that communities emerged in the late nineteenth century before the advent of the automobile.[7] Those streetcar suburbs and other neighborhoods are today some of the most attractive and sought-after living environments in North America, and include neighbor-

Figure 5.5 Alternative transportation modes: walking. Like most European cities, Verona has pedestrianized large portions of its downtown.

Figure 5.6 Alternative transportation modes: bicycling. Amsterdam has created many bike paths separated from vehicle traffic.

Figure 5.7 Alternative transportation modes: bus. Turkey has one of the world's best inter-city bus systems, which includes this hub in Ankara.

Figure 5.8 Alternative transportation modes: rail. Unlike most other North American cities, Toronto never tore out its streetcar system, and now has some of the best public transit on the continent.

Figure 5.9 Alternative transportation modes: informal transit. Small, privately operated vans are a primary source of transportation in the developing world.

hoods such as Dupont Circle, Adams Morgan, and Cleveland Park in Washington, DC, Jamaica Plain and Brookline in the Boston Area, North Oakland and Berkeley in the San Francisco Bay Area, and the Annex, the Beaches, and Kensington in Toronto.

Pricing strategies to reduce automobile use include raising the cost of driving (by measures such as gas taxes, registration fees, parking charges, road use tolls, or fees on total mileage driven) and reducing the price of alternative modes of transportation (by free or subsidized public transit, company rebates to employees who do not use parking spaces, and other means).[8] Strategies such as levying high gas taxes face fierce political opposition in the United States, but are the norm in many other countries, where gas typically costs around $3.50–$4 a gallon as opposed to $1.50 to $2 in the US and Australia. Higher parking charges and "eco-pass" programs (under which employee transit use is free or subsidized) are among the most effective tools to reduce automobile use and have been adopted by a number of municipalities.

The fierce reaction of Americans to any increase in the price of gasoline, no matter how minor, shows just how far we need to go to change public attitudes towards transportation. Driving is seen as a right by many people. The personal vehicle is a cultural icon, status symbol, and indicator of personal freedom. Meanwhile, the negative effects of excessive vehicle use are commonly ignored. Yet the days of unfettered adherence to the automobile culture are over, as shown by growing movements for walkable and bikable cities and opposition to road expansion, and much attention is being paid to strategies to reduce driving. Comprehensive "transportation demand management" (TDM) strategies may include not just the pricing, land use, and transportation alternative policies mentioned earlier, but also techniques such as providing better information to drivers about available garage parking (to avoid extensive circling in urban areas looking for a parking space), programs to promote car and vanpooling, and other forms of educational activity. None of these strategies alone can solve the problem of automobile dependency, but together they can help address this archetypal challenge of sustainability planning over the long term. (See Box 5.1.)

Box 5.1 Changing approaches to transportation planning

Traditional approach	Sustainability-oriented approach
Engineering perspective	More holistic perspective
"Traffic oriented"	"People oriented"
Focus on large-scale movements, often ignoring local trips (within zones)	Concern for local movements, small-scale accessibility
Automobile as the priority	Pedestrians, bicycles, and transit as priorities
The street as traffic artery	The street as public space with multiple uses
Economic criteria for decision-making	Environmental and social criteria as well
Increase road capacity to handle projected demand	Transportation demand management (TDM) programs to reduce demand
Consider road-user costs and benefits	Consider other costs and benefits as well
Focus on facilitating traffic flow	Calming/slowing traffic where necessary
Segregate pedestrians and vehicles	Integrate pedestrian and vehicular space where appropriate through careful design (e.g. boulevards, woonerven)

Adapted from Stephen Marshall, "The Challenge of Sustainable Transport", in Layard *et al.*

Environmental protection and restoration

An obvious priority for sustainability planning is to help communities coexist in far better fashion with the natural environment. This task contains several important directions. One is to protect and conserve existing wilderness, species, and ecosystems. Another is to actively restore those environmental components that have already been damaged. The conservation movement dating back to the late nineteenth century has had many successes at the first of these, for example in establishing national parks and wilderness areas, although many other areas have succumbed to development or despoilation. A related success has been in the reduction of pollution and certain toxic wastes. These objectives have been addressed fairly extensively by the environmental movement, especially in its post-1960s incarnation, through drives for legislation and regulation.

Environmental planners have helped create an extensive set of practices to minimize dumping or inadvertent releases of hazardous materials. An elaborate field of environmental assessment has arisen to review industrial processes with an eye on reducing pollution and waste. International standards known as ISO 14001, developed by the Swiss-based International Organization for Standardization and adopted in 1996, have helped standardize environmentally appropriate industrial processes. The US EPA and many other institutions in other countries have had active programs to work with companies around pollution prevention, and some industries have undertaken such programs themselves internally.

Somewhat more problematic are "polluter pays" strategies. These are intended to harness economic incentives to reduce environmentally destructive behavior, but are difficult to enforce if the relevant industries are powerful and politically well-connected. Extensive litigation has been required, for example, to track down polluters responsible for the Superfund sites in the US, and still in many cases federal or state governments have had to pay for clean-up.

The as-of-yet only partially glimpsed direction for environmental planning consists of restoration activities. These include revegetating previously degraded sites or ecosystems within existing cities, cleaning up "brownfield" industrial sites, unearthing culverted creeks and restoring wetlands, replacing asphalt with permeable paving that allows aquifers to recharge from rainfall, and many other related actions. The end goal is to connect every bit of existing communities with the natural landscape in some way, in the process making them far more attractive and livable. But active environmental restoration efforts, under which urban areas might be dramatically greened and large areas of land or ecosystems outside cities might be restored to something approximating an undisturbed state, are still in their infancy.

Figure 5.10 A channelized stream. The engineering approach to development has resulted in destruction of natural landscape elements, as illustrated by this channelized stream in Las Vegas.

Figure 5.11 A restored stream. Sustainable design instead respects and works with the natural landscape, in this case by unearthing a culverted creek in Berkeley and restoring a naturalistic, meandering stream channel.

As with other sustainability issues, many scales of planning can play a role. Global resource planning has had some successes, resulting for example in treaties to protect the Earth's ozone layer and to take initial steps to combat climate change. National environmental planning has resulted in environmental review frameworks and pollution control laws in many countries. Regional agencies have worked to enforce air and water quality. Local governments have at times regulated development to protect fragile local ecosystems. But in most cases vastly more needs to be done. Relatively new topics such as environmental justice have yet to be thoroughly integrated into environmental planning. Restoration activities are still very preliminary, and the ecological worldview overall still has far less influence on official decision-making than dominant economic perspectives. (See Figures 5.10 and 5.11.)

Energy and materials use

In an age of rapidly growing global consumption of energy and nonrenewable materials, a revised approach to resource use is another essential component of sustainable communities. In large part this will mean moving from open-ended resource systems, in which resources are taken from the earth, consumed, and then released as waste, to closed-loop resource cycles, in which "waste" materials are reused or recycled, and overall material consumption is substantially reduced.

The "Three Rs" – reduction, reuse, and recycling – form core strategies for sustainable resource use. "Reduction" is perhaps the most important of these and is brought about at the beginning of the materials use stream by lowering consumption, using less packaging, maintaining and extending the lifetime of existing products, and diminishing the industrial waste associated with making products. "Reuse" and "Recycling" take place after consumer use of resources. The former strategy simply keeps the product in the same loop of use and reuse (such as when glass bottles are collected, washed at a beverage distributor, and reused for drinks, as was done routinely in the US until the spread of plastic and aluminum containers in the 1970s). The latter approach involves a more extensive process of remanufacturing products from waste materials (as when paper is collected and reprocessed into new paper products).

Additional "Rs" include recovery (of energy or raw materials from waste disposal operations) and rethinking (of lifestyles and consumption patterns). Examples of recovery include producing electricity from waste incinerators or collecting methane from landfills. Examples of rethinking include individuals changing their definitions of what they need or what they see as the "good life."

Figure 5.12 describes the relation between different parts of the materials use stream.

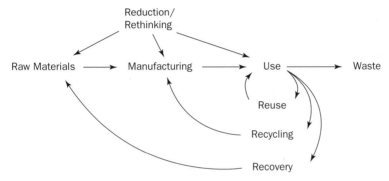

Figure 5.12 The urban waste stream. The traditional urban waste stream proceeds from raw materials through manufacturing to use by consumers and generation of waste. A sustainable resource cycle seeks to reuse, recycle, and recover waste materials, as well as to reduce consumption in the first place.

Green architecture and building

Buildings are one of the principal uses of energy and materials, accounting for some 40 percent of all energy used in the United States and a large proportion of the solid waste going into municipal landfills. They also profoundly affect our relation to the natural world, either insulating us from it in highly artificial environments, or providing us with a more integrative experience that takes into account natural light, ventilation, climate, and landscape as well as local culture and history. As pioneering theorists have been arguing for several decades now,[9] current architectural and building practices will need to be substantially rethought within a more sustainable society. Some architects have in fact been developing green design principles and practices for more than 40 years now, though these practices are not nearly in the mainstream, and state and local building codes have become substantially greener, for example in many places requiring improved energy efficiency as well as water-saving devices. But such efforts have only scratched the surface.

The architecture and construction of buildings may not seem like an appropriate topic for planners, who are usually concerned with the larger-scale functioning of neighborhoods, cities, and regions. However, buildings are of course a central feature of the urban

Figure 5.13 Ecological homes. Commissioned by the Dutch Environment Agency, the community of Ecolonia uses passive solar architecture in many of its 101 homes, as well as solar hot water heating and thick, heat-retaining walls. These and other strategies reduce gas consumption by 40 percent and water consumption by 20 percent.

environment, and the landscaping around them equally so. Collaboration between planners, architects, and landscape architects will be essential in further developing the field of ecological design, in particular to structure a larger context of regulations, guidelines, and incentives that can help greener buildings and construction practices emerge. Such topics are discussed more specifically in Chapter 13. (See Figure 5.13.)

Equity and environmental justice

One of the most disturbing trends around us is the widening of gaps between rich and poor, both globally and within individual nations, and the existence of widespread poverty and discrimination in most corners of the globe. These inequities have all sorts of secondary effects, ranging from the deforestation of countrysides in the developing world by poverty-striken families seeking fuel, to the inflated costs of housing in affluent jurisdictions where a monied professional class has chosen to live. Inequities in access to health care, housing, education, employment, and clean environments are also problems in most areas.

Within Western nations some inequities have been addressed through civil rights legislation and various government regulations against discrimination. But economic inequalities, which are among the most fundamental, have received relatively little attention. Indeed, since the neoconservative resurgence of the 1980s with the election of Ronald Reagan in the United States, Margaret Thatcher in Britain, and Helmut Kohl in Germany, many nations have tolerated much greater disparities in wealth. Progressive tax structures have been weakened and government safety nets for the least well off shredded, while handouts to upper classes through tax breaks, industrial subsidies, and give-aways of public resources have increased. As Kevin Phillips has shown, such policies have been typical throughout the history of the United States, but are now leading to an unprecedented concentration of wealth and power and serious questions about the future of American democracy.[10]

Since wealthy groups hold a great deal of political power within any particular country and control much of the media, addressing these fundamental issues of economic inquality is not easy. But there are opportunities for planners at many scales to increase equity through strategies such as increasing affordable housing within communities, promoting small, locally owned businesses, and ensuring that underrepresented constituencies are heard within local government. State and regional-scale planning offers other opportunities to equalize tax resources, educational opportunities, and school funding between communities. National planning provides an opportunity to set broad policy on questions of discrimination and access to resources, as well as to establish progressive tax structures and balanced economic development policies that will benefit lower-income groups rather than wealthy corporations and their stockholders.

Environmental justice, as previously discussed, can also be addressed by planners in many ways. Local land use and facility siting decisions are key, to protect disadvantaged communities from toxic threats. Beyond this is the more general goal of moving society away from reliance on industries that produce large amounts of pollution and toxic chemicals, which inevitably will harm some communities more than others. Larger-scale infrastructure planning is also important, for example through providing lower income communities with decent public transportation to jobs. Other steps such as improving public access to decision-making processes, the transparency of local government, and representation of various communites on boards and commissions can also be pursued.

Economic development

We live in a society in which economic rules and values largely structure the context around us. Rethinking those economic constructs will be central to the process of sustainable development. In particular, ideals such as economic growth as the main determinant of "progress" at either local or national levels need to be reconsidered. As writers such as Daly and John Cobb have pointed out,[11] the sheer volume of production of goods and services has relatively little to do with human happiness or quality of life, as well as overall social and ecological well-being. Yet growth in quantitative output remains the bottom line of economic progress. A number of groups have already proposed different measures of success. Daly and Cobb have developed an Index of Sustainable Economic Welfare that would take into account the costs of pollution, commuting, resource depletion, and environmental damage, while counting as positives public spending on health and education. The Redefining Progress organization has proposed a Genuine Progress Indicator that includes the economic contributions of household and volunteer work, but subtracts factors such as crime, pollution, and family breakdown.[12] Marilyn Waring has also proposed that national accounting systems be rewritten to include the value of unpaid work that women do, such as child care, elder care, and the running of households.[13]

Writers such as Hawken, Lovins, E.F. Schumacher, and Hazel Henderson have argued for revisions to current economic systems so as to better reflect values underlying sustainable development. Hawken's resorative economics calls for revised incentives to put the energies of entrepreneurship and economic creativity to work restoring environments rather than degrading them.[14] Lovins has argued that efficient use of energy and other resources is really in business' best interest in both the short and the long term.[15] In his influential books several decades ago, Schumacher called for "appropriate technology," especially within Third World development, and for small-scale, locally based solutions to development problems.[16] Henderson has argued that our current economies must eventually move to a renewable resource base,[17] and many other thinkers have advanced similar arguments as well.[18]

One basic alternate economic development direction is to promote a more locally-oriented economy. The idea is generally not to have a totally self-sufficient community – this is virtually impossible given the present state of global interconnection, and not particularly desirable either, since we all gain much from goods and services created elsewhere – but to have those products produced locally whenever this can reasonably be done. Advantages include businesses that are more closely linked to local resources and cultures, and that if owned locally keep their profits in the community. Small-scale businesses exert less of a homogenizing influence on local societies, and produce less of an overall concentration of economic power and wealth. The goal would be to also have them supply stable, decent-paying, long-term jobs for the community, which global corporations, with their highly routinized job specializations and constant threats of layoffs or plant relocations, often fail to do.

Various alternatives to the private corporation are also widely seen as having promise for ecomomic development within a more sustainable society. Cooperative businesses, which are jointly directed either by producers or consumers, represent one alternative. There has been a rich tradition of cooperative economic activity within most industrialized countries. In the United States, household products such as Sun-Kist oranges and Land-o-Lakes butter are marketed by producer coops, while many local cooperative grocery stores function as consumer coops. The general rationale behind coops is that they spread profits and responsibility among community members, promoting a more equitable and responsive society.[19]

Revised forms of corporate governance can also help make economic activity more socially and environmentally responsive. In the United States, the few small corporations at the time of the nation's founding received relatively limited licenses from states to perform a certain task such as operating a toll road or a canal for a limited period of time. Gradually, licensing requirements have been loosened to the point of near-meaninglessness, often by corporations playing one state off against the other to locate their mailing add-resses in those, such as Delaware, that allow the weakest charters. Another problem has been that the US Supreme Court granted corporations the same rights as persons in its 1889 Santa Clara decision, with few of the responsibilities. Addressing this imbalance through revised chartering, requirements for disclosure of community impacts, or other means is likely to be an eventual part of developing a more sustainable economy.

Different levels of government can take different sorts of steps to redefine the economic context in ways that promote sustainability. Such initiatives will be explored further in subsequent chapters. Given the entrenched structural forces promoting unsustainable economic activity, the process of moving to a more sustainable economy might be equated to turning around the Titanic. Such a change in course is likely to take a long time. But bit by bit it can be done, and as Meadows and others have suggested the sooner the changes can begin the greater the likelihood that crises can be averted.

Population

One of the thorniest challenges of all for sustainable development is population growth. Few if any elected officials or citizens want to touch this issue, in part because they are often then accused of being elitist or anti-immigrant since much of the most rapid population growth is occurring in the developing world or among immigrant groups. Within local government, "growth" is typically only addressed in terms of controls on future housing development, which therefore limit the number of future residents. These strategies often tend to be exclusionary, aimed at keeping lower-income residents or minorities out of communities. If similar controls are not placed on economic development, they also lead to an imbalance of jobs and housing, and higher housing prices.

Yet the sheer growth in population globally and within most countries is an issue with enormous implications for resource use, land use, local environmental protection, equity, and quality of life. No matter how well we plan for growth, such planning will inevitably be imperfect and population expansion will have unavoidable negative impacts on the sustainability of human communities. (A few countries, mainly in Europe, have stable or declining native populations, but this condition is often offset by the arrival of immigrant minorities.)

Seriously addressing population issues will mean focusing on the hot-button topics of family planning, immigration, poverty, and the status of women. Some of these issues can be dealt with at many levels of government through public health services, educational campaigns, and programs to improve educational and economic opportunities available to women, a topic which was a special focus of the World Summit on Population held in Cairo in 1995. Improving such opportunities has been shown in many countries to be one of the most effective ways to reduce birth rates. Needless to say, improving opportunities available to the female half of the population is an important equity issue as well.

The subject of immigration on the other hand has few clear policy handles. It is dealt with primarily at the national level through quotas and border controls. The basic intent of these policies is to limit population pressure on a given country and areas within it. Yet these measures are only partially effective. Enormous flows of refugees and illegal immigrants move in many parts of the world regardless of efforts to stop them. In the process, large amounts of human suffering are caused.

Addressing immigration will ultimately mean tackling the economic, political, and social causes that produce so much of it. Poverty, lack of economic opportunity, violence, oppression, and local overpopulation are at the root of many people's desires to leave their native localities or countries to seek new lives elsewhere. The lure of Western material culture and lifestyles, as broadcast around the world by television and advertising, is certainly a factor as well. Reducing these factors fueling immigration will mean fundamentally changing the global economic and political environment. As authors such as Samir Amin and Edward Goldsmith have argued for decades, it is likely to require an end to the current situation in which multinational corporations undercut local economies in developing world countries and drain profits back to the First World. As critics such as Noam Chomsky have also argued, it is likely to require an end to First World support of violent or oppressive regimes in developing nations. A complete analysis of steps that might be taken to improve conditions in the developing world, thus reducing population growth and immigration pressure on First World nations, is far too lengthy to be undertaken here. But these measures are no secret, and are being developed as a coherent political agenda by the anti-globalization movement.[20] The fact that they are not being consistently pursued to date, especially by nations such as the United States, is a testimony to the power of global capital, economic and political elites, corporations, and the mindsets that they have helped construct within industrialized nations.

* * *

These issues form much of the agenda of sustainability planning in the decades ahead. Good work is already being done in most areas, often by individuals and organizations on the fringes of political and economic power, although not infrequently by governments and corporations as well. Whether more substantial changes can be brought about will depend in large part on whether mainstream political parties, civic leaders, citizens, and business organizations can take more substantial action, and whether the enormous political and economic power of unsustainable development interests can be effectively reduced or countered. In any case, there are many roles for planners and other professionals at different institutional levels to address these central concerns. The following chapter will consider various tools and strategies that might be applied in different contexts.

6 Tools for sustainability planning

Given that sustainability planning aims to bring about major changes in a range of areas, how can this be done? What methods might be particularly useful to planners, political leaders, or activists? Possible options include many traditional urban planning strategies – if handled in creative and proactive ways – as well as some new tools that have been developed particularly with sustainable development in mind, such as sustainability indicators and ecological footprint analysis. Educational and consensus-building processes, as well as old-fashioned political organizing and coalition-building, are also important.

The "toolbox" available to planners is increasingly large and varied. New mechanisms are invented almost daily. Yet each has strengths and weaknesses. Rather than becoming too entranced with particular strategies for their own sake, what is most important is to pick and choose appropriate methods for each situation, and to innovate or combine methods when appropriate.

Planning processes

Plan-making

The most traditional planning tool, by definition, is the simple construction of a plan, that is, a well-supported collection of strategies for achieving desired results in the future. Plans can be constructed at any level of government, though at local scales they tend to include more details of land use, transportation, economic development strategy, and the like, and at larger scales they tend to take the form of broader lists of goals, shared principles, and commitments to action. Much literature exists on the history and nature of planning. There are various models of this process, which usually include research and information-gathering tasks, analysis, generation of alternative future scenarios, and selection of a preferred scenario. Recently, public participation has become a priority as well. Traditional models include the famous "survey–analysis–plan" dictum of Patrick Geddes early in the twentieth century, and the later comprehensive rational planning model which is often seen to include four steps: (1) defining goals, (2) identifying obstacles to these, (3) identifying alternative solutions, and (4) comparing the merits of these.[1]

Specific types of plans particularly useful for sustainable development can include national "Green Plans" setting out environmental and development policy, regional plans to manage urban growth and/or develop transportation systems, municipal plans to regulate development of a particular area or the city as a whole, and neighborhood plans to establish specific zoning and design standards for a neighborhood. Development approval processes, through which governments approve specific projects supposedly in accordance with these plans, and establishment of particular government programs are ways to implement the principles established by planning documents. Plan-making plus systematic implementation might therefore be seen as the essence of the urban planning field.

At the local level General Plans (also called Master Plans or Comprehensive Plans) establish an overall vision and policy framework for a municipality. US states typically require each city or county to have such a plan and to update it regularly, and often specify key elements that must be covered, such as land use, circulation (transportation), housing, conservation, open space, noise, and safety. Cities may choose to add other elements as well, for example, to cover urban design, economic development, environmental protection, and resource use. Though most of the plan consists of written goals, policies, and analysis of existing conditions, General Plans usually contain a land use map for the city specifying areas for different land uses and densities. Such a map is in essence a physical development plan for the city and forms the basis for more detailed zoning regulations.

There has been a continuing debate in academic planning circles about the usefulness of General Plans. Pushed strongly in the mid-twentieth century by figures such as Jack Kent at the University of California at Berkeley and Harry Chapin at the University of North Carolina as a vehicle to generate a consensus blueprint for a city, the General Plan has been attacked by others as a formalistic exercise producing vague language with little effect on actual urban development. Critics point out that cities routinely amend their General Plans to accommodate developers, and often fail to consult these documents when developing new programs. Originally oriented around specific visions of physical form, General Plans also frequently became more abstract and technocratic, with less power to galvanize public opinion or direct spatial development. However, in recent years such plans have begun to emphasize physical planning and urban design again, a hopeful sign for sustainability planning that depends heavily on a focus on physical place.

General Plans vary widely in quality and effectiveness. Some consist of boilerplate rhetoric with lofty goals but little linkage to specific policies and programs. Such plans gather dust on shelves and are rarely referenced within planning debates. Others do contain good specifics but are simply not followed in practice. State legislation requiring consistency between plans and development practices can help in this regard. But potentially General Plans can serve a powerful role in developing consensus around sustainable development directions, setting forth specific policies, and holding politicians accountable for achieving agreed goals.[2] As Michael Neuman (1998) points out in a classic article on the usefulness of the General Plan, strong physical plans can help to portray collective hopes about the future of a city or town, allow necessary political conflict to emerge, build social, intellectual, and political capital within communities, and set agendas for powerful public agencies.[3]

Specific Plans (also called Area Plans, Neighborhood Plans, Sector Plans, or Precise Plans) develop a planning vision for a particular area within a city, such as for a downtown, a transit station area, an older industrial district, or a neighborhood. Cities often develop these more focused plans with an intensive public process including workshops, meetings, and design charettes (workshops in which groups of participants develop potential designs for urban places). Specific Plans may include a detailed land use vision at a parcel-by-parcel scale, particular economic development strategies, and recommended zoning changes and urban design guidelines to help bring about desired forms of development. Within a sustainability planning framework Specific Plans provide a crucial mechanism through which cities or other public agencies can involve the public in developing a vision of how a particular place should develop. Although Specific Plans may cover broad areas within the city, increasingly cities are using much more narrowly focused

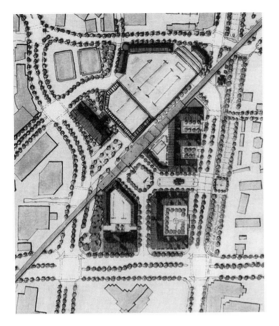

Figure 6.1 Plan for infill development at a transit station.

Figure 6.2 Perspecive view of transit station development. As Figures 6.1 and 6.2 shows this plan for the Pleasant Hill BART station in the San Francisco Bay Area, created in a public charette, envisions the creation of a dense, walkable village center. A series of area plans over 20 years has helped infill development occur around this station. Credit: Contra Costa County Redevelopment Agency and LCA Town Planning & Architecture, LLC.

versions to help guide revitalization and development of smaller areas, such as around transit stations or along key commercial corridors which can be retrofitted as more pedestrian-friendly, mixed-use areas. (See Figures 6.1 and 6.2.)

Functional Plans develop a planning vision for a particular issue or topic area within a city, offering an opportunity for the city to explore sustainability planning directions in depth. Some of these policy documents may eventually be incorporated into General Plans, or may amplify elements of existing General Plans. Functional Plans may be drawn up in a wide range of issue areas such as bicycle planning, transportation demand management, energy policy, urban design, recycling, economic development, or parks and greenspaces. Sometimes state governments mandate that local jurisdictions develop particular types of Functional Plans. The State of California, for example, requires cities to update housing elements of General Plans more frequently than other elements, and so these documents can be considered free-standing Functional Plans. States such as Oregon and Washington have required cities to develop growth management plans which establish UGBs; these may amend, replace, or augment portions of existing General Plans.

Implementation

Far too often excellent plans are prepared but sit for years without being implemented or acknowledged within decision-making processes. Plan preparation is only part of the battle; the most important work is often to get the goals envisioned in the plan brought into reality.

Whether or not a plan is implemented depends on many things: whether funding is available, whether staff time exists to work on the desired changes, whether different departments within city government or other agencies will cooperate with each other to get things done, whether elected leaders will pay attention or follow through, whether public interest can be sustained and opposition defused, and whether planners can keep the goals and commitments represented by the plan in front of everyone's eyes. Many steps taken during the plan-making process itself can help ensure that plans achieve their intended results. A plan that takes existing resources into account, that builds consensus between constituencies, that gets decision-makers and the public excited about future possibilities, and that includes specific budgeting or legally binding changes in city zoning code or laws stands a much better chance of being successful in the long run. Active efforts in such areas by planners, decision-makers, and members of the public will be necessary to ensure that sustainability oriented plans achieve their desired result.

Visioning

Even before the creation of planning documents, the development of long-term, wide-ranging, and creative visions of alternative futures is important to help decision-makers and the general public understand that there are choices in how we develop our communities and society. Going beyond the general goals typically set forth in plans, such visioning can take place through the production of vision statements or reports, through graphic images that illustrate alternative futures, through the manifestos of particular groups, through films, through design charettes and public workshops, or through various other means.

Many of the most influential planning movements historically have been stimulated by particular visions of ideal urban environments. Howard's *Garden Cities of To-morrow*

(1902 [1898]) presented a very carefully worked-out vision of alternative physical and social designs, together with diagrams that continue to be widely reproduced 100 years later. This vision influenced a whole Garden Cities movement in Britain, the Greenbelt New Towns planning in the US, and many individual planners and politicians. The modernist vision of architecture and community form, codified by the Charter of Athens in 1933, likewise developed an enormously influential set of principles and graphics promoting particular urban designs and policies.

In recent decades more environmental and humanistic visions have helped lay the groundwork for sustainable urban development. Ian McHarg's *Design with Nature* (1969) set out a vision of a more ecologically oriented landscape design process, Jane Jacob's *Death and Life of Great American Cities* (1961) reasserted a vision of traditional urban neighborhoods, and Dolores Hayden's *Redesigning the American Dream* (1984) developed a feminist vision of urban environments. The nonprofit New York Regional Plan Association has developed three visions (1927, 1965, and 1996) of development in the New York metropolitan area,[4] while Urban Ecology prepared a *Blueprint for a Sustainable Bay Area* in 1996 for the California region centered on San Francisco, Oakland, and San Jose. In the late 1990s Herbert Girardet and others developed a vision for a sustainable London. At the most visionary end, ecocity theorist Richard Register (1987) developed a 100- to 200-year vision for the future reshaping of Berkeley, California in his book *Ecocity Berkeley*, proposing that development be removed from many ecologically sensitive areas of the city through a program of ecocity zoning and placed instead in dense, walkable urban nodes.[5]

Vision documents can inspire others, inject new ideas into political discussions, educate the public, and spur planning reform. They can also help develop consensus on shared planning goals and how to implement them. Municipal General Plans usually start with a vision statement for a particular city or town, although this is typically quite general and often does not mention objectives such as ecological sustainability or social equity. Much stronger, more visual, and more specific visions will probably be required to inspire people with new ideas and motivate them to work for change. Of course, there is always the challenge of relating visions to reality. They must contain elements that respond to current problems and are broadly appealing, otherwise they may be ignored or dismissed as irrelevant. They must make clear the linkages between visionary goals and practical implementation, so that readers or viewers can see how such dreams might actually come about in practice. They should be written or presented in a way that is accessible to a broad audience and that does not turn people off with arcane language or bizarre concepts. It also helps if vision statements are timed to be maximally useful within particular decision-making processes or to achieve widespread distribution through the media.

Though they rarely consider themselves visionaries, planners can play an important role in making sure that such ideas are considered within debates. One prime place to do this is within Environmental Impact Statements or Reports (EIRs), dry documents to be sure, but processes which by law in the US must present alternatives to a proposed project. These alternatives are then reviewed and contrasted with the proposed project in terms of environmental impacts. EIR alternatives are often weak or trumped-up straw men, leading to much litigation. But strong, meaningful alternatives can help show decision-makers that different strategies exist for any particular project, and are a pragmatic way to insert alternative scenarios into the planning process.

Review of proposals and alternatives

Once courses of action are identified within planning processes, it is vital to review them and compare them with other alternatives. Environmental review is one way to do this. In some countries this process is known as Strategic Environmental Appraisal; in the United States it usually takes the form of Environmental Impact Statements (required by federal law), Environmental Impact Reports (required by some state environmental policy acts), or Environmental Assessments (a lower-intensity form of analysis that can be required by either). The heart of such documents is generally the comparison of a project's impacts with various alternatives, including a "no project" alternative. A sizable consulting industry has sprung up to assist government agencies and private developers or corporations in preparing environmental reports. State or federal legislation requires that these documents take a broad definition of environmental impact, including many social impacts. In addition to analysis of land use, air quality, water quality, and hazardous materials they may contain sections on traffic impacts, archeological resources, noise, cultural resources, and employment.

There have been efforts to promote the use of social impact statements as well. The World Bank, for example, uses a Poverty and Social Impact Analysis process to appraise projects, including distributional impacts and 'before and after' estimates of poverty in the relevant geographic area. Though the process is difficult, the International Association for Impact Assessment is attempting to develop guidelines on Social Impact Assessment that establish standards and models for this sort of analysis.[6]

Cost–benefit analysis is a traditional way of assessing the impacts of any course of action in economic terms. This approach is widely used in corporate and governmental sectors. However, cost–benefit approaches typically minimize or leave out environmental and social impacts; these must be quantified in monetary terms to be meaningful to such models. Environmental economists have worked mightily to incorporate these essentially noneconomic dimensions into cost–benefit frameworks. But the process is unlikely to prove satisfactory in the long run.

Perhaps the most promising strategy is to develop a comprehensive Sustainability Appraisal of any proposed action that takes social, environmental, and economic impacts all into account. The British Department of the Environment, Transport, and the Regions has taken some steps in this regard.[7] Methodology for this is in its infancy, and such appraisal will place substantial demands on staff and decision-makers, requiring them to be able to forsee and understand the full range of impacts of a project. Among other things, such ability may require a broad and sophisticated education, and substantial experience.

Best practices

One way to make visions real for people – and to prove that they are feasible and can actually be carried out in practice – is to publicize existing "best practices" within particular areas of planning. Consequently there have been concerted efforts internationally to develop databases, websites, guidebooks, and exhibits containing examples of best practices of urban and suburban development, often mixed with sample guidelines or linked to annual awards programs.[8]

The United Nations Center for Human Settlements and other organizations first coordinated an international exhibit of Best Practices at the 1996 Istanbul conference. Displays from scores of countries showed a wide range of design and policy ideas connected with urban sustainability. At the visionary end, Australian planners showcased

Adelaide's "Halifax Project," in which local nonprofit groups were spearheading redevelopment of an inner-city industrial block as a mixed-use, alternative energy-powered neighborhood. On a more pragmatic note, China exhibited models of very large social housing projects designed to provide livable, high-density residential environments. Other interesting projects included Vienna's plan for a regional greenbelt, Chattanooga's ecological clean-up and riverfront restoration programs, and initiatives for grassroots democracy in the city of Belo Horizonte, Brazil. Similar international exhibits have been staged at subsequent conferences, including Johannesburg in 2002. The government of Dubai funded a prestigious international award program beginning in 1995, entitled the Dubai International Award for Best Practices in Improving the Living Environment. Awards are presented every two years and a database has been established with more than 1150 best practices from 125 countries.[9]

In the United States, the federal Department of Energy's Center of Excellence for Sustainable Development has established an on-line database of local success stories (at www.sustainable.doe.gov), while the EPA has launched an extensive website detailing policies and case studies on the related topic of smart growth (www.epa.gov/livability). These government resources of course are dependent on political backing; one of the first actions of President George W. Bush's Secretary of Housing and Urban Development, Mel Martinez, was to discontinue that agency's Best Practices program. But the value of such examples means that the number and variety of such resources is generally increasing. Many nonprofit organizations such as the Local Government Commission, the Congress for the New Urbanism, the Sustainable Communities Network, and the Resource Renewal Institute have also publicized best practices of community development.

Sustainability indicators

One widespread initiative within sustainability planning has been the development of indicators that track progress towards sustainable development. More than 25 large US cities have developed such frameworks, including Portland, Seattle, San Jose, San Francisco, Santa Monica, Austin, Chattanooga, Jacksonville, Tampa, Indianapolis, Milwaukee, Boston, and Cambridge.[10] The use of indicators within planning is by no means new, but sustainability indicators tend to be more inclusive than many previous sets of performance measures, and focus more directly on showing trends concerning crucial environmental and social problems.[11]

The best-known prototype for sustainability indicators is the Sustainable Seattle process begun in the early 1990s. This grassroots effort initially drew together several hundred leaders from a range of organizations within the community with the intent of developing consensus on 20 key indicators of regional sustainability. The group published an initial list in 1993, which later expanded to 40 indicators in several main categories.[12] Though the Sustainable Seattle Coalition had no institutional authority to plan for the region, these indicators helped influence the City of Seattle's 1996 General Plan and the work of regional agencies, some of which developed similar lists of indicators for their own use. The Sustainable Seattle process also became enormously influential internationally as a model of a community-based indicators effort. Interestingly, after updating and expanding its initial set of indicators in 1995 and 1998 the Sustainable Seattle organization revisited its indicator approach in 2000 and decided to change its strategy. Local officials had complained that indicators such as "Wild Salmon Returning to Spawn" involved factors out of their control, such as logging practices throughout the watershed, dam construction, and global weather patterns. Consequently Sustainable Seattle refocused its work on smaller-scale indicators more closely related to neighborhood health and vitality.[13]

Other influential sets of sustainability-related indicators have been developed by many local governments including Jacksonville, Florida, whose Quality of Life Indicators were pioneered in 1986, Santa Monica, and the region of Hamilton-Wentworth in Ontario.[14] A leading citizen-based indicator set has been developed by more than 2000 residents of Calgary, Alberta, who have produced two "state-of-our-city" reports and agreed on 36 social, environmental, and economic indicators for that community. Over nearly ten years these indicators show some improvement in environmental quality, but much work remaining to be done in areas of equity, sustainable economy, and resource use.[15]

Indicators potentially have great power to demonstrate problems, motivate action, educate the public, and show the positive effect of sustainability policies. They can be helpful in monitoring program effectiveness and in guiding revisions to policy over time. Some indicators such as levels of air or water pollutants have direct public health implications and are tied to state or national policy. For example, the federal Clean Air Act mandates regular measurement of air pollutants such as carbon monoxide, nitrogen oxides, ground-level ozone, and certain particulates. Other indicators are good reflections of the health of particular species or ecosystems and are routinely monitored by environmental agencies under the Endangered Species Act or other legislation.

Yet indicators are a tool that must be used judiciously. If they do not have political commitment and buy-in by the relevant institutions and leaders, their development may prove to be merely a symbolic exercise. Creating a set of indicators for a region can be difficult, expensive, and time-consuming. Some indicators are much easier to quantify and maintain than others. Public agencies already routinely collect data on some, such as air quality, water quality, housing affordability, and public transit usage, but not on others, such as inequities of wealth, income, or tax base within metropolitan areas. Some indicators such as quality of life or sense of community are qualitative in nature and may be difficult to measure (the usual approach is to conduct surveys through which people are asked for their subjective ratings). Some indicators are much better than others at capturing the public imagination and encapsulating the health of complex systems within a single image. For example, Sustainable Seattle's criteria of "Salmon Returning to Spawn" worked particularly well because it used a charismatic species that served as a symbol and actual keystone element of regional ecosystems.

Once indicator frameworks have been set up, they must be maintained over time and updated regularly. If the staff and resources are not available to do this, indicators may quickly fall into disuse and undermine the entire effort. Perhaps most importantly, indicators must be linked to agencies or levels of government with the power to actually address the problems they measure. Preferably these groups should actually be involved in the creation of the indicators, so as to develop institutional commitment to them. If an indicator effort is not linked to implementation mechanisms and political power, it may serve little function other than short-term public education.

One last problem is that citizens may develop unrealistic expectations that sustainability indicators will show great improvement in response to changed policies. For some initiatives, such as air and water quality, such dramatic change may in fact be possible over relatively short periods of time. But for others, especially those involving complex urban social or environmental systems, change may be much slower and may depend on a range of interwoven policies being put in place over time. For example, no specific action is likely to quickly reduce automobile use in metropolitan areas. What is required instead is that a constellation of land use changes, pricing policies, improved transportation alternatives, and other steps such as an improved balance between jobs and housing within each community be put in place over an extended period of time. Gradually then people will drive less. But those who look for immediate reductions in a measure such as "per capita vehicle miles traveled" as a short-term result of smart growth policies may well be disappointed. For these reasons indicator efforts must be approached with caution. (See Box 6.1.)

Box 6.1 The Sustainable Seattle indicators

Indicators of Sustainable Community 1998

SUSTAINABILITY TRENDS

Declining Sustainability Trend

Solid Waste Generated and Recycled
Local Farm Production
Vehicle Miles Traveled and Fuel Consumption
Renewable and Nonrenewable Energy Use
Distribution of Personal Income
Health Care Expenditures
Work Required for Basic Needs
Children Living in Poverty

Improving Sustainability Trend

Air Quality
Water Consumption
Pollution Prevention
Energy Use per Dollar of Income
Employment Concentration
Unemployment
Volunteer Involvement in Schools
Equity in Justice
Voter Participation
Public Participation in the Arts
Gardening Activity

Neutral Sustainability Trend

Wild Salmon
Soil Erosion
Population
Emergency Room Use for Non-ER Purposes
Housing Affordability
Ethnic Diversity of Teachers
Juvenile Crime
Low Birthweight Infants
Asthma Hospitalizations for Children
Library and Community Center Use
Perceived Quality of Life

Insufficient Data

Ecological Health
Pedestrian- and Bicycle-Friendly Streets
Open Space near Urban Villages
Impervious Surfaces
Community Reinvestment
High School Graduation
Adult Literacy
Arts Instruction
Youth Involvement in Community Service
Neighborliness

Standards and benchmarks

A somewhat related set of tools to help bring about sustainable development consists of measures that establish standards for building or planning many elements of cities and towns. Rather than assessing problems or the effects of past policies, these benchmarks can help planners, architects, and many other professions improve their work in the future.

Historically, of course, building codes, road design standards, zoning frameworks, and other professional guidelines have often shaped urban development in directions that are unsustainable. For example, transportation engineering associations and state highway departments (later usually renamed transportation departments) established road design standards that institutionalized overly wide roads within suburban development. Even in residential neighborhoods 40- and 50-foot wide streets became the standard in the US, along with wide turning radii and other features appropriate for relatively high rates of speed.[16] Other agencies such as the federal Home Loan Administration also required particular types of street networks if new developments were to qualify for loans, in particular mandating subdivisions with a high percentage of cul-de-sacs (such neighborhoods were thought to be the modern ideal and likely to hold property values better than communities with old-fashioned gridded streets). The result of such standards was to enforce a particular model of low-density, automobile-oriented suburban development that we now see as profoundly unsustainable.

In many cases these old standards must now be reviewed and changed. Zoning codes constitute one of the most central set of standards for urban planning, and will need special attention. Minimum lot sizes in the US, for example, are commonly set at 5000 to 10,000 square feet by many cities, enforcing relatively low-density patterns of development. These might be reduced substantially, say to 2000 square feet for townhouses, and maximum lot sizes put in place instead to prevent wasteful use of land in urban areas.

In the past decade or two a number of new standards have been developed to promote sustainability. The federal Energy Star guidelines, for example, set standards for energy efficient appliances. Consumers often receive rebates from utility companies for purchasing appliances that receive Energy Star certification. Simply labeling appliances to show their energy efficiency has been an important educational technique. To take another example, the LEED (Leadership in Energy and Environmental Design) standards that were established in the late 1990s represent an extremely useful way to promote green buildings and energy-conscious architecture. Under these, buildings are rated on a variety of criteria and receive basic, "silver," "gold," or "platinum" certification (see Chapter 10 for more information on LEED).[17]

Despite the huge benefits that standards can have in terms of spreading sustainable design practices into the mainstream of urban development and consumer choice, they have disadvantages as well. Standards can often be too rigid, and have difficulty keeping up with changing technology and innovation. Initially they may prevent innovation, until codes are changed to accommodate new techniques (for example, straw bale construction was forbidden by building codes in most communities until these codes were amended in the 1990s). Firmly established standards may reduce creativity, in that design or development becomes a process of meeting established benchmarks rather than "pushing the envelope." Extensive formalized standards can also add cumbersome bureaucracy and paperwork if not developed carefully.

"Performance standards" represent one way to avoid rigid codes. Under these, a building or development must simply meet certain overall criteria, such as keeping energy use below a certain level or maintaining a certain species diversity in an ecosystem. The exact means are up to the developer or policy-maker, thus opening the door to creative new approaches. Within building construction, future performance can often be modeled by computer or analyzed by engineers. Within larger-scale urban development, care must be taken that impacts are measured over time and aspects of the development adjusted to achieve performance goals. A somewhat related mechanism is the performance zoning that Andres Duany, Elizabeth Plater-Zyberk, and other New Urbanists use to specify the character of new communities. These simple, graphic codes illustrate desirable building form, street design, and neighborhood layout. Advocates believe that they provide a simpler, easier-to-understand, and more intuitive approach to urban design than complicated written zoning codes and street standards.[18]

Ecological footprint analysis

One of the most intriguing methods to quantify the environmental impact of human communities is the ecological footprint model. Initially developed by William Rees at the University of British Columbia and Mathis Wackernagel at the Redefining Progress organization in Oakland, this technique seeks to turn various aspects of human resource consumption into equivalent amounts of land that would be required to produce such resources. Each individual or community is therefore assigned a "footprint" in terms of acres or hectares that represents their ecological impact on the planet.[19]

At a simple, intuitive level the ecological footprint model has a great deal of appeal as a way to dramatize the impacts of resource consumption or changes in materials use over time. Ordinary citizens can run on-line versions of footprint models to calculate the impacts of their own lifestyles,[20] and some analysts have attempted to calculate footprints for large urban regions or entire nations. Girardet, for example, calculates the footprint of the greater London area at 19,700,000 hectares (about 48 million acres, or 76,060 square miles), an area almost as large as Britain itself. This city, in other words, would take the equivalent of a land area 125 times its size to meet its resource needs.[21] Other studies have found that the average American requires an ecological footprint of more than 12 hectares, while the average Briton requires 6 hectares and the average Indian just 1 hectare.[22]

However, the usefulness of such statistics is questionable. What does one do with such figures? Moreover, such models involve a huge number of assumptions about how various forms of resource use or pollution translate into land area, and can become tremendously complex spreadsheet exercises. Also, many key elements of urban sustainability, especially involving equity, livability, and social well-being, are virtually impossible to incorporate into such a quantitative model. Ecological footprint analysis therefore seems a limited sustainability planning tool with applications more useful in public education than in specific policy-making.

Other research and analytic tools

Virtually all the research tools developed by the planning profession to date can be used in some way within sustainability planning. Their usefulness depends in large part on what initial research questions and processes are guiding them, and on how time- and resource-efficient they might be. Some tools, such as very complicated computer models of traffic generation, may fall relatively low on the scale of usefulness, or may be misused currently in ways that insert unquestioned assumptions into the decision-making process (such as that smooth-flowing traffic is always good), that obfuscate issues, or that distance the public from decision-making. These methods must be rethought, and perhaps laid aside in favor of others. But such instances are probably relatively rare, and even then the tool itself may not be to blame.

A small number of planning tools are relatively new and have been developed to assist in environmental or sustainability planning. A few of the best-known are listed below.

GIS and mapping

One of the most rapid areas of growth within the urban planning field has come through increased use of "geographic information systems," or GIS. These computer-based applications provide a sophisticated ability to map and analyze many different spatial layers – land use, topology, roads, rail lines, census data, hydrology, soils, slopes, fragile habitats, endangered species, and so on – across urban or rural regions. Such mapping can be used for many types of analysis to support growth management planning, environmental protection efforts, environmental justice analysis, and many other sustainability oriented endeavors.

GIS systems and computer modeling based on them have for example been used to support regional growth management planning in Minneapolis-St Paul, the San Francisco Bay Area, and Portland, Oregon. Planners and citizens have been able to study the implications of different metropolitan growth scenarios in terms of open space consumption, urban densities, population near transit, air quality, traffic generation, and other variables. Extensive GIS systems are also being used by the Nature Conservancy to map ecoregions in California and prioritize lands to be protected from development, and by the Greater Atlanta Regional Agency to analyze urban development, transportation, and air quality in that region. Academic scholars have used GIS to study topics ranging from less-educated job seekers' access to jobs in the Boston Metropolitan Area[23] to the emergence of suburban nodes of relatively dense housing around Seattle.[24]

GIS systems, though a useful tool for sustainability planning, are by no means a panacea. They can provide important information for planners and highly educational material for the public, but it can be time-consuming and expensive to set up such computer databases, and the data must be of high quality and well-maintained to be useful over time. Good design of GIS systems is very important so that they fit well with policy-making, contain the right information, and are understandable by the general public.[25] Some types of problems also do not lend themselves to quantitative spatial analysis, but require on-the-ground observation and work with local communities instead. The location of a particular building or activity, for example, should not just be decided on the basis of abstract analysis about proximity to other facilities and so forth, but on what the proposed site actually looks and feels like. GIS analysis should never take the place of common sense or hard-nosed organizing and action. With such a technology, the temptation is great to study problems at enormous length rather than taking action based on already-extensive data to address them.

Environmental assessment

Environmental assessment, also called Environmental Impact Assessment (EIA), is a science-based discipline that has emerged since the early 1980s in response to the passage of regulation regarding air quality, water quality, toxic chemicals, and other environmental threats. The focus is usually on assessing conditions and risks for particular facilities, industries, or watersheds. This assessment then becomes the basis for documents such as EIRs and for environmental management policies designed to reduce risk. Since its inception the field has evolved from measuring the effects of particular chemical pollutants on a single species to a more broad-based assessment of the impacts of multiple stressors on ecosystems.[26]

Environmental assessment relies on an increasingly standardized set of methods for assessing threats to ecological health. Agencies such as the US EPA's National Center for Environmental Assessment, the Canadian Environmental Assessment Agency, and Britain's Environment Agency have established guidelines for conducting ecological risk assessments.[27] This discipline is particularly important to certain professions such as real estate, which must know the status of certain properties, and to industries that fall under extensive federal or state environmental regulation.

Environmental impact reporting

A formal process of researching environmental and human impacts of new development projects is now legally required by many nations and states under varying titles that might be collectively labeled "Environmental Impact Reporting." The US Environmental Impact Statement (EIS) process, required under the National Environmental Policy Act of 1970 (NEPA), was one of the first such pieces of legislation; it established the principle that the impacts of any federal action "significantly affecting the quality of the human environment" should be studied, and alternatives to this action considered that might have less impact. Most American states subsequently passed their own versions of this requirement in "mini-NEPA" legislative acts. The California Environmental Quality Act, for example, requires Environmental Impact Reports to be prepared on major projects. State courts have extended this requirement to private sector as well as public sector projects having significant environmental impacts. Briefer environmental assessments or "findings of no significant impact" must be prepared for smaller projects.

Other nations have similar requirements. Canada requires environmental assessments on projects that are authorized, funded, or carried out by federal agencies. The European Union has required member states to adopt environmental impact assessment processes in line with the United Nations' Convention on Environmental Impact Assessment, which entered into force in 1997.[28] As mentioned previously, such assessments can both provide information on environmental inputs and help evaluate alternatives to a proposed action.

Development path analysis

One method used, especially in Europe, to assess how various paths of economic development might lead towards sustainable development is development path analysis (DPA). Under this technique activities are categorized into one of six different potential development paths, depending on their impact on the environment. The paths range from business-as-usual to dramatically different types of economic activity. This method has been used particularly by the European Union to allocate Structural Funds within its Building Sustainable Prosperity Program.

Institutions and policy mechanisms

The role of institutions

The institutions of planning – boards, commissions, processes, guidelines, regulations, and the like – are among the most important tools at our disposal. These entities can help shape the physical, social, and economic landscape around us in more sustainable directions, although they can also hinder change and become enormous obstacles if poorly designed or captured politically by particular interests. The process of carefully improving such institutions is essential to sustainable development.

Among the most important institutional tools are agencies of government, which may seem static and immutable but in the longer term can be and have been changed dramatically. Progressive-era innovations in the early twentieth century introduced notions such as open meetings, a professional civil service staffing local governments that was supposed to be immune from politics, and citizen commissions to advise city councils on policy and administer zoning. Regional agencies are another innovation that first arose in many metropolitan areas in the 1960s, but that still need considerable strengthening and refinement in most places. New metropolitan, watershed, or bioregional agencies may be particularly important to sustainable development, since many issues of urban growth and environmental protection are now seen to transcend the boundaries of past local government institutions. Election reforms are a somewhat different type of institutional change that is also essential, in particular to reduce the influence of big-money donors on government. These reforms are needed in order to ensure that equity and environmental interests have the same sway within government as economic constituencies.

Mechanisms through which institutions determine and set forth policy (that is, agreed principles, strategies, and standards) are open to change as well. The notion of an urban General Plan was an innovation in the first half of the twentieth century, as were zoning, street standards, and the like. Some of these mechanisms may now be proving counterproductive (zoning for example is often seen as too rigid and enforcing an inappropriate separation of land uses), and may need to be revised. New mechanisms may also be needed. Urban design codes and guidelines have proliferated in recent years, for example, as planners and citizens have realized that the public sector needs in many cases to more proactively shape the physical form and character of communities.

Consistency provisions

One main problem with current planning, which governments have not yet thoroughly addressed, is the need for consistency between planning tools. If specific area plans and zoning do not follow city master plans or state growth management goals, for example, then these broader planning frameworks become meaningless. Also, if local code enforcement officials or city commissions do not uphold officially approved plans and codes, or grant numerous exceptions to them as frequently happens, then these planning tools become meaningless as well.

Consistency problems have arisen as various planning tools have been put into place in piecemeal fashion over the past century, often with somewhat half-hearted commitment by political leaders to the overall concept of planning. And certainly, the idea of a single, legally binding planning framework between different scales of government can be a frightening one. But on the other hand a lack of consistency between these levels can defeat the whole purpose of planning. So ways must be found to improve consistency and enforceability while also making the framework responsive to changing local conditions. Usually this is accomplished by setting broad goals and policy directions at higher levels of government, with more detailed policy and implementation occurring locally.

A number of US states have begun to adopt consistency requirements mandating that different levels of plans be consistent with one another, and giving the public the legal power to sue to enforce consistency. This provision essentially puts teeth into planning. States such as Florida and Oregon have been in the lead in this regard. City governments can also pass ordinances or adopt charters requiring consistency between different scales of planning. Such provisions are still weak in many locales, however.

Intergovernmental incentives and mandates

One of the most important toolsets related to sustainability planning has to do with "carrots and sticks" by which various levels of government can encourage each other to plan for sustainability.[29] Typically, higher levels of government establish general goals and provide incentives and mandates for lower levels to take action implementing these, but the process can sometimes work the other way as well, for example if particular cities or states adopt stringent environmental requirements that are then copied at the national level. In the current era in which citizens are often skeptical of government, it is unlikely that any particular level of government by itself will be strong enough to adopt and implement comprehensive sustainability policy. Strong regional agencies, for instance, have historically often been seen as the solution to metropolitan problems, but the fact is that with very few exceptions American urban regions have failed for more than 50 years to bring such institutions into being. So what is needed instead is for our existing, imperfect institutions to reinforce one another, working together to establish inter-governmental frameworks for sustainability planning.

States such as Oregon have relied extensively on an interlocking framework of incentives and mandates for decades, with generally positive results. That state first developed a set of statewide land use and environmental planning goals in the late 1960s and early 1970s, and then has systematically worked with regional and local governments to encourage them to meet those goals. Such efforts have included direct grants to local government for planning and implementation that meets state goals, technical assistance through staff and data, and strong mandates and timelines requiring that local governments produce results.[30] The state of Maryland likewise has developed both incentives and mandates to encourage local governments to implement its Smart Growth program. The state will only fund infrastructure within "Priority Funding Areas" designated by local government consistent with state criteria established under the Smart Growth and Neighborhood Conservation Act. These criteria emphasize planned urban growth in areas where infrastructure is already in place and preservation of the state's "rural legacy."

In contrast to the traditional model of top-down planning common to many European countries and some US cities in the past, such a decentralized framework of inter-relationships between levels of governments seems more likely to succeed in the North American context. Moreover, much of the world is following the American model, with regional agencies losing power (those governing metropolitan London, Barcelona, and Copenhagen were dissolved in the 1980s), and an increasing emphasis on participatory, consensus-based, or market-based planning rather than top-down control. So like it or not, the new political model in many places is one of flexible governance in which a variety of public and private sector institutions work together to get things done. It is this environment within which most sustainability planning will take place.

Education, communication, and consensus-building

Among the most important tools for long-term change are strategies of education, communication, and consensus-building. These include the whole range of activities that make up Friedmann's "social learning"[31] – action-oriented practice, education, group decision-making, and cognitive growth. But this set of strategies also includes more recently publicized processes of consensus-building, meeting facilitation, and networking which often are included under the "communicative action" label.

One common denominator behind such strategies is the recognition that for political or social change to occur people's beliefs, knowledge, values, and paradigms of thought must also change. This understanding has given rise to a wide range of educational activities by public agencies and NGOs, including extensive newsletters and publications, websites, planning workshops, design charettes, creation of demonstration facilities, and the like. Another foundation to this approach is the recognition that "social capital," – defined by Putnam as "features of social organization, such as trust, norms, and networks, that can improve the efficiency of society by facilitating coordinated actions"[32] – is necessary for institutions to work well. Building social capital has thus become an important goal of many sustainable development advocates.[33] Many of the same strategies apply to this objective as well. Particularly important are structured community processes through which participants can get to know one another over time, build trust, establish a common base of knowledge and data, and develop networks that can bear fruit in the long run. Examples include urban design charettes and the coordinated resource management programs (CRMPs) that have been undertaken in some western US states to bring all stakeholders in a watershed together to determine how best to protect resources in the long run.

As noted earlier, caution is in order when contemplating extensive consensus-building processes as a tool for sustainable development planning. Such processes are often lengthy, expensive, and time-consuming, and their outcomes depend on the willingness of all parties involved to participate in good faith and forgo unfair power advantages. "Participation" in general has been a mantra within planning circles for several decades now, and certainly is extremely important, but also has problematic aspects. Deferring too much to the desires of local communities may not be wise, if these groups are bound to a narrow viewpoint or limited self-interest (exemplified in many NIMBY battles within local government). One particularly vocal minority may hijack public meetings, or assert an interest that runs counter to the interests of broader constituencies or underrepresented groups. Planners must use discretion in structuring such processes and taking their input into account, and must seek opportunities to educate both themselves and others about the full range of groups and goals involved in any given planning debate.

Organizing and coalition-building

Although sustainable communities advocates can use many new and traditional planning mechanisms to bring about change, and may seek to restructure institutions and build social capital to create the context for sustainability planning, at some point they must also confront realities of political and economic power. They may then need to develop alternative sources of power. Perhaps this can be done by winning over current politicians and corporate leaders, but developing coalitions, advocacy groups, political parties, and public events calling for change may also be necessary.

Since organizations calling for progressive social change are often small and limited in resources, coalition-building has been a necessary skill in many places. Natural alliances are often formed between environmentalists, social justice groups, religious organizations, public health advocates, educators, affordable housing builders, and some labor unions. Themes such as "livable communities" can draw these and other groups together. In the Portland, Oregon region, such organizations banded together to form a Coalition for Livable Communities. Beginning in the 1960s a broader, de facto alliance also came about in that region between environmentalists, family farmers, and progressive developers that resulted in statewide growth management legislation. In the San Francisco Bay Area, some 46 different local organizations banded together to form the Transportation and Land Use Coalition, dedicated to reforming the region's transportation and growth policies. At a national level, the Surface Transportation Policy Project and the National Clean Air Coalition have been examples of coalitions that have succeeded in passing or altering significant pieces of legislation.

Progressive political parties are a somewhat different mechanism for organizing political power. Green parties in many countries have helped push a broad agenda (typically ten main points) that is very closely linked to sustainable development. Because of the nation's strong two-party system, Greens have had a hard time making headway in the United States. But in nations where political institutions allow a greater role for minority parties they have been more successful. The German Greens stunned that nation's political establishment in 1979 by winning seats in the Bundestag with more than 5 percent of the nation's vote, the level under the German system of proportional representation at which parties may be represented in the national government. Ever since then the Greens have exerted a significant influence on that nation's environmental and social policy. Green parties have also been active in France, England, and many other nations. In New Zealand, a Values party was formed in 1972 that achieved more than 5 percent of the national vote in 1975 and pushed for sustainability-oriented policies. This constituency has evolved into that nation's current Green Party.

Coalition-building and political organizing is often less daunting at a neighborhood or local government scale, where just a few individuals can form an organization and a few organizations can represent a political movement. Sustainable community groups have arisen in a great many locales. Key challenges for these organizations are to figure out creative ways of gaining local press attention (a few media stories have a powerful multiplier effect for small political movements), gaining access to politicians (which at a local level can often be had just by asking), and maintaining a consistent presence over time. Attracting involvement by individuals or businesses with substantial resources is also an important task. Finally, a core task of sustainability organizing at any level is to establish a positive and proactive agenda, rather than simply opposing bad projects. A constructive agenda can have greater power to inspire people in the long run, and can accomplish much more than the draining process of fighting rear-guard actions against inappropriate projects initiated by existing forces.

7 International planning

We turn now to look at possible directions for sustainability planning at many different scales of activity. "Planning" is often thought of as an activity that occurs at a local government scale through specific land use and development decisions within cities. But many forms of planning occur at much larger scales as well. International agencies, national governments, and states or provinces all seek to shape development contexts for the future. Coordination of goals starting at the broadest possible scales is essential to the process of sustainable development. Large-scale policies and programs can help smaller scale planning happen more effectively, and can ensure that minimum standards of environmental protection and human well-being are met even if local governments are particularly resistant to providing them.

The challenge, in short, is not just to "think globally and act locally," as activists began to say in the 1960s, but to "think at many scales and act wherever possible." This chapter focuses on the broadest level of planning – the global scale – and following chapters will address increasingly smaller institutional purviews. A central question throughout is how to create an interlocking framework of sustainability planning efforts at all different levels. Since no single institution or level of government has the power to address the multitude of sustainability issues, that is the way a more sustainable society is most likely to come about.

Institutions

Consciousness of global interdependence is one of the hallmarks of recent ecological thought and efforts for sustainable development. But institutionalized actions to co-ordinate development goals internationally are still relatively recent, having taken place mainly since the establishment of the United Nations at a series of events in San Francisco in 1945, and the World Bank and the International Monetary Fund at a conference in Bretton Woods, New Hampshire in 1944. (The League of Nations, founded in 1919 after the First World War, did not last long enough to accomplish much in this regard, and previous efforts by colonizing powers such as Great Britain and Spain to spread their models of public administration around the world did not necessarily take joint needs and goals into account.) More than 50 years after the founding of these institutions, international coordination is still difficult, and the role of the World Bank at least has been extremely problematic in terms of sustainability planning. Moreover, the recent emergence of the World Trade Organization and its related agencies with their economic agendas seems to be in direct conflict with the UN institutions with their social and environmental goals. Yet despite such worrisome developments, the gradual accumulation of global institutions and treaties as well as the rise of global consciousness are hopeful signs that our species can learn to coordinate its actions internationally to meet the challenges it faces on a small planet.

Dominant cultures have always exported their models of how to build cities and towns. The ancient Greeks and Romans laid out gridded cities throughout Europe and the Mediterranean, from Egypt to England, creating a standardized form of urban development that can still be seen today in many places. In similar fashion the Spanish planned cities throughout the New World starting in the 1500s, developing each new town according to principles that King Philip II codified as The Laws of the Indies in 1573. Like the Roman military towns, these laws called for a gridiron street pattern around a central plaza. Lots were reserved for a church, a hospital, and a governor's house. Subsequent urban growth was to be symmetrical and contiguous to this initial grid.[1] More recently, the British in the nineteenth and twentieth centuries imposed their methods of city planning on many nations that they colonized in Africa, Asia, and North America. British administrators simply applied standards of the British Town and Country Planning Act to these foreign locations, even encouraging use of British building materials such as brick in colonial contexts. In the process a great deal of local architectural and cultural tradition was lost.

In a somewhat less direct fashion specific urban design ideas such as that of the "garden suburb" – pioneered in England and the United States in the late nineteenth and early twentieth century – have gained influence on all continents. More recently, American models of automobile-oriented suburbs have been widely emulated, especially since Third World governments and developers frequently hire American or British consultants to design new communities. First World corporations and media spread images of suburban life and related material products worldwide, shaping cultural desires regarding ideal urban environments. International development agencies such as the World Bank have typically promoted western models of development as well, including freeways and other large-scale infrastructure. The result currently is that unsustainable development models are spreading across the Third World in a twenty-first-century form of cultural imperialism.

In contrast to these forces for unsustainable urban development worldwide, in recent decades there have been more conscious and systematic attempts to build global consensus on alternative development directions, expressed primarily through international treaties, UN conferences, some of the more progressive international development agencies, and the work of NGOs. Often downplayed or disregarded within the United States, international agreements are the foundation underlying many sustainability initiatives. The Geneva Conventions on the rights of prisoners of war, first formalized in 1864, the Charter of the United Nations, and the Universal Declaration on Human Rights (adopted in 1945 and 1948 respectively) were among the first of these statements of global values. Although few nations have systematically implemented all their provisions, these declarations nevertheless began the process of laying out the intrinsic rights of human beings and establishing basic norms of national behavior.

International attempts to develop consensus on principles of sustainable development date mainly from the 1970s. The 1972 Stockholm Conference on the Human Environment – the original "Earth Summit" – was a catalytic event helping to develop global understanding of development crises and stimulating the first use of the term "sustainable development." The 1976 Vancouver Conference on Human Settlements ("Habitat 1") represented the first global attempt to consider desirable goals and directions of urban development. This conference approved a declaration that stressed a moral imperative to plan for poor residents of the world's cities, strongly attacked land speculation in urban areas, and called for national steps to tax the "unearned increment" of land value resulting from speculation.[2] Not surprisingly, this call was ignored in capitalist nations in which a large fraction of personal wealth has historically been built on the rapid escalation of land values.

Twenty years later a second major round of United Nations conferences was considerably more successful at focusing international attention on sustainable development directions. These conferences included:

- The 1992 Conference on Environment and Development (the "Earth Summit" held in Rio de Janeiro).
- The 1993 World Conference on Human Rights (held in Vienna).
- The 1994 International Conference on Population and Development (held in Cairo).
- The 1995 World Summit for Social Development (the "Social Summit" held in Copenhagen).
- The 1995 World Conference on Women (held in Beijing).
- The 1996 Conference on Human Settlements (the Habitat II "City Summit" held in Istanbul).

All these events developed declarations on their respective topics, usually lengthy and convoluted documents agreed to through laborious consensus processes.[3] Yet these statements are a first step towards global principles concerning many development issues. These conferences were also notable in that they gave a major boost to the global growth of "civil society" – the nongovernmental sector characterized by nonprofit organizations that try to provide services or shape public policy, generally in humanistic or environmental directions.

The Rio Earth Summit is the best-known of the 1990s UN conferences. Although it did not address urban development per se, this event nevertheless helped initiate much planning for sustainable communities. Chapter 28 of the *Agenda 21* document adopted in Rio requires that

> each local authority should enter into a dialogue with its citizens, local organizations and private enterprises and adopt "a local Agenda 21". Through consultation and consensus-building, local authorities would learn from citizens and from local, civic, community, business and industrial organizations and acquire the information needed for formulating the best strategies. The process of consultation would increase household awareness of sustainable development issues. Local authority programmes, policies, laws and regulations to achieve Agenda 21 objectives would be assessed and modified, based on local programmes adopted. Strategies could also be used in supporting proposals for local, national, regional and international funding.[4]

Following these guidelines, many nations began "Local Agenda 21" (LA21) planning processes. In countries such as Australia, Bolivia, Denmark, Finland, Japan, Netherlands, Norway, the Republic of Korea, Sweden, and the United Kingdom national governments helped coordinate LA21 efforts. In other nations, local governments took initiative by themselves. By 1996 more than 1800 local agencies in 64 countries reported being involved in LA21 activities.[5] By 2002 that number had grown to 6200 LA21 efforts worldwide, although 5000 of these were in Europe.[6] UN agencies, bilateral development agencies such as those run by Canada, the Netherlands, and Denmark, and NGOs have sought to support such actions. Nonprofit groups such as the International Council on Local Environmental Initiatives (ICLEI) have been particularly active in working to see

LA21 documents implemented (indeed, ICLEI originated the LA21 idea the year before the Rio Summit). The follow-up World Summit on Sustainable Development, held in Johannesburg, South Africa, in 2002 also under UN auspices, gave additional impetus to local efforts to implement Agenda 21 and share best practices of local sustainability planning. Although overall results from this conference were highly mixed,[7] it did result among other things in substantial agreement on strategies to tackle water and sanitation issues within the world's cities and towns.

The 1996 "City Summit" in Istanbul represented the first time that the world's nations had gathered to specifically share strategies on urban development and display best practices. The resulting charter avoids certain topics strongly opposed by powerful nations such as the United States, including land use and speculation taxes. But it does endorse a universal goal of "ensuring adequate shelter for all," calls for changing "unsustainable consumption and production patterns, particularly in industrialized countries," and embodies international agreement on "land-use patterns that minimize transport demands, save energy, and protect open and green spaces."[8] Although such principles are widely ignored by national and local governments, the Habitat Declaration represents a first step towards at least setting forth global principles of sustainable urban development that may some day be linked to more effective implementation or incentives.

The establishment of an international network of World Heritage sites and Biosphere Reserve sites is a further way to plan on a global scale. The former list includes more than 754 sites worldwide of great natural or cultural value, with candidate sites nominated by national governments with the understanding that they will be protected under the 1972 Convention for the Protection of World Cultural and Natural Heritage.[9] The Biosphere Reserve site concept was initiated in 1974 by the United Nations Educational, Scientific, and Cultural Organization (UNESCO)'s Man and the Biosphere program to designate places that conserve biological diversity, promote economic development, and maintain cultural values.[10] Although such areas are designated by national governments and do not necessarily now enjoy strong protection, the concept of a consistent framework of sites that are valuable to the entire human species, or to the Earth itself, is potentially a very powerful one.

The European Union and its associated organizations have been a prime forum for developing continent-wide sustainability policies. In 1992 the EU's Fifth Environmental Action Plan, entitled "Towards Sustainability," attempted to address the failure of regulatory approaches to meet environmental standards by adapting the method of the Dutch National Environmental Plan, which combined regulatory, voluntary, and market approaches.[11] The 2001 Sixth Environmental Action Plan focused action on four target areas: climate change, biodiversity, health and the environment, and sustainable resource use and waste management. This plan establishes specific environmental objectives to be achieved by 2010 as well as longer-term goals. For example, it establishes a target of having 12 percent of energy derived from renewables by 2010, as well as a longer-term abolition of subsidies for fossil fuels.[12]

The 1997 European Union Treaty of Amsterdam also calls for sustainable development. This document establishes common policy on a wide range of subjects from immigration to the protection of animals, but within its environmental policy section emphasizes sustainable development. Among other actions, the treaty calls for prices to be amended to take into account social and environmental costs; for all major legislative proposals to assess environmental, economic, and social costs and benefits; and for action on climate change, public health, and other subjects.[13]

Through these and other vehicles, the EU has promoted environmental analysis of agricultural, economic, and regional policies, as well as the use of environmental impact assessments.[14] It also facilitates cross-border planning and the transfer of knowledge concerning environmental best practices between jurisdictions throughout Europe. At times the EU has had the power to change national institutions, as in Britain, which has had to codify certain civil rights, adopt wildlife protection legislation, and improve water quality standards in order to be in compliance with European courts.[15]

The Council of Europe, a separate organization bringing together spatial planning ministers from many countries, developed agreement on Guiding Principles for Sustainable Spatial Development of the European Continent in 2000.[16] This document expressed agreement on topics ranging from an efficient continent-wide transport system to the need to preserve ecological and cultural landscapes. The European Council then adopted a European Union Sustainable Development Strategy at a summit meeting in Gothenburg in June 2001. This policy platform includes many specific measures related to reducing carbon dioxide emissions, pollution, and nonrenewable resource use. Although implementation of such initiatives remains slow, as of 2001 eight EU member nations had adopted carbon taxes (designed to reduce CO_2 emissions) and nine had enacted taxes on waste disposal.[17]

Multinational development organizations – particularly the World Bank and the United Nations Development Program (UNDP) – took up the subject of sustainability very actively in the early 1990s, and altered their urban development policies and programs based on it. The UNDP has adopted "sustainable human development" as a primary goal, and prepares a Human Development Index that is a leading global indicator of progress toward this end. The World Bank appointed Egyptian-American architect Ismael Serageldin as Vice President for Sustainable Development in 1989, and initiated a series of annual conferences on Environmentally Sustainable Development in the early 1990s. After years of criticism from activists, significant efforts were made to improve environmental and social impact evaluation of Bank projects. The extent to which these initiatives affected World Bank operations is debatable, however, and activists continued to call for outright dissolution of the agency, believing that many of its projects undermine the cause of sustainability in developing countries.

Many bilateral development agencies – coordinating aid between individual countries and developing nations – went further. The bilateral development agencies of Canada, Denmark, Sweden, and the Netherlands have been particularly active at working to develop sustainability planning worldwide, funding and providing technical assistance to local programs in areas such as alternative energy, public health, appropriate transportation, and local self-sufficiency.

Along with national, multinational, bilateral, and other official international agencies, much action on the international scale is undertaken by the institutions of civil society. Various environmental organizations, religious groups, trade associations, networks of local officials, and other entities have become highly active on the international scene, both in educational and program-implementation roles, and as legitimate players within decision-making. A number of international networks of cities participating in sustainability planning efforts were established or expanded during the 1990s. These associations include the European Sustainable Cities and Towns Campaign, the Sustainable Communities Network, the International City Management Association, the Sustainable Cities Program (a project of the United Nations Center for Human Settlements (UNCHS) and the United Nations Environment Program (UNEP)), and the Urban Environment Forum (a UNCHS-affiliated network of cities). Participants at the European Conference on Sustainable Cities and towns in Aalborg, Denmark, in 1994 approved a particularly influential document known as the Aalborg Charter. This conference also gave birth to the first organization mentioned above, which as of 2002 had been joined by more than

1000 local authorities representing more than 100 million people in 36 European countries.[18] In addition to local efforts, the Campaign worked with the European Union in the early 2000s to secure commitments to using environmentally related taxes and incentives, stricter targets for greenhouse gas emissions, and planning for more sustainable European transportation systems.

NGOs are playing a major role internationally around specific issues and techniques. For example, ICLEI has conducted extensive environmental impact assessment trainings with staff from 17 African cities in six sub-Saharan countries.[19] The aim has been to increase the capacity of local officials, politicians, and members of the public to recognize and respond to environmental problems. Oxfam has done an enormous amount of work to promote sustainable agricultural practices in many countries. Doctors Without Borders has been active not just in responding to public health emergencies, but in pushing for more sustainable long-term solutions to health care, including improved human rights conditions, better access to medicines, and improved distribution systems for food. Many NGOs have also been active in pushing for stronger international treaties in creating and promoting best practice examples of sustainable community development.

One of the most ambitious efforts to establish international consensus on sustainable development principles is represented by the Earth Charter, a document developed by an international network of NGOs and international agencies starting in the 1990s. This document aims to establish global consensus on fundamental ethical principles underlying sustainable development, stressing global interdependence and acknowledgement of shared responsibility for global well-being.[20] Its exact usefulness in the future will depend on how it is viewed and used by governments, NGOs, and ordinary citizens, but it joins the Global Declaration of Human Rights and other United Nations documents as a pioneering attempt to establish a foundation of common global values on which to base action.

Such networks and programs help counterbalance global expansion by corporations and business-oriented agencies such as the World Trade Organization (WTO). However, their power is not strong in relation to these economic interests and the national governments that frequently support them. The risk is that economic globalization by itself will bring about the opposite of sustainable development – a world in which corporations are relatively free to exploit local people and resources, circumventing local laws and regulations by appeals to international trade agreements.[21] In this world corporations would dominate local, national, and international politics, standardized products would displace those of local cultures, values of materialism and individualism would trump those of cooperation and collective welfare, and public sector action to promote environmental protection, social equity, or local community would become extraordinarily difficult. Such troublesome dynamics might take place under cover of an atmosphere of tolerance and modest reform on the surface, as long as these did not threaten the fundamental processes of the accumulation of wealth and power.

To counter such a prospect, a sustainability planning agenda needs to strengthen international agreement on social and environmental goals, and to back this up with agreements on specific targets and funding for implementation. Those working internationally can also increase the sharing of ideas and best practices concerning sustainability planning. Dedicated efforts by citizens, NGOs, international institutions, the media, and progressive politicians will be necessary to counterbalance the weight of business and international capital and to make sure that sustainability objectives are the focus of planning. This organizing effort began in the 1990s as activists connected with the International Forum on Globalization (based in San Francisco) and *The Ecologist* journal (based in London) helped mobilize opposition to world trade conferences in locations such as Seattle (1998),

Genoa (2000), and Cancun (2003). Public protests at such events, reported by the world's media, helped let vast numbers of people know that opposition to economic globalization does in fact exist.

At the same time a series of alternative events has helped develop an alternative view of a more sustainable, locally based world economic system. These events have included The Other Economic Summit, first organized in the mid-1980s by the New Economics Foundation and held concurrently with the annual Group of 7 summit of major industrialized countries, and the World Social Forum, initiated in the early 2000s in Pôrto Alegre, Brazil, and held concurrently with the World Economic Forum, which annually gathers public and private sector capitalist leaders in Davos, Switzerland. These alternative events bring NGOs, academics, progressive politicians, and others together to share information and develop strategies for more sustainable development internationally.

International planning issues

Sustainability issues at an international scale are numerous and many by now well-known. The Kyoto Accords and other international agreements surrounding global warming are one area in which the world's nations and other institutions can be said to be planning jointly, although without universal assent or a great deal of measurable success so far. These Accords, agreed on in December 1997, form an international treaty calling for developed countries to cut their production of global warming gases to an average of 5 percent below 1990 levels by 2012. So far progress is not encouraging. By 2000 the US was 14.2 percent above 1990 carbon dioxide emissions, and the administration of George W. Bush subsequently renounced any affiliation with the agreement.[22] Germany, on the other hand, has successfully moved towards meeting its Kyoto target, reducing emissions in 2002 to 19 percent less than in 1990, close to its goal of 21 percent.[23] Britain reduced its emissions 12.8 percent between 1990 and 2000, seemingly on course towards a target of 20 percent reductions.[24] The Kyoto agreement failed to set targets for developing nations, and with China and India rapidly industrializing, overall global emissions are likely to continue growing.

Action to address depletion of the Earth's ozone layer has been more successful. The 1987 Montreal Protocol (amended in several subsequent sessions) was eventually ratified by 183 countries. This agreement called for nations to phase out production of ozone-depleting chemicals such as chlorofluorocarbons (CFCs), halons, carbon tetrachloride, and methyl chloroform by the year 2000 (2005 for methyl chloroform). Gradual reductions in methyl bromide were subsequently negotiated as well. Industrialized countries also established a Multilateral Fund, administered by the United Nations Development Program and United Nations Environment Program, to help developing countries meet Montreal Protocol targets. The result has been stabilization of the size of the hole in the layer that appears each fall over Antarctica. Models predict that the ozone layer will gradually recover over the next century.[25]

Other topics have yet to be seriously addressed by international mechanisms, though nonspecific statements of the desire for change in many areas are contained in documents such as Agenda 21. These issues include loss of biodiversity (a Convention on Biological Diversity has been drafted and many programs are underway, but these have as of yet done little to stem the problem[26]), depletion of fossil fuels and other nonrenewable resources, damage to the Earth's oceans, overpopulation, global inequities, and various forms of violence and warfare. Each of these is potentially a crisis of worldwide proportions, but efforts to plan for a more sustainable future in these areas are still in the early stages.

Other topics are less than global in scope, but transnational in the sense that they affect several nations or a part of the world, crossing national boundaries. These "subglobal" issues include acid rain, other forms of air pollution, water use, and ecosystem protection in transnational areas such as the Great Lakes, the Mediterranean, or the Amazon. Some treaties, clean-ups, and other actions have been negotiated in these instances. For example, the US, Canada, several states and provinces, and a variety of business and environmental groups have collaborated on a series of Great Lakes Water Quality Agreements, which, despite slow and uneven implementation, have substantially reduced discharges of many pollutants and toxic chemicals into these shared bodies of water.[27]

As international institutions, trust, traditions of cooperation, and understanding become better developed, more can be done to plan for solutions to global issues. Meanwhile, some existing institutions will need to be rethought. The Bretton Woods institutions (the World Bank, International Monetary Fund, and associated agencies), set up after the Second World War to promote international development and stability, have often promoted forms of development now seen as unsustainable. These institutions have made attempts at reform, but have a great institutional momentum and attachment to mainstream agendas of capitalist economic expansion; much more substantial reforms are likely to be required. New institutions specially set up to promote sustainable development are still relatively weak, and new mechanisms may need to come into existence to deal with issues that are not presently being well addressed, such as declining global biodiversity and rising inequity.

Sustainability issues in industrialized vs. less-developed countries

Cities and towns the world over share many problems because they are being shaped by similar forces, such as the spread of automobiles, technology, and materialistic lifestyles. It makes less and less sense to talk of "developed" and "underdeveloped" countries, First World and Third World, South and North. All of these distinctions, to the extent that they were ever meaningful, have become blurred by recent changes. A growing number of nations occupy a middle ground or illustrate a great range of development within themselves. Countries such as Turkey, Brazil, South Korea, and Mexico have become highly urbanized, technologically sophisticated, and affluent in many ways, even while large segments of their populations remain poor and isolated. In Turkey, for example, cities such as Istanbul and Ankara are relatively modern and well tied in to global finance and business networks. The country's highly efficient intercity bus system is far better developed than similar ground transportation in the United States. Yet many other parts of the country remain without electricity or telephones, and literacy nationwide is only in the 40–50 percent range. Even in its large cities much of Turkey's housing is informal – built illegally by immigrants from the countryside, as occurs thoughout the developing world. So this country like many others illustrates a complex mixture of characteristics that are both "developed" and "less developed."

Conversely, many communities within the "developed" United States suffer from enormous poverty, pollution, inequality, and deteriorated infrastructure. Immigrants and poor residents create a variety of informal housing types in this affluent nation – most dramatically the large "colonias" developments along the Texas/Mexico border – and homelessness is a major problem in most large cities. Much of the nation's development is also clearly unsustainable in terms of its environmental or social impacts. So the "developed" label is problematic also, and does not necessarily mean that nations are consistently or well developed.

Still, many basic challenges vary between industrialized and less-developed countries. Cities in less-developed nations face difficult problems by virtue of their extremely rapid growth, poverty, and level of technical and administrative development. In particular, such countries often feature rapidly growing "megacities" – cities of more than eight million persons that have doubled, tripled, or quadrupled in size within a few decades. Cities of such size never existed until the middle of the twentieth century, and as late as 1975 no city in the world had as many as 20 million residents. However, by 2015 the United Nations Population Fund estimates that at least half a dozen will be reaching this threshold, and some 23 will have exceeded 10 million. (See Box 7.1.)

Box 7.1 The growth of megacities (population in millions)

1975		2000		2015 (expected)	
Tokyo	(19.8)	Tokyo	(26.4)	Tokyo	(26.4)
New York	(15.9)	Mexico City	(18.1)	Mumbai	(26.1)
Shanghai	(11.4)	Mumbai	(18.1)	Lagos	(23.2)
Mexico City	(11.2)	São Paulo	(17.8)	Dhaka	(21.1)
São Paulo	(10)	Shanghai	(17)	São Paulo	(20.4)
		New York	(16.6)	Karachi	(19.2)
		Lagos	(13.4)	Mexico City	(19.2)
		Los Angeles	(13.1)	New York	(17.4)
		Kolkata	(12.9)	Jakarta	(17.3)
		Buenos Aires	(12.6)	Kolkata	(17.3)
		Dhaka	(12.3)	Delhi	(16.8)
		Karachi	(11.8)	Metro Manila	(14.8)
		Delhi	(11.7)	Shanghai	(14.6)
		Jakarta	(11)	Los Angeles	(14.1)
		Osaka	(11)	Buenos Aires	(14.1)
		Metro Manila	(10.9)	Cairo	(13.8)
		Beijing	(10.8)	Istanbul	(12.5)
		Rio de Janeiro	(10.6)	Beijing	(12.3)
		Cairo	(10.6)	Rio de Janeiro	(11.9)
				Osaka	(11.0)
				Tianjin	(10.7)
				Hyderabad	(10.5)
				Bangkok	(10.1)

Source: United Nations Population Fund, *The State of the World Population 2001.* Available at www.unfpa.org/swp/2001/english/tables.html.

These gargantuan urban places, most of them in the Third World, must cope with the planning challenges of providing basic infrastructure and services such as sewage, water, electricity, roads, public transit, public health, and education. They are also frequently faced with massive transportation congestion and growing automobile use within street systems that are often totally unprepared for such needs. Bangkok is a good example; its streets and alleys were developed primarily for foot traffic and do not connect in ways that help diffuse automobile traffic across the city. Basic emissions controls and other pollution regulation are also frequently lacking in Third World cities, as is efficient government and basic information on land tenure (legal ownership of land). Poverty, hunger, and health care are pressing needs.

These challenges – in many cases faced by industrialized nations in the nineteenth century – are often overwhelming in the age of megacities. Ensuring that the population has access to food, water, shelter, and sanitation often takes precedence over questions of how to reduce pollution and restore urban ecosystems. The pressing need for action and limited governmental resources can sometimes lead to highly creative and sweeping innovations, such as Mexico City's system of allowing cars to drive only on odd- or even-numbered days, depending on their license plate number, or the dramatic land use planning and social programs found in Curitiba, Brazil. In this latter city, widely trumpeted as a best practice of urban development, planners unable to afford a subway system organized the city's land development around a highly efficient bus network instead (as well as undertaking many other creative planning endeavors). They provided the private bus companies with dedicated lanes, elevated tube-like loading platforms to speed boarding, and rapid transfer facilities. Meanwhile planners ensured that a large amount of relatively dense development was built along the transit lines, making the city one of the world's premier examples of transit-oriented development.[28]

One common characteristic in Third World cities is the presence of large amounts of informal housing. Individuals or entire communities flood into the city from the countryside and construct housing for themselves with whatever materials are available on land that they do not own (typically public land or large parcels controlled by absentee owners). Such "invasions" are often coordinated by middlemen, who demand a price for their services. The result is that large shanty towns emerge literally overnight. The nature of these towns varies enormously. Some may be constructed of cardboard or plywood, but others may consist of multistory reinforced concrete apartment buildings. Without any regulation or oversight, informal settlements often feature winding, narrow streets in an "organic" urban form reminiscent of medieval cities, as well as very dense and mixed-use development. Over time, these communities become formalized. Buildings and services are improved. Local governments often cannot evict residents, since there is nowhere else for them to live and they may represent a powerful political force. Also, 50 percent or more of a city's residents may be housed this way. Instead, local politicians often extend city services to residents and eventually legalize the settlement. Decades after initial settlement the result may be an urban neighborhood that is highly compact, mixed-use, and tightly knit socially and economically. From a sustainability point of view these neighborhoods may in fact be superior to the sterile, modernist social housing blocks that governments have often built for their citizens, or the lower-density, automobile-dependent suburbs created for the middle and upper classes by private developers. (See Figure 7.1.)

Another distinguishing feature of many Third World cities has been the relative lack of automobiles and the presence of a wide range of creative transportation alternatives (though this situation is changing rapidly). People in such cities depend heavily on foot transportation or bicycles – both modes that planners in North America are desperately trying to reinvigorate. Also, these cities are served by informal systems of vans, jitneys, or taxis, in addition to public transportation. These privately owned services are often highly efficient and may be a better illustration of "free" markets at work than transportation in the United States, where cheap gasoline, limited public transportation options, and a failure to factor the externalities of automobile use into the price of driving have skewed the transportation system heavily towards automobile dependence.

Figure 7.1 Informal settlements. Much housing in the developing world is built in "informal" settlements such as this one outside Istanbul, where residents lack title to the land and basic services are often missing. Sustainability planning in such locations must take into account basic needs for clean water, sanitation, health services, food, and employment.

Much of the suffering of Third World cities has arisen because of the lingering effects of centuries of colonization or other forms of exploitation. Frequently wealthy elites in developing countries have literally stolen the benefits of development for themselves through outright theft as government officials, or have gained them legally through close association with western corporations, banks, and political leaders, who have reinforced each others' interests. Often colonization set up a class of westernized elites, who garnered control of much of a country's land, resources, political machinery, and military, and have remained in power with varying groups of allies ever since. In those few cases in which populist revolutions have been successful, such as in Cuba and Nicaragua, governments have been undermined in almost every way possible by capitalist interests in the United States and elsewhere.

Despite their differences, cities in developed and Third World countries face many similar problems. Planners in most countries must seek to preserve or create compact urban form, avoid sprawl, reduce automobile dependency, lower resource use, improve equity, promote locally-oriented businesses and employment opportunities, and adopt green architecture and landscaping practices. These needs are nearly universal. Even the old, compact cities of Europe such as Amsterdam, Paris, or Florence are now surrounded by sprawling suburbs. The Japanese with their bullet trains face rising automobile use and traffic. Countries that are supposedly committed to equality face huge problems in assimilating minority populations – Algerians in France, Turks in Germany, Indians in Britain, Koreans in Japan, and African-Americans and Latinos in the United States – as well as growing disparities between rich and poor. Equity issues also concern the status of women everywhere. In developing nations, women can play a central role in bringing about many sustainable development practices – from the initiation of locally-oriented micro-enterprise businesses to the use of less polluting stoves – and improving education and employment opportunities for women is seen as a central way to reduce population growth. In developed countries, the degree to which women feel comfortable and safe in public environments is one of the prime indicators of urban quality of life.[29]

Although their problems may be somewhat different, "developed" and "less developed" nations can learn from one another. The United States can certainly learn from the compact urban form, efficient transportation systems, and less materialistic lifestyles of other countries. Meanwhile, many other nations are copying elements of US environmental regulation and planning. A diffusion of sustainability planning ideas between both industrialized and Third World countries – aided by international institutions and NGOs – can be a central part of the process of creating more sustainable communities.

8 National planning

Although their influence may be slowly waning as both global and regional institutions gain in power, nation-states are still the dominant political force in the world, and a constituency whose leadership is essential for sustainability planning. Those of us in the US do not tend to think of our federal government as being engaged in planning activities, especially in the traditional sense of guiding local land development. The idea that the national government had a specific land use policy would be anathema to many Americans, and "planning" is not a word that is widely used on Capitol Hill. However, the federal government's actions do fundamentally shape the physical and economic landscape around us, and have for decades. For example, national spending on highways, federal home loan guarantees, and tax deductions for home mortgage interest have promoted suburbanization,[1] while federal spending on military research and development has promoted economic development in the sunbelt states at the expense of the "rustbelt" states of the upper Midwest and Northeast.[2]

In many other countries national governments play a much more overt and conscious role, establishing policy on land use, spatial planning, housing, public transportation, and other topics that shape the physical environment of cities and towns. Britain, the Netherlands, and Sweden have had active programs to build new towns; South Korea and Britain established large greenbelts around their capital cities; and the national governments of France and Japan made pioneering investments in high-speed rail networks, which many other countries are now emulating. Many nations in the developing world have directed large infrastructure and industrial investments towards certain "growth pole" cities that were expected to leverage broader development in their surrounding regions. Rather than physical planning, the main national leadership of the US government has been in environmental and civil rights legislation which other nations have reproduced to varying extents.

Whatever the emphasis of particular administrations in the past, the challenge ahead seems to be to more explicitly and consistently relate the actions of national governments to sustainability objectives. For example, this might occur through programs implementing global treaties on global warming, environmental protection, and human rights, or through programs of consistent support for state, regional, and local action in cases where these scales are the most appropriate level for planning.

Who plans at a national scale?

The range of institutions that engage in planning at a national scale is substantial. The process of setting the national agenda starts with constitutions, which may have sustainability-oriented principles written into them. National constitutions written in the post-Second World War period often provide a broader range of human and environmental rights than those documents created earlier, since these values had become part of the international political discourse by that time. Constitutions for nations such as Italy, Germany, and Japan were largely imposed by the victorious powers in that war, and may include more progressive principles than found in the US founding document itself. Ironically, with the oldest constitution among the major industrial nations the United States has a more limited and conservative set of guiding principles with which to work, as well as a deeply entrenched national reluctance to amend this document. Nations such as Great Britain that have no national constitution represent another extreme, relying instead on "common law" – a large collection of precedents and court decisions that serves much the same function, but is more open to interpretation and amendment in some ways (for example land policy) while being bound by tradition in other ways (such as by perpetuating the anachronistic institution of the House of Lords).

Constitutions affect urban planning directly in large part because they help spell out the relation between citizens and land. Those frameworks that prioritize individual property rights and treat land as a commodity make the job of sustainability planning far more difficult than those that view land as having social value and intrinsic value. The American Constitution, for example, places great weight on private property and under the Fifth Amendment asserts that "nor shall private property be taken for public use, without just compensation." This clause allows individuals to sue for "takings" by local governments if the development value of their land is diminished without clear justification. In contrast, the Swedish Constitution provides relatively weak protection for property rights and did not explicitly provide for compensation for property owners until 1994, when the country incorporated into Swedish law the European Convention on Human Rights, which requires that the right to compensation for state interference with private property be established.[3]

Legislative branches of government establish more specific policy in many areas, with legislation typically proposed and/or signed by executive branch leaders. Congresses or parliaments also approve budgets that determine the ability of public agencies to implement this policy. Meanwhile, national court systems play an extremely important role in interpreting and enforcing legislative or executive actions. At times, courts also establish policy themselves through their interpretation of constitutional language. This role has been particularly important in establishing civil rights in the US Other equity-related issues, such as a public right to relatively equal educational facilities and fair housing opportunities within particular municipalities, have also been enforced by courts at various levels, though unevenly and without always a high degree of success. In a famous set of Mount Laurel decisions in New Jersey, for example, the state Supreme Court ordered the town of Mount Laurel to "provide realistic opportunities for the construction of affordable housing constituting a fair share of regional need."[4] This community had been sued by the National Association for the Advancement of Colored People for its exclusionary zoning policies, which ensured that only relatively expensive single-family housing could be built within its borders. However, although the Court issued its initial decision in 1975, the municipality continued to resist change, in part by appealing the decision to the US Supreme Court. It took a second, even stronger decision by the state Supreme Court in 1983 and passage of a state Fair Housing Act in 1985 to begin increasing the amount of affordable housing in the jurisdiction, and even then the city's foot-dragging continued.[5]

Executive branch agencies implement and enforce policy, and carry out substantial planning activities in their own right. Housing, urban development, and interior agencies often affect city planning directly; environmental and social welfare agencies are also major players. Staff in such agencies has the ability to shape public debate through regulation, incentives, or events that promote education and dialogue. For example, the US EPA under the Clinton administration played a significant role in supporting nation-wide Smart Growth efforts, joining with other nonprofit and governmental organizations to create a Smart Growth Network and underwriting educational efforts on this subject. For a time the EPA also made Sustainable Development Challenge Grants available to local and regional agencies to support innovative sustainability planning efforts. The EPA, Department of Transportation, and Department of Housing and Urban Development (HUD) also help finance urban infrastructure and make many other forms of grant funding available to local government. The British Department of Transport, Local Government, and the Regions plays an even more active role by issuing planning guidance documents that in effect establish national policy on urban development and planning.

Special national commissions and task forces represent an institutional mechanism to develop debates on planning issues at the national level. The President's Council on Sustainable Development (PCSD) in the US was instrumental in promoting dialogue on sustainability during the mid-1990s, and released a report, *Sustainable America*, in 1996 that attempted to build national consensus on this topic.[6] The Rogers Commission in the United Kingdom was influential through its 1999 report *Towards an Urban Renaissance*.[7] In Canada, parliamentary legislation in 1994 established the National Roundtable on the Environment and the Economy, which has commissioned extensive research, produced numerous reports, and involved many stakeholders in sustainable development-related discussions.[8]

Nongovernmental organizations are an additional type of institution that can very sub-stantially influence national planning efforts. In Britain, the Town and Country Planning Association has been an extremely important force since its founding by Ebenezer Howard in 1899. Leading members such as Frederick Osborne authored a seminal 1944 White Paper on "The Control of Land Use," which led to perhaps the country's most sweeping piece of planning legislation, the 1947 Town and Country Planning Act. Groups such as the Civic Trust (which has the Prince of Wales as its patron), the Royal Town Planning Institute, and Friends of the Earth have also been highly influential. In the United States, environmental legislation and land use policy (at least for park and wilderness areas) have been strongly influenced by NGOs such as the Sierra Club, the National Audubon Society, the National Wildlife Federation, the National Parks and Conservation Association, and the Wilderness Society since the early 1960s, when these groups follow-ing the lead of the Sierra Club under its then-director David Brower initiated large-scale national lobbying efforts. More recently, a broad coalition of organizations called the Surface Transportation Policy Project helped rewrite national transportation policy through the Intermodal Surface Transportation Efficiency Act of 1991, and has helped preserve these gains in subsequent reauthorizations of this bill.

In other nations the Wuppertal Institute (Germany), the Canadian Urban Institute (Canada), and Friends of the Earth (in many countries) have also been influential NGOs in the areas of urban and environmental planning. Quasi-autonomous NGOs (QUANGOs) are a further institutional mechanism set up by governments to meet specific programmatic needs. However, although nongovernmental entities have been successful in promoting action at the national level in some areas, especially environmental policy and certain areas of civil rights, their influence is limited in many other fields. They have as of yet had only modest success in reforming energy and resource policies in many countries, in creating stronger land use planning and growth management incentives, and in changing the basic inclination of many national governments to favor large corporations and wealthy investors in their economic policies.

British Commonwealth countries occasionally use the institution of a royal commission or task force to develop policy or consensus on important topics. For example, the Rogers Commission report has influenced debates on urban revitalization in the UK. This report emphasized the role of good urban design in bringing about the regeneration of cities and towns, and called both for a national urban design framework to disseminate key design principles and for demonstration projects to illustrate the nature of these techniques.[9] The Rogers task force also called for 65 percent of transportation expenditures to be allocated to projects giving priority to walking or public transport. In the area of land use, it called for Urban Priority Areas to be established with streamlined development permitting, expanded compulsory purchase powers by local government, and fiscal incentives for infill development. Such recommendations may be stronger and more unfettered than could be produced by a government body itself, and can expand the boundaries of a debate or focus public attention.

Finally, incentive grants and competitions sponsored by national agencies can be an effective way to stimulate local action. The US EPA's Sustainable Development Challenge Grant program helped stimulate local planning initiatives in the late 1990s. "Best Practices" awards by the US HUD Department also helped publicize and reward successful local initiatives. And in 2000, the Federation of Canadian Municipalities and the consulting firm of CH2M Hill began co-sponsoring an annual Sustainable Community Award, which received submissions from 52 municipalities all across the country, likewise stimulating and rewarding local interest in the subject.[10]

National sustainability frameworks

Stimulated in part by the Brundtland Commission Report, Agenda 21, and other international activities, a number of countries have prepared national sustainability planning frameworks. Most of these efforts are still in the early stages and have yet to systematically affect national policy. Many nations have also focused initially on traditionally "environmental" topics such as resource use and pollution rather than more fundamental questions of urban development, transportation, land use, and equity. But still, substantial progress is being made, and the idea of a broad-based sustainability plan is an important one.

Comprehensive "green plans" have been adopted by nations such as the Netherlands, Sweden, Canada, and New Zealand.[11] The Dutch government first prepared an "Urgent Policy Document" on the Environment in 1972, and has created several iterations of its National Environmental Policy Plan (NEPP) plan beginning in 1989. Intended to be a single, comprehensive, ecosystem-based policy, the NEPP integrates several hundred policy initiatives and requires each of the nation's 12 provinces to do a more detailed plan for its own jurisdiction. The fourth revision of the NEPP was approved in 2001 and adopts a relatively long-term time horizon extending to 2030. This plan focuses on problems of biodiversity, climate change, nonrenewable resource use, threats to human health and safety, and quality of life. Various versions of the NEPP have stressed voluntary agreements or "covenants" between government and business, and seem to have achieved significant results. For example, since such planning first went into effect industry has reduced its waste production by 60 percent, recycling has increased 70 percent, and sulfur dioxide emissions have fallen 70 percent.[12] In addition, Dutch spatial planning has been oriented around profoundly environmental objectives, in particular the country's effort to preserve a "Green Heart" of agricultural land from urban development within a large circle of cities including Amsterdam, Leiden, The Hague, Rotterdam, Dordrecht, and Utrecht. The Green Heart concept has been criticized by some as a misleading metaphor when this central agricultural area is not homogeneous and contains some urban development,[13] but is supported by others who argue that the concept has helped a wide variety of Dutch efforts to plan compact new urban development and to protect the already intensively used countryside.[14]

Swedish land use and transportation planning has historically sought to meet sustainability objectives such as compact urban form and transit-oriented development. Like Copenhagen, the city of Stockholm has been developed since the late 1940s around a finger-like structure of commuter rail lines, initially with a strong focus on suburban new towns such as Vallingby and Farsta, but more recently with an emphasis on channeling urban growth into denser, more mixed-use infill neighborhoods such as Stockholm Sodra. The nation also has a long track record on many other environmental issues. It has decreased water consumption substantially since the 1960s, held energy consumption constant despite a growing economy, cut waste generation by 50 percent in many cities, and expanded use of renewable energy.[15] By 1998 virtually all of the country's 288 municipalities had decided to initiate Local Agenda 21 planning processes, following a period of active encouragement from the national government.[16] Between 1999 and 2001 the Swedish government enacted Environmental Quality Objectives legislation designed to set specific targets and deadlines in fifteen areas of environmental policy. Substantial new funding was allocated as well.

Although like many nations it started adopting environmental legislation in the late 1960s, Germany began explicit sustainability planning efforts in the early 1990s at the time of the Rio Earth Summit. Chancellor Helmut Kohl established a National Commission for Sustainable Development in 1991. Three themes that formed the basis of the nation's sustainability policies were the precautionary principle (that environmental risks and damage should be avoided from the very outset), the polluter pays principle, and the principle of cooperation. The Federal Environmental Agency produced a Sustainable Development progress report in 1997, and a national Green Plan was drafted in the late 1990s that emphasized reducing global warming emissions, ecosystem protection, resource use, human health, and environmentally sound transportation.[17] A set of indicators measuring progress in these areas was also developed. One of the country's main successes has been in the reduction of solid waste, in part achieved through the Packaging Ordinance of 1991 that requires companies generating wastes (including packaging and used products) to take them back for recycling, and to integrate the costs of doing so into purchase prices. Even German automobile manufacturers have become used to recycling used automobiles. To reprocess waste in this way German corporations set up a private recycling system, the Duales System Deutschland, which recycles goods marked by a green dot for which the producer pays a fee. A number of German provinces (lander) also have strong sustainability-oriented programs.

British planning since the early 1990s has explicitly emphasized sustainable development, and Local Agenda 21 programs are underway in most British municipalities (the Labor government proposed a 100 percent LA21 involvement target in 1997). A UK Round Table on Sustainable Development formed after the 1991 Rio Earth Summit moved forward national discussions of this topic, and a 1990 White Paper on the Environment and a 1994 UK Strategy for Sustainable Development helped develop national policy. The UK government has set some ambitious targets for specific changes, such as for an 88 percent reduction in global warming emissions between 1990 and 2050.[18] Unlike some other nations the United Kingdom has emphasized land use and transportation as key elements of sustainability planning, in addition to more traditional environmental topics such as resource use and pollution. In 1994 sustainable development became a featured goal within Planning Policy Guidance (PPG) 13 on planning and transport, while 1995's PPG 6 moved the government away from promoting exurban sprawl development, especially of retail stores.[19] In 1997 the Blair administration called for all local governments to prepare Agenda 21 strategies, and three years later established a Sustainable Development Commission to further national action on this topic.[20]

Britain's Department of Transport, Local Government and the Regions (DTLR) prepared a 2001 *Green Paper* on reforming the nation's planning, beginning with the premise: "We need good planning to deliver development that is sustainable and which creates better places in which people can live and work."[21] While simplifying many aspects of the planning framework – including eliminating most county planning – the *Green Paper* requires local governments to create Community Strategies oriented around the Three Es and a long-term vision:

> Local authorities have a new duty to prepare Community Strategies, which they develop in conjunction with other public, private and community sector organisations. Community Strategies should promote the economic, social and environmental well-being of their areas and contribute to the achievement of sustainable development. They must have four key components:
>
> * a long-term vision for the area which focuses on the outcomes that are to be achieved;
> * an action plan identifying shorter-term priorities and activities that will contribute to the achievement of long-term outcomes;
> * a shared commitment to implement the action plan and proposals for doing so; and
> * arrangements for monitoring the implementation of the action plan, for periodically reviewing the community strategy, and for reporting progress to local communities.[22]

Such local strategies must be in accord with national goals and Regional Planning Guidance developed by the eight regional agencies within the country. They must also set out a formal "statement of community involvement" stating arrangements for involving the public. The *Green Paper* has been criticized for eliminating existing county planning processes, but may have potential to assert a clearer set of national policies oriented around sustainability and to improve public involvement.[23]

In the late 1990s the Blair government also took steps such as establishing a national goal of reaching 60 percent infill development by 2008 in order to reduce suburban sprawl,[24] and adopted transportation planning policy emphasizing transportation–land use coordination and accessibility by public transport, walking, and cycling.[25] A far-reaching 2003 strategy entitled *Sustainable Communities: Building for the Future* then called for increasing the quality and quantity of affordable housing, improving the design of cities and towns, reducing homelessness and social exclusion, protecting green belt areas, and designating four areas within the country for intensive growth (the Thames Gateway, Milton Keynes and the South Midlands, Ashford, and the London–Cambridge corridor).[26] Some £2.5 billion was designated for these purposes.

New Zealand has initiated some of the most comprehensive national environmental planning of any country, beginning with its 1991 Resource Management Act, which set a sweeping agenda for managing the nation's air, water, soils, ecosystems, and natural resources. In contrast to many other nations' approach, no one is allowed to discharge pollutants unless specifically authorized by a permit. A follow-up 1994 Environment 2010 plan called for integrating economic, social, and environmental planning, which is to be achieved through decentralized, participatory decision-making rather than the top-down regulation that marked earlier environmental policy. Business and trade groups have been relatively proactive in adopting standards conforming to these plans, and certain regions and cities in the country, such as Waitakere and Christchurch, have become models of sustainable community planning.[27]

Australia adopted a National Strategy for Ecologically Sustainable Development in 1992, in conjunction with the Earth Summit.[28] This strategy grew out of the deliberations of nine different working groups, combining industry, environmental, scientific, community, and government interests, over two years. It established policy in areas of reducing greenhouse gas emissions, protecting forests, managing waste, preserving biodiversity, and managing rangelands. All three tiers of the nation's government, Commonwealth National, State and Local, approved this strategy at a 1992 conference of heads of government. Other pieces of national legislation have sought to implement this strategy, such as the 1999 Environment Protection and Biodiversity Conservation Act. The nation has also adopted a set of 24 sustainability indicators that will be updated regularly.

Japan's Basic Law on the Environment, adopted in 1993, instituted sustainable development as a central national priority and mandated preparation of local environmental plans.[29] Environmental strategies were also integrated into the country's five-year economic plan and a "New Earth 21" plan launched to develop environmental technologies.[30] In keeping with national tradition, implementation of such environmental initiatives is primarily through voluntary compliance, loans to assist businesses with conversion, and consensus agreements mediated by various levels of government. The nation has also set very ambitious goals for wind energy, with a target of 3 million kilowatts by 2010.[31] Interestingly, there is a strong case that Japan was one of the world's more sustainable societies during the 300-year Edo period, from the early 1600s to the late 1800s.[32] During this time the nation was closed to the outside world, was self-sufficient, had a low rate of population growth, and recycled almost all materials. No foreigners were allowed to land and no Japanese to depart. This period came to an end with the 1853 arrival in Tokyo (Edo) Bay of an impressive US military force under Admiral Perry, sent with the explicit intent of negotiating a treaty opening the country to the outside world.

In the mid-1980s the work of the Brundtland Commission stimulated the Canadian government to establish a National Task Force on Environment and Economy, involving business, government, environmental groups, labor, academic experts, and representatives of indigenous peoples. This task force and the federal Environment Canada agency produced a national Green Plan in 1990. However, subsequent national administrations failed to back this plan and it has been in large part ignored. The federal government in Canada is relatively weak in any case, and much oversight of environment and development rests with the strong provincial governments. Particular Canadian cities have been leaders in sustainable development, especially Vancouver, one of the best examples of compact, high-amenity urban development anywhere, and Toronto, which has been known for progressive transportation, urban design, housing, and greenway planning, though suffering from many other problems such as suburban sprawl, automobile traffic, and air pollution.

In the United States our federal government has so far done relatively little to promote sustainable development. The most explicit national initiative has been the PCSD, active between 1992 and 1999. Appointed by President Bill Clinton and co-chaired by Jonathan Lash, President of the World Resources Institute, and David T. Buzzelli, Vice President of Dow Chemical, this 29-member commission produced the landmark 1996 report *Sustainable America*. This document proposed a set of ten national goals for sustainability planning, including general principles such as stewardship, equity, economic prosperity, international responsibility, and civic engagement. The commission suggested sustainability indicators in each of these areas. Although the report mentioned the need for physical planning reforms to stop sprawl, its recommendations on this front were limited to general suggestions for local government action. Instead the commission focused on resource use topics, recommending greater regulatory flexibility, use of market forces, pollution taxes, extended product responsibility by manufacturers, and collaborative planning processes.[33] Despite the vague nature of many of its recommendations, if the Clinton administration or Congress had taken *Sustainable America* seriously it could have served as the basis for national sustainability planning. But unfortunately little implementation of the PCSD's ideas has taken place.

Other federal initiatives have been undertaken by the US EPA through its Sustainable Development Leadership Grants, and the US Department of Energy, which established a Center for Excellence in Sustainable Development mainly focusing on green building practices. A wide range of NGOs in the United States is also working to promote sustainability-oriented planning policies at federal as well as lower levels of government. Such organizations include the Surface Transportation Policy Project, the Smart Growth Network, the Sustainable Communities Network, the Sierra Club, the Natural Resources Defense Council, the US Green Building Council, and many others.

The passage of a wave of federal environmental legislation in the late 1960s and 1970s marked the active entry of the federal government into environmental planning. The National Environmental Policy Act (1970), Clean Air Act amendments of 1971, Clean Water Act of 1972, Resource Conservation and Recovery Act of 1974, and Superfund Act of 1980 established the legal basis for regulating environmental quality in and around cities. At roughly the same time civil rights and housing initiatives helped address social and equity dimensions of urban sustainability, though measures such as the Fair Housing Act and the Non-Discrimination Act were just a beginning. This first wave of environmental planning established strong but piecemeal federal regulation in a number of key areas related to community sustainability.

The passage of the 1991 federal transportation bill – the Intermodal Surface Transportation Efficiency Act of 1991 (ISTEA) – marked another a milestone in that many urban regions were given flexibility to reprogram federal transportation money for transit, bicycle and pedestrian planning, and even transit-supportive land use planning. This approach was continued under the bill's 1998 successor, the Transportation Efficiency Act for the 21st Century (TEA-21). Under the Clinton administration HUD adopted much improved policies relating to public housing, in particular converting many sterile, modernist housing projects to more human-scaled townhouse neighborhoods through its HOPE VI program.

The US federal government has been very reluctant to develop policy related to many other dimensions of sustainable development, including land use, energy consumption, resource use, greenhouse gas emissions, equity (apart from civil rights legislation), or urban growth management. Land use and spatial planning in particular are seen as local government concerns, with great deference given to private property rights. We have no equivalent to Britain's 1947 Town and Country Planning Act, which established the principle of public control over the development of land, required large-scale plan-making by counties, and subjected these plans to approval by a national minister. In the United States planning tools such as zoning do not rest on any such national legislation, but rather on the vague concept of "police power," a residual legal ability of government to pass laws to protect public health, safety, and welfare.[34] The last significant bill attempting to establish national land use planning policy failed in the 1970s.

The situation in the United States, in fact, has been one of strong national policy that often undermines the cause of urban sustainability. US refusal to abide by the Kyoto Protocols on global warming emissions and its disregard of *Agenda 21* and other United

Nations initiatives are particular black marks on the nation's sustainability planning record. Much more active leadership is needed from the national level if the United States is to move towards developing more sustainable communities.

The starting point for a sustainability planning agenda at the national level, then, is to actively recognize global goals and treaties, and to make good-faith efforts to follow up on these through domestic policy. Anything less essentially amounts to anti-social behavior on a global scale. The development of explicit urban planning guidance and policy at the national level is important for no other reason than to ensure that national expenditures on infrastructure, housing, transit, and other categories do not undercut local sustainability efforts. The British PPG statements and the national green plans of countries such as the Netherlands and New Zealand are an example of such national policy frameworks. But nationwide sustainability planning should ideally go far beyond this to provide active support of local sustainability planning through planning grants, targeted infrastructure funding, tax policy incentives, and environmental policy. Environmental and civil rights goals in particular are often necessary to establish at national level, since local governments may resist setting policies in such areas or at best will be inconsistent in doing so. Even in countries where planning policy-making is more decentralized, such as the United States, federal incentives for sustainability planning can be extremely useful. Historically funding such as the Section 701 planning grants, Community Development Block Grant (CDBG) funds, and Congestion Management and Air Quality (CMAQ) grants has been pivotal in bringing about new local planning and programs; as we will see, the state Smart Growth programs being developed in many parts of the country rely heavily on such incentives as well. Where no stronger policies or programs are politically possible, incentives for lower-level action represent the least a national government can do.

In recent decades national governments have had a tendency at times to devolve powers to more local levels of government. The trend towards devolution was particularly pronounced under Nixon, Reagan, and the two Bush administrations in the United States, and under the Thatcher administration in Britain. Green Party platforms also back devolution of power to promote grassroots democracy, calling for decisions to be made at the lowest feasible level. Devolution has been justified as reducing national-level bureaucracy, promoting the operation of markets, and enhancing local responsibility and democratic decision-making. However, its downside is that it tends to weaken national standards for civil rights, environmental protection, and a host of public welfare issues. It also undermines the rationale for planning at the large scales at which it is often needed – for example to establish national land use planning goals, to coordinate national transportation spending with land use and environmental policies, to manage energy and resources effectively, and to establish affordable housing policies. The effect of devolution is often to weaken planning across the board, since it reduces or eliminates the framework of upper-level policies and incentives supporting planning. Therefore a balance must be struck between levels of government that maintains the active involvement of those higher institutional levels that can best take a big-picture view, and that are most immune from the parochialism and occasional corruption of local government.

National planning issues

National sustainability planning issues of greatest urgency tend to be those in which lower levels of government do not have the perspective, resources, jurisdiction, or political will to effectively bring about change.

Addressing civil rights and other equity-oriented concerns at the national level is particularly important, since state and local governments tend to vary widely in their ability or willingness to develop such policies on their own. Without a national mandate, privileged communities have little incentive to accommodate less affluent citizens and minority groups, or to share resources with them. In the same way that international norms such as the Universal Declaration on Human Rights and the Geneva Conventions place moral, legal, or persuasive pressure on national governments to conform to higher human rights standards, national frameworks help raise standards regarding equity and civil rights among more local governments.

National steps to equalize access to financial resources across state and local jurisdictions also represent an important equity measure. Otherwise, jurisdictions with a low tax base and high need for public services, such as central cities and impoverished states, are at a disadvantage compared with more affluent jurisdictions (whose benefits in turn often rest on federal spending for infrastructure, research, or the military). This problem occurs both within states and nationally. Since the early 1970s various "revenue sharing" programs in the US have provided a modest amount of unprogrammed block grant funding from the federal government to cities across the nation, but this program has diminished in size in recent decades. In Britain, the national government serves a stronger tax equalization function by collecting all business taxes at uniform rates and redistributing revenues according to population.[35] Such a strategy also helps eliminate pressure on state or local governments to give tax breaks to businesses in order to attract economic development, since such tax rates are set nationally.

Action on global issues such as climate change, ozone depletion, deterioration of marine environments, nonrenewable resource use, and endangered species protection virtually requires a national commitment as well. Lower-level governments alone are able to undertake only piecemeal action on such issues, though such small-scale efforts are certainly significant. Local and regional agencies have little incentive to think about global issues, since almost all political pressures on them are local. In the absence of a national mandate for action, such large-scale topics are unlikely to be dealt with. Other broad areas of environmental policy also typically have been the purview of national governments. National environmental policy laws, endangered species legislation, air and water quality legislation, and other big-picture environmental topics have generally been addressed at this level. State governments frequently take action on these subjects as well, so what results is a landscape of overlapping policy frameworks from these higher levels of government. Typically the state actions expand on the national goals, providing more stringent regulations to meet state and regional needs. For example, California has adopted much stricter air quality regulations than the US federal government because of severe air quality problems in the Los Angeles basin and other parts of the state.

Policy on large-scale infrastructure such as transportation systems is also typically handled at the national level of government. Large-scale road or public transit systems require large amounts of funding, which is often not available at the local level, and their planning often crosses local, regional, or state lines as well. National commitment to high-speed rail by continental European countries has greatly boosted rail transit networks in recent years, for example. In contrast, neglect of the British national rail system under the Thatcher administration proved a major setback. National commitment to bicycling is bearing results in the Netherlands and Germany. Between 1978 and 1996 the former country more than doubled its mileage of bike paths separated from automobile traffic, while between 1976 and 1995 the latter almost tripled its bike path network.[36] US national preference for funding motor vehicle-related infrastructure during the post-Second World War period, rather than passenger rail systems or bicycle and pedestrian facilities, has had profound effects on urban and regional development nationwide, providing the infra-structure for suburban sprawl and shifting the majority of freight transport onto public roads rather than railway lines.

A final area in which national policy is particularly vital concerns land use, even though this is seen in many places as a local prerogative. By establishing the basic legal relations regarding real property, national constitutions, legislation, and court decisions set limits on what local planners can do. If it is presumed that owners are entitled to a relatively unfettered ability to develop land, or else must be paid high levels of compensation by local governments, then public regulation of development is limited. Tools such as zoning and urban design regulation become suspect, and cautious politicians avoid using them proactively. Such is the situation in much of the US. If, on the other hand, endangered species, ecosystems, historic cultural landscapes, and traditional communities are acknowledged to have value in themselves, then the balance may tilt more towards a more conducive context for sustainability planning. To date some US initiatives such as the Endangered Species Act have begun to move in this direction. But such steps are limited, and rethinking national values regarding land development is likely to be a very long process.

9 State and provincial planning

State and provincial governments form an intermediate level of authority that is present within many medium or large-sized countries such as the United States, Canada, Australia, Germany, Italy, Mexico, India, and Brazil. These nations either came together from a collection of quasi-independent regional entities, or possess a sufficient scale of territory as to make a layer of provincial governments below the national level desirable. Going beyond this, some nations have internal regions with their own cultures or quasi autonomous institutions of government; Great Britain has Scotland, Wales, and Northern Ireland, and Spain has Catalonia and the Basque country. Many of these large cultural territories might easily be their own nations, or once were. States, on the other hand, form a consistuent tier of government between national and regional or local scales, usually without such pretensions – a level that may prove best suited to some dimensions of sustainability planning.

States control a substantial amount of political power and in federal systems are the primary source of power; national governments derive their powers from a voluntary relinquishment of these by the states. An extreme model of a state-centered politics is Switzerland, where the cantons are the primary locus of decision-making. Although not decentralized to nearly this extent, the United States and Germany also have systems in which the American states and the German lander both hold significant power. There is substantial popular fear of too much centralized authority in both nations, for different historical reasons.

For the purposes of sustainability planning, state or provincial governments often have key roles to play in overseeing land use planning, transportation systems, environmental protection, equity, and the formation of municipal governments. Land use and growth management planning in particular may be best handled at this level, as state or provincial governments are still relatively close to local territories and cultures, but far enough removed from local politics to be less affected by the parochialism that often afflicts city, town, and county government.

Certain relatively progressive states, or states with particularly serious problems, can at times go much further than a national government in developing policy and programs. California, for example, has led the United States in developing automobile emissions regulations, while states such as Vermont and Oregon have been leaders in growth-management legislation. States also at times offer a smaller and more manageable political arena in which to bring about change than national government. Such considerations argue for viewing state governments prominently as a locus for sustainability planning.

Who plans at the state and provincial scale?

State governments in the United States are organized much like the national government, with two houses of a legislature, an executive (the governor), and a court system. Various agencies of the executive branch handle different program areas related to planning. Particularly important in terms of physical planning are transportation agencies, which are the dominant road-building organizations in the country (the federal Department of Transportation channels funds to the states, rather than building infrastructure itself). These state departments of transportation (DOTs) are often dominated by old-line engineering attitudes, and tend to see large new construction projects as a way to justify their own existence. It has frequently been a challenge to get them to instead prioritize alternative modes of transportation, and to establish funding programs for local cities and towns to carry these out. In the end these agencies may need to shrink and redefine their mission, as transportation planning moves away from its past agenda of road expansion.

State departments of housing and/or community development are perhaps the most directly focused on urban development issues. California's Department of Housing and Community Development, for example, administers more than 20 grant programs related to affordable housing, alleviating homelessness, public facilities, and infrastructure. Such programs can be used to leverage additional action at the local level. State environmental protection agencies and public health agencies can also play leading roles in sustainability planning. Minnesota's Office of Environmental Assistance, for example, sponsors the state's Sustainable Communities Network, which conducts education and networking statewide. Officials from several agencies in that state have also developed the Minnesota Sustainable Building Guidelines that will be applied to all new state buildings.[1]

Statewide NGOs are increasingly effective at shaping policy, especially in areas of growth management, environmental protection, and education. "1000 Friends" groups have proliferated in many US states, inspired by the success of 1000 Friends of Oregon, and chapters or field offices of groups such as the Sierra Club, the Audubon Society, the Surface Transportation Policy Project, and the Nature Conservancy are very active as well. Public Interest Research Group (PIRG) organizations, originally inspired by Ralph Nader in the 1970s, are found in many states. Teachers' unions, good government groups such as Common Cause, associations of public health professionals, and labor unions are frequently influential at the state level. Potentially such constituencies can be organized into political coalitions supporting sustainable development initiatives.

State or provincial planning issues

Land use and growth management

With the federal government in the United States essentially abdicating any substantial role in land use planning, sustainable resource policy, housing provision, or many other areas of sustainable development, some state governments have stepped in to fill the void. Vermont, Oregon, Florida, Hawaii, New Jersey, and Rhode Island pioneered the first wave of state growth management policy in the 1970s, sometimes referred to as the "quiet revolution" in state land use planning. Vermont, for example, established a regional permitting process for large developments; Florida required both state and regional review of large projects and later mandated that local plans be consistent with regional and state planning goals, although local resistance undercut this objective.[2] Since then some states have developed second- and third-generation growth management frameworks. In the late 1990s Maryland became the national exemplar of Smart Growth planning with a number of initiatives under governor Parris Glendenning. Under its 1997 Smart Growth Act the state enacted policy to restrict infrastructure funds available to cities that do not place new development within locally determined Priority Funding Areas. The state also set up a fund to acquire open space and forest land, and has adopted policy to site state facilities within existing urban centers. Such a framework represents a stronger and broader form of statewide growth management than that adopted by most states in the 1970s, in that it not only attempts to protect land from sprawl, but provides incentives for good development and rethinks state investment priorities.

New Jersey has long been a leader in statewide growth management planning. Its legislature first passed a Statewide Planning Act in 1985 that established a State Planning Commission and Office of State Planning. This agency then created the first State Plan in 1992 and an updated plan in 2001 with sustainable development as an explicit theme. Although local jurisdictions retain authority over land use, the New Jersey State Plan attempts to orchestrate agreement on development principles that emphasize the revitalization of existing communities and the preservation of open space.[3] State policy establishes a procedure known as "cross-acceptance" through which local, regional, and state plans are made consistent with one another. Sustainability planning in New Jersey has been advanced considerably owing to the efforts of New Jersey Future, a statewide nonprofit organization that released a Sustainable New Jersey plan in 1999.[4]

Oregon's statewide growth management framework is particularly well known and has evolved substantially over more than 30 years (see Box 9.1). Indeed, the state is an example of the benefits of incrementally improving policy and institutions over time.[5] In several works, planning historian Carl Abbott has documented ways that Oregon has developed a culture of planning that stresses continued innovation and improvement, something that he calls "the Oregon planning style."[6]

Minnesota has likewise taken the lead in a number of sustainability-oriented planning measures since the 1960s, including authorizing establishment of regional government in the Minneapolis-St Paul area and the adoption of revenue sharing to reduce tax base disparities between local jurisdictions in the Twin Cities region. Although the state has not been nearly as strong as some others at setting land use and growth management goals, it has focused on other aspects of sustainability planning. In particular, former Governor Arne Carlson launched a Minnesota Sustainable Development Initiative in 1993 as a collaborative effort of business, government, and civic interests. Through a number of reports and guideline documents this initiative has sought to develop consensus on more sustainable resource use, land use, energy consumption, and community development. The State Legislature added to the effort by approving 11 planning goals in its 1997 Community-Based Planning Act. A 2000 state report provides a model ordinance for local governments on how to implement sustainability planning, focusing on land use planning (including use of UGBs and transfer-of-development-rights frameworks), urban design, energy efficient buildings, water supply, sewage treatment, and economic development.[7] The Metropolitan Council of Minneapolis-St Paul has established a Livable Communities grant program that won a 2003 Smart Growth award from the US EPA. This program has awarded 292 grants totaling nearly $100 million to 106 local jurisdictions, which have used the money to revitalize brownfields, create mixed use town centers, and to provide affordable housing in rural, suburban, and urban settings.[8]

Box 9.1 Oregon's growth management planning

The state of Oregon has a more than 30 year history as a leader in statewide planning in the United States. After an initial attempt to establish statewide planning policy with a Land Use Planning Act, S.B. 10, in 1968, a much stronger bill, S.B. 100, in 1973 established a set of statewide planning goals.[a] These included the requirement that cities establish urban growth boundaries, and goals of protecting forest, farmland, and coastal resources. S.B. 100 also created a statewide Land Conservation and Development Commission to oversee application of these planning goals, and a Department of Land Conservation and Development to provide state assistance and resources to local governments.

The state continued to expand its planning role in subsequent decades. A 1991 state Transportation Planning Rule required cities and towns to plan to reduce automobile use. This regulation requires metropolitan areas to demonstrate that vehicle miles traveled (VMT) per capita will be reduced 10 percent within 20 years and 30 percent in 30 years, and directs local and regional governments to consider a host of physical planning initiatives to reduce automobile use. In the 1990s the state also undertook to develop a set of Oregon Benchmarks that would serve in effect as sustainability indicators. The result of these and other initiatives has been a multi-tiered planning framework that helps lead the state in many sustainability directions.

[a] For a detailed description of the Oregon land use planning system, see Abbott *et al.* (1994).

Environmental regulation

Most US states have duplicated many aspects of federal environmental regulation, passing their own environmental quality acts modeled on the National Environmental Policy Act as well as air and water quality legislation. California's Environmental Quality Act is substantially stronger than the federal version, and has been extended by the courts to regulate private development as well as publicly funded projects. That state has also seen its tough automobile emissions standards copied by New York and other northeastern states.

Many states have taken the initiative on issues related to energy policy. Just over half of state governments have incorporated energy efficiency requirements into their building codes, resulting in large savings in energy use for home heating and cooling. California, for example, established Title 24 of its building code in the early 1980s to raise requirements for energy efficiency, and through this conservation initiative now saves the energy equivalent of the production of several large power plants. However, many other states, including Arizona, New Jersey, Texas, Illinois, and Michigan, have yet to adopt energy efficiency codes.[9] States regulate public utility companies that provide electricity and natural gas, and thus are in a position to require these companies to pursue energy conservation programs and renewable sources. However, the process of deregulation in the mid-1990s drastically reduced attention to these programs. The infamous problems with deregulation in California and elsewhere in the early 2000s has made it clear that states will always need to play a strong oversight role vis-à-vis public utilities and private energy suppliers to ensure creation of more stable, consumer-focused, and sustainable energy systems.

States have taken the lead on programs to promote reuse, recycling, and waste reduction. Bottle deposit laws in states such as Vermont, Oregon, New York, Michigan, Iowa, Maine, and California have helped promote recycling of beverage containers. Most of these states instituted a $0.05 refund per container in the 1970s. Although this amount has not been raised in decades, it has promoted recycling and resulted in substantially less roadside broken glass and litter. Actual reuse of glass bottles was the norm in the United States until the 1960s. Residents routinely brought soft drink bottles back to retailers for a deposit, and the bottles were then washed and refilled at the regional bottler. Although a bewildering variety of plastic bottles and aluminum cans has replaced standardized glass bottles in the United States, this model of reuse is still the norm in many other nations. Perhaps standardized, refillable bottles will one day return, and state incentives or requirements will be used to promote this. State requirements that local governments reduce the volume of solid waste produced by their residents have been essential in bringing about local action on recycling and reuse of materials. In 1989 the California legislature, for example, required that the state's cities reduce their solid waste by 50 percent by the year 2000. Actual reductions totaled 47 percent in that year, very close to the goal.[10]

In 2001 New Jersey established the first state-supported Sustainable Development Science Institute in the US, a cooperative effort with a number of universities in the state, based on the model of the Dutch Sustainability Science Institute. This project aims to conduct a public process to set sustainability goals and indicators for the state, which the state government will then consider how to implement. The Institute is one result of a Sustainable State Project originated in 1994 when a delegation of state legislators and cabinet officials traveled to the Netherlands to observe urban planning there. Most unfortunately, initial funding for this Institute was taken from the defunding of the state's Office of Sustainable Business. The state of Oregon has also established a government board to examine, promote, and codify sustainable practices. Initiated by the state legislature in 2001, the eight-member Sustainability Board is directed to propose sustainability incentives and legislation and issue a biannual report on the state's progress toward sustainability.[11]

Transportation

Although it is tempting for politicians to announce major roadbuilding programs to provide jobs during a recession, or to equate progress with new freeway projects, state governments are gradually diversifying their transportation planning away from the traditional focus on road construction. However, most commuter rail and bus transit systems are developed regionally, so direct state involvement in alternative transportation modes is often limited to a small handful of intercity rail projects. State agencies pass most other non-highway transportation money through to regional metropolitan planning organizations (MPOs). But they can provide small, targeted grant programs to encourage these other modes. "Safe Routes to School" grant programs now offered in California and New Mexico illustrate such creative grant-making.

A far greater impetus for sustainable development would be to link state transportation investment with good local land use planning. Localities that did not plan for new development near public transit stations, or that allowed automobile-dependent sprawl to continue, might not receive funds for local road maintenance, for example. Such policies would help ensure that investments in new transportation systems were cost-effective by ensuring land use patterns that would support them. It makes little sense to invest in new rail transit systems in particular unless the local governments along each line will allow intensified development near stations – building ridership and ensuring that the line is well-used. It also makes little sense to build or expand freeways if local governments allow sprawling, automobile dependent development to flood them with additional commuters. The threat of withholding state transportation funding – or conversely, incentives for good local planning in terms of increased state transportation money – can be a powerful lever for change. Such conditionality will be fiercely resisted by local governments, and has yet to be used widely. But as previously mentioned, the state of Maryland has instituted some conditions on its infrastructure money, as part of its statewide Smart Growth program, making funding conditional on completion of local plans designating Priority Funding Areas.

Housing

State governments can play a major role in enforcing fair share housing distribution among local governments, so that each city or town approves a reasonable number of units affordable to those making less than 80 percent (for low income) or 50 percent (for very low income) of the area median income. Again, state infrastructure funding could be made conditional on local governments complying with fair share targets. Or incentive grants could be made available to those that exceed their requirements by a certain amount.

Some states currently offer low-interest loans or other assistance to nonprofit affordable housing providers, often in the form of tax credits, through which wealthy investors are enticed to lend funds to build affordable housing. Other state actions to assist housing production can include brownfield cleanup assistance, legislation to reduce the risk of litigation by NIMBYs against housing projects, and incentive grants to municipalities that exceed their previous performance in constructing affordable housing (as has been tried in California).

Equity

States can play a crucial role in promoting equity, and have already done so in many parts of the United States, albeit sometimes pushed by the courts. Fair housing policies have already been mentioned, and programs to provide partial equalization of local school funds are another main area of activity. Some states also have relatively progressive tax systems, for example, with steeply graduated income taxes that help reduce disparities in wealth. Hawaii's tax schedule, for example, rises from 2 percent at the bottom bracket to 10 percent at the top. A few other states such as Oregon have persistently avoided instituting a sales tax, widely viewed as the most regressive form of taxation in that it falls most heavily on the poor, for whom basic purchases of household goods and transportation are a higher percentage of income. Many states exempt food, shelter, clothing, and medical expenses from sales taxes for equity reasons.

In the future, states will offer an arena for addressing the serious inequities in tax resources that exist at the local level – and the intense pressure for "fiscal zoning" by local governments. Under the fiscal zoning scenario, cities and towns zone for as much commercial development as possible in an attempt to gain sales and property taxes, but in so doing merely steal tax base from their neighboring local governments and promote suburban sprawl. The benefits of this misguided local tax system typically go to relatively wealthy fringe suburban jurisdictions with ample vacant land and few existing urban problems, while older suburbs and central cities suffer from the flight of their taxable businesses. By reforming tax structures statewide to collect revenue through relatively equitable income and property taxes and then distribute it to cities on the basis of population, state governments could vastly improve both equity and local land use decision-making.

Economic development

Most states or provinces have active economic development efforts, and routinely court large corporations or industries to locate in the state. Governors may travel internationally, seeking to attract new investors, and may offer substantial incentives to interested parties. States typically also promote tourism very heavily, since it is a major source of revenue that can build on the state's natural advantages. The desire to compete successfully with their many peers often leads states to reduce their tax rates, adopt anti-union legislation, lower workers' compensation requirements, or take other steps aimed at a "business-friendly" reputation. Needless to say, these efforts do not necessarily produce sustainable development. Too often they produce few new jobs, many of which may be low-wage and temporary (if footloose corporations relocate elsewhere again). If they are in fact successful, traditional economic development strategies may produce overly rapid growth that strains public services and fuels suburban sprawl. These "business-friendly" policies may also reduce much needed state revenues and deprive workers of decent wages and benefits, if state labor policies are compromised in a "race to the bottom" with other states.

Corresponding state efforts to promote green economic development are few and far between, but there are some likely possibilities for action. Many states have historically been dependent on particular natural resource industries, which in many cases are exhausting their resource base. Coal, oil, timber, mineral, and fisheries industries fall into this category. State governments can take proactive steps to plan for a transition to more sustainable harvesting methods in many of these cases, or else for a transition to alternative products. Legislators can also plan so that the state budget does not take an inordinate hit when taxes from such resources run out.

States can also invest in human capital rather than in corporate recruiting incentives, a strategy that is likely to produce a capable, educated workforce in the long run that can help a variety of local businesses be successful. Investments in state university systems often bear handsome dividends, especially as graduates start new businesses. Similar investments in K-12 education are perhaps even more important, and play a substantial role in improving human welfare as well.

Along with programs to develop human capital, quality-of-life related investments by state governments are likely to help produce a more diversified, sustainable economy in the long run, attracting creative professionals and high-wage companies. Rather than creating a low-tax, low-amenity environment, the most sustainable economic development strategy may be to invest in education, parks, the arts, public health, and community facilities. This may not generate the quick jobs that a single large manufacturing plant might create, but such a strategy may instead lay the groundwork for both better-quality jobs and a better-quality living environment in the long run.

* * *

As with most of the rest of sustainability planning, examples of statewide action are still in the early stages in the United States. Yet initiatives have been increasing in areas such as growth management, open space protection, and regulation of pollution. Examples of state-level sustainability planning can be found in certain locations internationally as well. Within the United Kingdom, Scotland is pursuing an Action Plan on Sustainable Development that includes a cabinet-level commission, periodic public forums, a grant program for sustainability-oriented projects, and a sustainability indicator program.[12] In Canada the Ontario Roundtable on Environment and Economy played a major role in promoting community work on sustainability topics in the province of Ontario, until dismantled by a more conservative provincial government. In Spain, the Autonomous Government of Catalonia has adopted an Agenda 21 program that aims for improved land use, transportation, environmental, and natural resource planning.[13] In India, the state of Kerala is well known for progressive policies relating to social welfare and environmental issues, resulting in highly efficient distribution of food, low consumption of resources, a male life expectancy 10 years longer than the Indian average, and a female life expectancy 15 years longer.[14]

State and provincial governments, then, can occupy a crucial middle ground between national and local levels in terms of sustainability planning. The sustainability planning agenda at the state level often focuses on issues larger than individual cities, towns, or metropolitan areas can handle, particularly in the areas of urban growth policy, environmental protection, transportation, equity, and resource planning. But state planning frameworks are likely to be more specifically tailored for the area than national policies, and may in fact be quite detailed and stronger than their national equivalents. Their foundation is often a set of broad goals for the state, with much of the implementation of these principles necessarily still local. Incentive grants and conditions attached to state funding can be powerful incentives for better local implementation.

10 Regional planning

One of the paradoxes of planning is that many social and environmental problems are best approached at a regional scale, but this is the weakest level in terms of government institutions and public understanding. Issues such as air quality, water quality, transportation planning, suburban sprawl, and tax base inequities overlap municipal and county boundaries and virtually require a regional planning approach. In the postmodern landscape cities and suburbs have expanded so much that it makes sense to think in terms of "the regional city," as Peter Calthorpe and William Fulton (2001) have argued in their book by the same name.[1] Los Angeles, for example, often seen as a prototypical late-twentieth-century urban region, stretches for nearly 100 miles in several directions and includes hundreds of cities and parts of four counties. The City of Los Angeles itself has become a relatively small part of the urban region.

Yet although many development problems are best dealt with at a regional scale, most politicians and members of the public do not think regionally. Daily life occurs at a more local scale – in neighborhoods and cities – while national and state governments are strong and established institutions that dominate public attention. Regional planning issues tend to fall through the cracks. Still, many opportunities exist to promote sustainable development at this scale of planning.

In recent years, a growing movement calling for regional solutions to sustainability-related problems has emerged that might be termed a "New Regionalism."[2] This movement includes political scientists and sociologists concerned about equity within metropolitan regions,[3] environmentalists and urban designers concerned about growth and suburban sprawl,[4] and economic analysts who see urban regions as increasingly important economic actors in the global economy.[5] While few participants hold out much hope for strong new regional governments to tackle such topics, many instead call for flexible and sophisticated governance using existing state and regional agencies as well as ad hoc partnerships of regional stakeholders to bring about change.

Types of regions

The "region" has been defined in many different ways within the urban planning literature. In the early twentieth century a number of planning pioneers took a broad and holistic approach to the study of regions. Scottish biologist and polymath Patrick Geddes, often viewed as the father of regional planning, proposed a "synoptic vision"[6] of regions combining geographical, economic, social, and political dimensions. This vision was to be based on firsthand observation, quantitative data, extensive drawing and mapping, and historical study of the region's evolution. Criticizing the "artificial blindness"[7] of book-learning, Geddes (1915) proposed that planners survey the region from vantage points similar to his

Outlook Tower in Edinburgh, walk through it, and prepare multi-media exhibitions illustrating regional evolution. Only after planners and the public had gained a thorough understanding of the region's character, he believed, should plans be developed.

Howard's Garden City proposals (1902 [1898]) were equally holistic, including physical planning and urban design concepts, economic mechanisms for local industry and cooperative ownership of land, and provisions to maintain fair rents and promote equity. According to Howard, "a town, like a flower, or a tree, or an animal, should, at each stage of its growth, possess unity, symmetry, [and] completeness, and the effect of growth should never be to destroy that unity, but to give it greater purpose."[8] Although his proposals were focused particularly at a town scale, Howard's vision integrated these towns with their greenbelts into a larger, carefully organized metropolitan region.

Following the lead of these British thinkers the Regional Planning Association of America (RPAA), especially through its members Lewis Mumford and Benton MacKaye, sought to develop an "ecological regionalism" in the United States in which city and countryside, industry and nature were viewed as a whole. Mumford and others developed a vision – unfortunately never extensively implemented – integrating economic development, management of natural resources, transportation, large-scale physical planning, and humanistic architecture and site design.[9] As with Howard, the unit of development was to be the garden city. The Great Depression curtailed plans to build a demonstration model, and only a small portion of it was actually built in Radburn, New Jersey. Rather than serving as a model for new regional development, Radburn laid the groundwork for a new form of suburban subdivision emphasizing cul-de-sacs and an interior greenspaces network.

The objective of all these early regionalists was to decentralize population from the overcrowded, unhealthy nineteenth-century industrial city into a more balanced regional development pattern. Unfortunately, these regional thinkers wrote before the era of the automobile and did not foresee that cars would allow an extreme version of decentralization to occur with devastating ecological and social impacts. They also did not foresee that the planning profession would become increasingly pragmatic and dominated by modernist science during the twentieth century. Their holistic and idealistic approach to the "region" was disparaged by later generations of pragmatic planners who focused on more narrow and technocratic specialties such as transportation planning, economic development, and the administration of zoning.

Another approach to regionalism arose in the 1930s and 1940s, based in large part at the University of North Carolina in Chapel Hill. Sociologists there such as Howard W. Odum and Harry Estill Moore analyzed cultural regions such as the American South and were known as "cultural regionalists."[10] A prime motivation for this school was to preserve the unique social values and traditions found in such regions. Such cultural regions were typically very large and difficult to define, since social groupings and traditions overlap considerably.

Since the late 1940s many academic regionalists have defined "regions" as fields of economic activity, and have focused on areas such as Southern California, Silicon Valley, the Tennessee Valley, the Emilia-Romagna area of northern Italy, or the even larger Sun Belt across the southern United States. Economic analysis and scientific method have been dominant within this group, rather than the physical design visions and normative values of earlier regionalism. The field of "regional science," founded by Walter Isard at the University of Pennsylvania in the 1940s, has concentrated almost exclusively on economic analysis. In their classic 1964 volume *Regional Development and Planning*, John Friedmann and William Alonso referred to the region as an "economic landscape,"[11] and Friedmann wrote that regional planning was concerned mainly with "problems of resources and economic development."[12] References to real places disappeared almost entirely within this analysis. Perhaps the ultimate extreme of modernist regionalism was Mel Webber's early-1960s concept of the "non-place urban realm."[13] In his view place-oriented urban community would disappear entirely within highly mobile, automobile-oriented regions, and individuals would somehow establish "community without propinquity." Although this vision did in fact describe much twentieth-century development, it gave the shivers to increasing numbers of individuals in the latter part of that century who viewed such a landscape as environmentally, socially, and culturally disastrous.

In the 1960s and 1970s neo-Marxist regional economic geographers and sociologists such as David Harvey and Manuel Castells brought a new analysis of power to the study of regions, looking at the way economic capital, social movements, elite groups, and "growth machines" dominate urban and regional development. The region for these observers became a terrain of power – economic, political, and social. Simply developing technical analyses without an understanding of power dynamics, in this view, was point-less for planners. Instead what was important was understanding the structural changes needed in order for planning goals to be effectively reached.

More recently there has been a resurgence of interest in regional physical and spatial planning – addressing land use, infrastructure, urban design, ecosystem planning, and equity primarily within metropolitan regions. This reframing of the field has come about because of the rapid growth of urban regions worldwide, a new wave of environmental planning emphasizing bioregions and landscapes, and concern about growing income and wealth disparities between central cities and suburbs. To some extent this new regionalism represents a return to more holistic early twentieth-century definitions of the "region."

This form of regional planning concerns itself particularly with metropolitan regions, watersheds, bioregions, and other physical landscapes rather than more abstract economic or cultural space. The first of these categories, metropolitan regions, refers to collections of urbanized cities and counties in geographic proximity. Since the 1960s federal trans-portation and housing funding has required that metropolitan planning organizations (MPOs) be set up to coordinate efforts across these jurisdictions. There is sometimes argument over the exact extent of metropolitan regions – whether a region, for example, should include its entire commuteshed, with some communities more than 100 miles away from the original core city. The Los Angeles region might or might not be seen as including the San Fernando Valley, Riverside, and other areas far removed from the core city of Los Angeles. (The Southern California Council of Governments does in fact include these areas.) Metropolitan regions have also expanded dramatically in recent decades, meaning that the borders of regional institutions have needed to evolve. In the mid-twentieth century the San Francisco Bay Area, for example, saw itself primarily as a five-county region. Now the Association of Bay Area Governments (ABAG) consists of representatives from nine counties. In another decade or two it may have to include as many as 14 counties as the region's development sprawls outwards into California's Central Valley.

Watersheds are ecological regions that may or may not overlap with urban areas. In contrast to the changing borders of metropolitan regions, a watershed is a naturally defined area rooted in the characteristics of particular drainage systems. However, since rivers and streams branch extensively there can be said to be watersheds within watersheds (fluvial geomorphologists speak of ten orders of waterways), and different scales of watershed may be useful for different sorts of planning. For example, Rock Creek within Washington, DC, forms a watershed draining dozens of square miles, with important issues of water quality, hydrology, habitat, and interface with recreational park uses. But

Rock Creek empties into the Potomac River, whose much larger watershed encompasses hundreds of square miles in Maryland, Virginia, West Virginia, and the District of Columbia. The effort to clean up the Potomac was a major environmental campaign in the 1960s and 1970s. The Potomac in turn drains into the Chesapeake Bay, whose watershed covers some 64,000 square miles in parts of six states (adding Pennsylvania, New York, and Delaware). In 1983 a Chesapeake Watershed Partnership including most of these states plus the US EPA and the Chesapeake Bay Commission began one of the nation's largest planning exercises to protect and restore the entire ecosystem.[14] Over two decades the Partnership has developed a comprehensive set of goals and programs affecting land use, pollution control, wetlands, forests, recreation, urban development, transportation, and certain aspects of local economies – a wide range of sustainability-related issues – throughout the entire watershed region. This extremely large scale of watershed planning is vastly different from planning for more local watersheds in the same area.

Watershed programs have a long history that illustrates changing values within regional planning. Regionalists of the 1920s and 1930s promoted watershed planning to manage natural resources and to meet social objectives of reducing poverty and providing employment. The crowning achievement of this period in the US was the Tennessee Valley Authority, initially formed to provide electric power, employment, and social development for a large, impoverished area of the American South. Unfortunately social objectives were soon forgotten, and this project proved successful mainly at building large-scale infrastructure – dams, highways, power plants, and a nuclear power laboratory. As such it fell victim to modernist impulses to tame and manage the environment rather than to coexist sustainably with it. Comprehensive river basin planning in the Western US during the mid-twentieth century likewise created dams, reservoirs, and power plants at great environmental cost. (The drowning of Glen Canyon by Hoover Dam, viewed by former Sierra Club executive director David Brower as one of the environmental movement's greatest defeats, was just one of the more disastrous examples.)

However, beginning in the 1960s and 1970s watershed planning shifted in a more ecological direction. People realized that large dams, levies, and artificially stabilized channels, combined with urban development in floodplains, often dramatically increased flood damage rather than reducing it. Other negative effects of dam-building worldwide became well known. Even the Army Corps of Engineers, responsible for much of the twentieth century's dam-building and river channelization, adopted a mission of sustainable development in the 1990s and undertook projects such as restoring the oxbows and meanders in the Kissamee River in central Florida, which had been channelized by a previous Corps project 30 years before.

Bioregional planning often overlaps with large-scale watershed planning and has been an attractive ideal to many environmentalists since the 1960s. However, there are as of yet few political institutions corresponding to this scale. Bioregions are probably best defined in terms of distinctive plant and animal communities that form a typical natural mosaic of ecosystems. Not coincidentally, they are also often characterized by particular human cultures with their roots in specific natural resource industries or forms of agriculture. Yet, as with watersheds, bioregions are difficult to define. Is San Francisco's bioregion, for example, the area centered around San Francisco Bay itself, the watershed of the Sacramento and San Joaquin rivers that feed the Bay (which would include much of central California but not necessarily the coast), or all of Northern California? Local bioregionalists have generally taken the latter approach, naming the area the "Shasta Bioregion" after Mount Shasta, a 14,000-foot extinct volcano in the northern part of the state. But this landmark is very remote from most areas within Northern California, and ecosystems vary enormously throughout this area. The redwood and Douglas-fir forests of coastal Northern California bear little resemblance to the grasslands of the Sierra foothills or the semi-arid chaparral of the many inland valleys. And to many residents of the Northern California coast, San Francisco bears more similarities to Los Angeles than to their own communities. In such places clear boundaries of a bioregion are hard to come by, and so it is difficult to develop ideas about how to plan for such an elusive region.

Airshed planning is yet another form of regional planning whose borders are to some extent defined by the natural landscape. Airsheds are typically created by mountain ranges or hills that trap air and create local pollution, although long-distance transport of pollutants can add to local problems. A large metropolitan area in a hilly region may contain several discrete airsheds that may be grouped together for planning purposes. However, rather than focus on each air basin separately, air quality planning agencies tend to adopt metropolitan regional boundaries as determined by the politics of the region.

Finally, "landscape planning" is a growing field within the discipline of landscape architecture, pioneered by figures such as Ian McHarg and Richard Forman, that has a strong regional component. Landscape ecologists are concerned with distinctive natural landscapes that may cross the jurisdictional borders of cities, counties, and even states. During the last two decades of the twentieth century this field developed a sophisticated terminology with which to describe landscape structure. Landscape ecologists speak of "patches" of habitat, "corridors" between them, "edge" and "interior" ecosystems, a background "matrix," and an overall "mosaic" of these forms across a broad area. Landscape planning then becomes a task of managing these ecosystem elements to meet goals such as biodiversity.[15] Again, few institutions may exist corresponding to landscape regions, but landscape ecologists may be able to use many different levels of government to meet their planning objectives.

These, then, are many of the types of regions that are important for planning purposes. Despite definitional difficulties, the "region" for any given planning problem can usually be defined in practice. The main problem is that strong governmental institutions do not exist for many sorts of regions, or don't have significant power to affect land use, environmental planning, distribution of fiscal resources, or other areas of interest within planning. Regional institutions have gradually been growing in strength – regional environmental planning agencies, park districts, and transportation commissions, for example, have expanded greatly in the last 50 years. But the process of institution-building is slow and incremental. Regional planners must therefore often make do with whatever institutional tools are available, while seeking to improve these in the long run.

Who plans regions?

A large number of often overlapping agencies exist to plan at many of these regional scales. Metropolitan Planning Organizations (MPOs) exist in every US metropolitan area as the agencies responsible for coordinating transportation decisions and dispersing federal transportation money to cities and counties. Since they handle large amounts of funding, these MPOs have considerable power over how the region develops. But they usually have no control over land use, which handicaps them in terms of managing regional growth. MPOs may or may not be synonymous with regional Councils of Government (COGs), which are voluntary associations of local governments in a metropolitan region generally set up in the 1960s or 1970s. These latter agencies collect information about the region, provide a forum for local cities and counties to coordinate strategies, and frequently provide technical support services to municipal governments.

Their membership is entirely voluntary, and they do not have legal authority over land use or other important planning subjects.

Britain has seen a long history of regional planning, in particular through the "county councils" that coordinated public housing and other planning needs beginning in the nineteenth century. These were abolished under the Thatcher administration, but the Blair government established Regional Development Agencies once again under a 1998 Act. Each agency was directed "to contribute to the achievement of Sustainable Development in the United Kingdom where it is relevant to its area to do so." However, this mandate must be balanced with four other goals, and so the overall focus of these regional authorities is less than completely clear.

France, the Netherlands, Sweden, and many other nations have a strong tradition of regional planning, often to coordinate development of large metropolitan areas such as Paris, the Randstad (the area including Amsterdam, The Hague, Rotterdam, and Utrecht), Copenhagen, and Stockholm. In many cases these agencies have established plans for the spatial development of regions. The "finger plan" designs of Copenhagen and Stockholm, in which development is channeled along rail transit corridors, are well known; regional authorities directing development around Paris have created a series of new towns and development poles, including areas such as La Defense, Marne la Vallée, Val Maubuée, Evry, Sénart, St Quentin en Yvelines, and Cergy Pontoise.

Actual metropolitan governments with statutory authority over land use planning are extremely rare in the United States. Portland, Oregon's, Metro Council is the best-known example. Since the 1970s this three-county regional government has gradually grown in power and respect, and is in charge of implementing Oregon's statewide land use planning goals in the Portland Area (see Box 10.1, later in this chapter). A similar Metro Council in the Minneapolis-St Paul area, with members appointed by the governor, has somewhat less power over local land use, although it is able to review large regionally significant projects and has overseen a regional tax-sharing framework since 1972. Consolidated city-county governments have also been created in metropolitan areas such as Nashville, Indianapolis, Jacksonville, and Miami as a way of providing more effective regional coordination. However, such regional planning entities have generally been more concerned with providing efficient public services such as police, fire, sanitation, and public health to metropolitan areas than with rethinking patterns of development.

Single-purpose regional agencies exist in most metropolitan areas and have authority under state or federal law to develop policy and programs in fields such as air quality, water quality, parks, public transit, and public utility planning. In addition to its Association of Bay Area Governments, the San Francisco Bay Area, for example, has a Metropolitan Transportation Commission (the MPO), an Air Quality Management District, a Bay Conservation and Development District, a Regional Water Quality Control Board, and numerous water, sewer, and park districts covering portions of the region. Some single-purpose regional agencies are well funded and powerful (the transportation MPOs fit into this category). However, they do not necessarily have a mandate to coordinate with other agencies or local governments around the region, or to "think regionally" in any comprehensive sense. Rather, these agencies focus on a single, narrowly defined function. Regional branches of state or federal agencies (such the US EPA and HUD) likewise concentrate on carrying out particular policies of these higher levels of government within the region, but may not have a mandate to coordinate with other agencies on broad regional policy.

A final notable force in planning urban regions has been NGOs. These citizen groups have grown in number and influence in many areas and can play a significant role by developing plans themselves, influencing plans developed by official agencies, or lobbying for change at the regional level. The Regional Plan Association of New York and New Jersey is the oldest continually active US organization of this sort. Formed in the 1920s by civic leaders and the Russell Sage Foundation, this Association prepared regional plans for the New York metropolitan area in 1929, 1964, and 1996. Although they lacked governmental authority, these visions greatly influenced the work of cities and agencies throughout the region. An even earlier example of citizen-based regional planning was the Plan of Chicago prepared by Daniel Burnham in 1909. This pioneering document proposed regional park systems, transportation systems, and a network of grand boulevards and public spaces. Of these, the lakefront park system was the most consistently implemented.

More recently, citizen organizations and networks have become a powerful force for regional growth management planning and environmental protection.[16] For example, the Chesapeake Bay Foundation has led efforts to protect the Chesapeake watershed since 1967, and boasts more than 100,000 members and 200 staff.[17] The work of 1000 Friends of Oregon and allied groups has been crucial in supporting regional growth management and transit plans in the Portland area. Sister 1000 Friends groups now exist in more than a dozen states nationwide, including Iowa, Washington, Florida, Minnesota, Wisconsin, New Mexico, Maryland, and Hawaii. Starting with the efforts of three University of California faculty wives concerned about Army Corps of Engineers plans to fill much of San Francisco Bay, the Save San Francisco Bay Association succeeded in getting state legislation passed in the 1960s protecting the Bay from development and establishing a regional agency to monitor its health, the Bay Conservation and Development Commission. In a more rural context, property owners and environmentalists in the Lake Tahoe area, led by the League to Save Lake Tahoe, have spearheaded the development of a regional plan to protect water quality and other environmental aspects of that large lake basin in the Sierra Nevada Mountains.[18]

Regional sustainability issues

Those groups wishing to promote sustainability-oriented planning at a regional level typically focus on types of certain sorts of issues best coordinated at this scale, including transportation planning, large-scale land use and growth management planning, watershed and environmental protection, air quality planning, regional equity planning, and regional economic development. The following subsections describe these subjects in greater detail.

Transportation

Ask people about their greatest concerns, which pollsters regularly do, and "transportation" is likely to top the list. Traffic congestion is a fact of daily life for millions and is worsening steadily in most metropolitan areas. Discerning strategies to reduce automobile usage; promote walking, bicycling, and public transit use; and reduce the amount that we all need to travel everyday is a key challenge for more livable and sustainable communities. Although all levels of government must be involved in so complicated a challenge as reducing automobile use, much transportation planning is coordinated at the regional scale. Regional transportation planning policies are also a crucial determinant of metropolitan land use and growth.

Regional agencies can play a role in all three of the areas of transportation planning reform mentioned in Chapter 5: better alternatives, better land use, and better pricing. In the United States MPOs typically prepare Regional Transportation Plans (RTPs) every two or three years to identify projects to be funded over a 20–25-year time frame. These plans help funnel federal and state money into regional transportation infrastructure and transit operations. Until the passage of ISTEA in 1991 such plans were almost exclusively focused on roads and large transit systems. However, this federal legislation gave regional agencies in areas suffering from air pollution increased flexibility to allocate monies formerly used for highways to transit, bicycle, and pedestrian planning and to programs adding amenities and urban design improvements in locations that would support transit. This bill was a breakthrough in that it acknowledged the interdependency between transportation and land use, and allowed transportation money for the first time to be spent to encourage transit-supportive development and urban design. Through such programs, regional agencies can not only fund measures to improve walking and bicycling, but can also provide incentives for better local land use planning, in particular for transit-supportive, pedestrian-friendly urban form, and for regional smart growth visions that will help coordinate local planning.

American urban regions such as St Louis, Atlanta, Portland, Washington, DC, the San Francisco Bay Area, San Diego, Los Angeles, and Dallas have built major new rail transit systems in the past 30 years. Agencies in some of these areas have also begun offering incentives to local governments to plan for new development near transit lines, minimize sprawl, and emphasize pedestrian and bicycle planning. The Atlanta Regional Council's Livable Centers Initiative[19] and the Bay Area Metropolitan Transportation Commission's Transportation for Livable Communities program[20] are two examples of such incentive programs. Both give grants to local governments for the planning and construction of new amenities near transit, and the latter also offers grants of up to $2000 per unit for new housing near transit.

As with state infrastructure funding, regional transportation funds can be used to leverage better land use by making grants to local government conditional on the adoption of good local land use plans that will reduce automobile use and sprawl. This step is more difficult politically and has yet to be taken by US regional agencies, although the US EPA has temporarily frozen transportation funding to metropolitan areas such as Atlanta and the Bay Area that failed to achieve compliance with clean air laws. But bills establishing this sort of linkage have been introduced into the California state legislature, and many observers believe that such "sticks" are needed as well as incentive "carrots" to induce local governments to change their ways.

Regional agencies can help coordinate local action in other areas of sustainable transportation policy as well, such as creating pedestrian- and bicycle-friendly streets, regulating parking charges, adopting transportation demand management programs, and ensuring transit- and pedestrian-friendly urban form. TDM programs are especially significant, and as discussed earlier, may consist of a wide variety of inducements for people to drive less. These incentives may include increased parking changes, Eco-Pass programs, preferential parking and rebates for car pools, new shuttle vans, bike facilities, "guaranteed ride home" programs in the event of family emergency for workers who carpool or take transit, and many other innovative services. Although no one TDM program is likely to greatly reduce driving by itself, many such initiatives combined with land use changes and better transportation alternatives can begin to reverse the unsustainable growth in automobile use.

Land use, growth management, and regional design

Stopping sprawl and revitalizing central urban areas through new development are major concerns in most metropolitan regions, even in relatively rural regions where low-density exurban development may be occurring around small towns and rural highways. Although they rarely have direct authority over land use, regional institutions can still play a major role in coordinating growth management and smart growth by bringing local governments together, facilitating consensus on better growth directions, providing incentive funding for local governments to implement regional growth goals, and potentially withholding funding from local governments who refuse to take regional needs into account.[21] As previously mentioned, transportation funding is a particularly important lever that regional planning agencies have to shape metropolitan growth. The extension of water and sewage systems also supports suburban sprawl. Regional decisions about where to build such infrastructure help determine the growth and form of a region. Building new suburban-serving freeways is likely to encourage further sprawl, as many metropolitan areas have found out. Conversely, regional decisions to forgo highway expansion and promote public transit serving existing urban areas may help limit sprawl.

Outside the US, regional agencies can potentially take a more overt and ambitious approach to shaping regional land use. During the past 100 years there have been several waves of efforts to design metropolitan regions in ways that manage growth; coordinate transportation, land use, and housing; and preserve open space. The Garden City visions of Howard Unwin and others, first conceptualized in the late nineteenth century, eventually bore fruit in the 1940s and 1950s in plans for many European metropolitan regions such as London, Copenhagen, Stockholm, and Paris. In South Korea, the nation's government established a broad regional greenbelt about 10 kilometers wide around Seoul in the post–Second World War period, and undertook construction of several large new towns outside the greenbelt, following the British model.[22] In the early 1960s regional planners in the Washington, DC area developed a "star" plan for development in that region, calling for future development to occur in radial corridors along transit lines. However, unlike the European examples, Washington had no strong agency to bring the vision about, and this plan was neglected by local governments. (See Figure 10.1.)

Beginning in the 1990s many planners even in the US have again focused attention on regional design.[23] New Urbanists such as Peter Calthorpe have been particularly active in this regard,[24] and have actively consulted with regional agencies in places like Portland, Toronto, Salt Lake City, and Minneapolis. Many of these visions are strikingly reminiscent of earlier garden city concepts, in that they cluster transit-oriented communities along rail lines throughout a metropolitan area, often with lower-density areas or open space between these. This basic regional design concept has proven remarkably resilient and attractive to different generations of planners.

Portland, for example, has designated nodes of density along stations of its new MAX light rail system, and intensified corridors of development along heavily used Tri-Met bus lines. These transit lines radiate out from downtown Portland, creating a traditional star-shaped pattern.

A wide range of regional planning documents helps implement such visions. These plans help coordinate land use with transportation systems, create regional park and greenspace systems, ensure economic development, and/or meet affordable housing needs across the region. Examples during the 1990s included the Region 2040 plan prepared by Portland's Metro, the Vancouver regional plan prepared by the Greater Vancouver Regional District, the Livable Metropolis plan prepared by Toronto's Metro Council, and the New York Regional Plan prepared by the Regional Plan Association of New York.

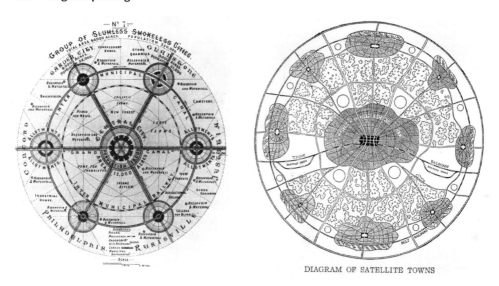

DIAGRAM OF SATELLITE TOWNS

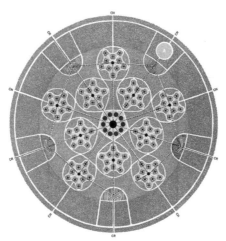

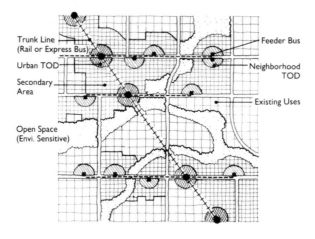

Trunk Line
(Rail or Express Bus)

Urban TOD

Secondary
Area

Open Space
(Envi. Sensitive)

Feeder Bus

Neighborhood
TOD

Existing Uses

Exploration into ideal regional design values was stimulated in the 1960s and 1970s by MIT planning professor Kevin Lynch, whose writings sought to explore the characteristics that make regions work well for people. In his most influential book, *A Theory of Good City Form* (1981), Lynch analyzed normative urban design values that help produce a "good" urban region. He first conducted a sweeping historical review to identify motivations influencing urban form throughout history, including maintaining property and tax values, increasing access, improving defense, maintaining political control, making profit, enhancing safety and physical health, and preserving desired environmental characteristics. He then prioritized a number of key urban design values that he believed appropriate for contemporary cities:

> So what is good city form? Now we can say the magic words. It is vital (sustenant, safe, and consonant); it is sensible (identifiable, structured, congruent, transparent, legible, unfolding, and significant); it is well fitted (a close match of form and behavior which is stable, manipulable, and resilient); it is accessible (diverse, equitable, and locally manageable); and it is well controlled (congruent, certain, responsible, and intermittently loose). And all of these are achieved with justice and internal efficiency.[25]

These concepts may sound abstract and academic, but all have real and tangible implications. "Legible" urban form, for example, refers to city design that can be easily understood by the average resident, that can be "read" without undue difficulty. Grids and other relatively regular, connected street patterns tend to assist with this characteristic. Meandering, fragmented suburban street patterns are often not particularly legible. "Well-fitted" urban form is one in which buildings, block sizes, and other physical dimensions of the city match human needs and activities, typically through a relatively fine-grained, human-scale urban environment. Enormous modernist buildings and "superblocks" would not meet this criterion; contemporary malls and housing tracts might not either. "Accessible" urban form refers to buildings and spaces that can be entered, used, and even modified by people – in which space is not rigidly controlled, such as in gated communities or many office parks, but is open to public use and can be adapted to the needs of different users over time. (Likewise, in his 1994 book *How Buildings Learn* Stewart Brand argues that the most interesting and livable buildings in the long run are ones that can be adapted and modified over the years by a wide range of inhabitants.)[26]

Lynch's city form values are certainly not the last word in terms of a normative urban form philosophy. Most of his work, for one thing, took place before the full impact of the environmental movement had reached the design professions, and environmental dimensions of sustainability are largely left out. Some of these design values are also overly abstract or formal. Yet Lynch played an enormous role in inspiring designers to rethink urban and regional form in ways that lay the groundwork for more sustainable development. More than anything, he pointed out the importance of discussing ideal urban form values, studying evidence about the actual effectiveness of different urban design from past cities and towns, and figuring out ways to achieve urban design values in new development and planning.

Figure 10.1 (opposite) Garden City regional design concepts. Similar regional design concepts have occurred time and again since Ebenezer Howard's famous Garden City vision of 1898, typically clustering development around transportation systems.

Another relevant framework of design principles, prepared at about the same time as Lynch's, was the "pattern language" developed by architect Christopher Alexander and colleagues (1977) in the architecture department at the University of California at Berkeley. These writers identified a range of design characteristics that they saw as relatively universal in good urban areas around the world. In terms of regional design, particularly important themes included a "agricultural valleys," "city country fingers," "scattered work[places]," a "mosaic of subcultures," "density rings," and a "web of public transportation."[27] In the view of Alexander and his coauthors (1979), urban development was best pursued through an organic, incremental process in which each builder followed these time-tested patterns and considered the entire context surrounding his project before beginning development.[28] This thoughtful, Zen-like approach is quite at odds with contemporary development, in which extremely rapid building is the norm. Yet it does point towards an alternative urban design philosophy, at the regional scale as well as that of the neighborhood and individual buildings, that can potentially be more responsive to cultural, ecological, and historical contexts.

A major question concerning regional design plans is how they are to be implemented. Such action requires a national, state, or regional agency with considerable leverage over local land use decisions. Generally such plans have been most successful in areas where strong higher-level government authority can help bring about regional and local land use changes, such as in the State of Oregon. National governments have played this role for metropolitan regions such as London, Copenhagen, Stockholm, the Randstad in the Netherlands, Paris, and Seoul. But in the absence of such higher-level leadership, extensive collaboration between local governments could potentially develop and implement regional design as well. Given the strong tradition of local government land use authority in the United States and resistance from federal and many state governments to overriding this, what will probably be necessary in this country is a combination of local and regional coordination with a limited amount of guidance and incentive from higher levels of government. Over time such a decentralized institutional framework may be able to build consensus around new regional growth directions. Of course, a stronger and more proactive role by federal and state governments should be sought as well, and in the long run would greatly assist in the regional planning process. (See Box 10.1.)

Open space, watershed, and environmental planning

It makes sense that plans for park systems, ecologically sensitive natural areas, and air and water quality should be prepared on a regional scale, since watersheds, airsheds, and ecosystems are regional in nature rather than limited to the bounds of any municipal jurisdiction. Some metropolitan environmental planning was done as early as the nineteenth century, for example, resulting in Frederick Law Olmsted's "Emerald Necklace" system of parks for Boston. Turn-of-the-century "City Beautiful" planning also resulted in metropolitan open space plans that might be termed "regional." These helped to produce Chicago's waterfront parks, Washington, DC's system of parks, monuments, and public spaces, and the extensive set of parks in Buffalo, New York. In England, greenbelt planning helped create impressive rings of open space around metropolitan regions in the 1940s. London's Greenbelt, set forth in the 1944 Abercrombie Plan, remains one of the most significant regional open space achievements. But most systematic efforts at regional environmental planning – especially with habitat and ecosystem management goals in mind – are relatively recent, dating back at most to the 1960s.

Box 10.1 Regional planning in Portland, Oregon

Agencies in the Portland, Oregon metropolitan area have been refining regional planning visions for more than 40 years. In 1966 the fledgling Metropolitan Planning Commission prepared three alternative scenarios under the heading "How Shall We Grow?" The options as planners saw them then were for the region to form a "Lineal City" with development in a broad north–south corridor down the Willamette River valley, "Regional Cities" with growth dispersed into 25 separate and relatively self-contained cities, and "Radial Corridors" in which development clustered around mass transit in several corridors radiating from the central city, with surrounding land preserved as open countryside.

As required by state growth management legislation, the strengthened Metro regional government approved an Urban Growth Boundary in 1979 around the entire metropolitan area (except for that portion that had already sprawled across the Columbia River into the state of Washington). During the 1970s planning began for the MAX regional light rail system, and new parks, public spaces, and infill development projects revitalized downtown Portland.

In the 1990s the Metro Council developed a highly detailed Region 2040 plan emphasizing transit-oriented development, infill within existing urban areas, creation of a regional greenspaces system, and improved design of new neighborhoods. Regional planners then began a long process of working with local governments to change zoning codes and other standards to reflect the regional vision.

For more information, see the websites of the Metro Council (http://www.metro-region.org/), 1000 Friends of Oregon (http://www.friends.org/), or Abbott (1997) and Abbott *et al.* (1994).

A number of both rural and metropolitan regions have developed regional open space plans that seek to meet ecological objectives such as maintaining wildlife habitat, creating wildlife corridors between existing parks, and restoring watersheds, in addition to managing urban growth and providing recreational amenities. In rural areas in particular such planning is widespread as federal, state, or regional park districts – or consortiums of agencies – seek to enhance ecosystem and recreational values across large areas. The Sierra Nevada Ecosystem Project (SNEP), for example, was a comprehensive assessment of the health of the Sierra Nevada region in California and Nevada, requested by Congress in 1992 and conducted by regional offices of the US Forest Service working with a variety of local organizations and researchers. Assessing social and economic systems as well as the region's environment, SNEP produced a final report in 1996 that has influenced other more local planning throughout Sierra Nevada foothill counties, national forest lands, and mountain parks.[29] In Florida, the Comprehensive Everglades Restoration Plan produced one of the world's largest regional ecosystem restoration plans in 2000. Covering sixteen counties and 18,000 square miles, it focuses on improving natural water flow though this fragile habitat. The plan was jointly developed by the South Florida Water Management District and the US Army Corps of Engineers – ironically the very agency that by channelizing and diverting rivers created many of the region's environmental problems in the first place.

In metropolitan areas coordinated open space planning is also on the rise. For example, in 1992 Portland's Metro Council adopted a Metro Greenspaces Master Plan that provided the blueprint for acquiring fourteen key natural areas. The regional agency subsequently developed a Water Quality and Floodplain Protection Plan in 1996 and crafted a Fish and Wildlife Habitat Protection Plan in 2002. These plans fit together with the region's overall Region 2040 plan to help coordinate environmental planning and urban growth throughout the three-county, 1.8-million-resident region. The New York Regional Plan Association (RPA) proposed a "greensward" network of open spaces as a major element of its 1996 regional plan for the New York metropolitan area.[30] The plan calls in particular for the creation of eleven Regional Reserves to conserve waterways and working landscapes in the region, for increased investment in urban parks, and for a network of corridors and greenways providing recreational opportunities as well as benefits for wildlife. Although advisory in nature, the recommendations in this RPA plan, as in 1929 and 1968 versions, are likely to be taken very seriously by local governments in the region. Finally, in the vicinity of the Telford new town west of Birmingham, England, regional authorities have established a "Green Network" covering some 6000 acres. The former site of mining for coal, ironstone, clay, and limestone, this region is one of the world's oldest industrial landscapes and has been designated a World Heritage Site by UNESCO. Now officials are seeking to restore forests, meadows, and fens, in part through the planting of six million trees and ten million shrubs, while creating 25 miles of footpaths and creating or preserving a mosaic of wildlife habitats.[31] (Also see Box 10.2.)

Regional air and water quality plans are usually developed in response to state and/or federal legislation by regional agencies specifically set up to handle those issues. In the area of air quality planning, many regional plans have been prepared to implement the 1972 federal Clean Air Act and various subsequent state legislation. Such air quality plans aim at meeting threshold criteria for ozone, nitrogen oxides, carbon monoxide, sulfur oxides, and particulate emissions established under this legislation, and can regulate motor vehicle use, smokestack emissions, dry cleaners, use of painting solvents, backyard

---

Box 10.2 Open space protection in the Vancouver region

The Vancouver, British Columbia area has taken a number of successful steps to protect its stunningly beautiful natural landscape from development (see Tomalty, 2002). Many of these initiatives have been spearheaded by the Greater Vancouver Regional District (GVRD), established in 1967. This agency approved its first Livable Region Plan in 1975, and a successor Livable Region Strategic Plan in 1996. Based on a Compact Region Scenario, the latter plan included a Green Zone component protecting areas of great social or ecological value from development, and establishing an Urban Growth Boundary for the urban area. The GVRD required local cities to designate lands to be protected under this framework. By 1999 the agency had set aside some 440,000 acres in this Green Zone. Many of these lands are additionally protected through public ownership, park designation, or inclusion in the region's Agricultural Land Reserve.

The Vancouver region is also trying to manage urban growth so as to limit the overall urban footprint, with mixed success. The core city of Vancouver has gained substantial population in recent years, but the largest proportion of growth is taking place in outlying municipalities such as Surrey and Richmond. Other goals of regional planning include the requirements that "complete communities" be built instead of single-use subdivisions, and the support of alternative modes of transportation. For more information, see www.gvrd.bc.ca/index.html.

barbecues, and many other sources of pollution. In the Los Angeles area, for example, the South Coast Air Quality Management District, established in 1977, prepared regional air quality plans in 1982, 1989, 1991, and 1997 setting out hundreds of "control measures" to reduce emissions.[32] These measures included cleaner vehicle fuels, retrofits to bus engines, bans on burning garbage, a program of emissions trading among industries, and many other steps. Although still only partially successful, the agency's plans and other state regulation of motor vehicle emissions have resulted in a substantial decline in health alerts and emissions violations. If regions remain out of attainment with federal air quality standards because of motor vehicle use, the EPA has the power to freeze federal transportation funds that might be spent on highway projects. Environmentalists can also sue to force regional agencies to comply with federal or state air quality law.

Regional water quality initiatives attempt to implement standards set by the federal Clean Water Act and a variety of other legislation. Since the original 1972 Act these efforts have focused particularly on better sewage treatment facilities, since historically many North American municipalities have dumped poorly treated sewage into waterways, and such pollution still often occurs when wet weather causes systems to overflow in cities that do not have separate sewage and storm drain systems. Water quality efforts initially also sought to regulate "point sources" of pollution (pipes or contaminated sites discharging pollution into watersheds), and somewhat belatedly came to recognize the importance of "non-point sources," including pollutants that drain off roads and parking lots, that wash out of the air during rainstorms, that drain off agricultural fields, golf courses, or construction sites, or that likewise do not have a specific, discrete source.

Collaborative efforts in recent years have sought to develop consensus on some of the most large-scale and difficult water quality challenges in North America. These efforts include the Chesapeake 2000 plan for the Chesapeake Bay Watershed,[33] the 1994 Ecosystem Charter for the Great Lakes-St Lawrence Basin area,[34] and the CALFED Bay Delta Estuary Process for the San Francisco Bay estuary system.[35] In the Great Lakes region, efforts to combat pollution – which led to Lake Erie being declared virtually dead in the 1960s – began with the first Great Lakes Water Quality Agreement signed by agencies of the US and Canadian governments in 1972, and continued with work by many cities, states, provinces, regional branches of federal agencies, and collaborative forums such as a Great Lakes Commission consisting of officials from throughout the region.[36] The CALFED project was an enormous multi-year consensus-building effort between 16 government agencies and many private stakeholders including farmers and environmentalists. Initiated in part because of the threat of unilateral action by the US EPA Region IX office, the process achieved a historic 2001 Record of Decision that set forth agreement on many basic principles, such as expanded water conservation programs, an additional 300,000 acre feet of water annually to improve ecosystem health, and policy that new water projects should be paid for by the users rather than government. Implementation will undoubtedly be a continuing struggle given California's fierce politics around water. Still, CALFED does represent one of the most far-reaching water quality plans prepared anywhere in the world to date.

Regional equity planning

Since the rise of the environmental justice movement in the 1980s and (to a lesser extent) the smart growth movement in the 1990s, growing attention has been paid to questions of equity within the metropolitan region. Planners and public agencies have increasingly considered the degree to which lower-income neighborhoods and communities of color are exposed to toxic chemicals, pollution, and locally unwanted land uses (LULUs). Some organizations and political leaders have also begun working to address inequities in regional transportation planning and growing disparities in income and wealth between central cities and suburbs. However, equity initiatives to address all these problem areas are still in the early stages.

At the regional scale the widening gap in income, wealth, and tax base between suburbs and central cities is one of the most dramatic and rapidly growing forms of inequity. Also, within the suburbs themselves, a gap is widening between more affluent or rapidly growing jurisdictions and older, inner-ring communities that have a stable, working-class population and aging infrastructure. Essentially, lower-income and minority populations are penalized as businesses and upper-income residents move to more affluent or rapidly growing suburbs on the urban fringe. Central city groups in particular may no longer have access to many decent-paying jobs, and their city governments have fewer tax dollars with which to meet service and infrastructure needs. Authors such as Myron Orfield, David Rusk, and john powell argue that these disparities demand action.[37] One approach towards reducing such disparities, typically resisted by higher-income communities, is to ensure that each municipality zones and plans for a variety of job and housing opportunities suitable for all income groups. Court decisions in many states have also mandated that state governments provide funding to reduce disparities in educational funding between rich and poor jurisdictions. However, despite these "equalization" programs disparities remain, and are made worse by the fact that parent groups in affluent neighborhoods are often able to contribute substantial time and money to improving the quality of their children's education.

Another basic strategy to address resource inequities has been to advocate for regional tax sharing, which helps even the tax base between rich and poor communities, and also has the benefit of reducing incentives for local governments to zone for high-tax-generating (and sprawl-inducing) land uses such as regional malls and automobile dealerships. Under regional tax-sharing frameworks, local governments contribute all or some portion of sales or property tax receipts into a regional pool, which then reallocates revenues based on population or need. Currently the Minneapolis-St Paul area is the only metropolitan region in the US that does this on a large scale. Since 1972 40 percent of new sales taxes generated by local development in the Twin Cities area has gone into a regional pool distributed by population. State legislator Myron Orfield credits this mechanism with reducing disparities in the commercial and industrial tax base between rich and poor communities of substantial size from 18 to 1 to 5 to 1.[38]

Of course, affluent communities are likely to fight tax-sharing tooth-and-nail. Orfield recommends a "metropolitics" strategy in which central cities and declining older suburbs form a political coalition against wealthy new suburbs to pass regional tax sharing in state legislatures. But needless to say, this will not be easy. Another, potentially more workable strategy would be for state or federal governments to step in and reform the tax system so that sales and perhaps property taxes are collected by higher-level governments and redistributed to localities through revenue-sharing formulas based on population. Nixon's federal revenue-sharing program begun in the early 1970s provides something of a model of how this might be done.

Regional equity has been sought in the realm of transportation planning by groups fighting to ensure that investment does not favor affluent suburban communities at the expense of central city residents. Often this debate occurs over whether funding is going preferentially to suburban-serving freeways and commuter rail extensions rather than to central city transit. In a 1993 lawsuit settlement, for example, the Bus Riders Union of Los Angeles won a promise from the Metropolitan Transportation Agency not to raise fares and to purchase 500 new clean-fuel buses. The agency had previously proposed to raise fares and cut service in part to pay for its expensive new Metrorail system. In 1998 an environmental justice coalition in Atlanta sued the Atlanta Regional Commission over its policy of emphasizing suburban road-building, specifically exempting "grand-fathered" suburban road projects that had supposedly been approved in the past from Clean Air Act restrictions on road-building. The coalition won a settlement eliminating 44 of 61 grandfathered road projects and freeing up money for transit projects.[39] Such transportation justice debates are just beginning in many metropolitan areas. (See Figure 10.2.)

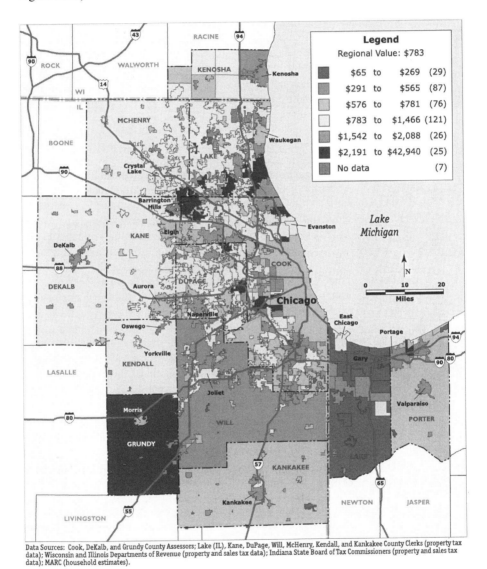

Data Sources: Cook, DeKalb, and Grundy County Assessors; Lake (IL), Kane, DuPage, Will, McHenry, Kendall, and Kankakee County Clerks (property tax data); Wisconsin and Illinois Departments of Revenue (property and sales tax data); Indiana State Board of Tax Commissioners (property and sales tax data); MARC (household estimates).

Figure 10.2 Inequities in a metropolitan region. A map of tax capacity per household in the Chicago area shows wealthy areas concentrated to the north and impoverished areas to the south of the city. From Myron Orfield's *American Metropolitics*, published by The Brookings Institution (1997).

Regional economic planning

A final area of regional planning concerns the field of economic development. Ironically, for decades regional planning within academia has focused on questions of economic development and economic geography, but real-life regional institutions and planning mechanisms for economic development are few and far between. Most economic development planning is handled at municipal or state levels instead. Moreover, scholarly regional economic research has not focused on sustainable development, but rather has tended to take for granted that conventional forms of economic growth are desirable and that the goal of regional planning should be to encourage these. Nevertheless, new directions in regional economic planning are possible if local governments coordinate their actions. Important steps are likely to include encouraging greater regional self-sufficiency in agriculture and key industries, developing regional programs for industrial recycling and sustainable materials use, coordinating investments in education so as to ensure a skilled regional workforce, and ensuring fair regional wages and hiring policies.

Regional economic planning is often a more pressing concern in developing nations than in industrialized countries such as the United States. In Third World countries national governments often play a greater role in developing economic policy, channeling funding to particular regions, encouraging businesses to locate there, and developing educational systems and infrastructure to promote regional development. In the United States, in contrast, the national government avoids specific spatial planning within economic development policy, with noted exceptions such as the establishment of the Tennessee Valley Authority in the 1930s and programs to reduce poverty in Appalachia in the 1960s. Nevertheless, other federal decisions have had profound and largely unacknowledged regional implications. For example, decisions to expand the military budget in the 1980s promoted the growth of sunbelt states – especially Southern California, Arizona, Texas, and Georgia – where these industries were located. Likewise the traditional federal subsidy of water in the West contributed enormously to the boom of California, Arizona, Nevada, and other states, as well as the decline of agriculture, industry, and population in many eastern and Midwestern communities.

Many of the most controversial regional debates in recent decades, such as over logging in the Pacific Northwest, have focused on "jobs vs. the environment." As sustainable development with its "Three Es" philosophy became a more widespread movement in the 1990s more people came to understand the falseness of this dichotomy. More and better jobs can be available in the long run through a sustainable economy. While traditional exploitative industries will suffer in the short term during the transition, opportunities will emerge in new fields such as sustainable forestry, pollution abatement and control, energy efficiency, renewable resource use, recycling, public transportation, ecological manufacturing, and ecological education.[40]

Whether traditional concepts of economic growth will be part of a regional sustainability paradigm is doubtful. Urban regions that have experienced rapid economic expansion in recent decades have often paid a high price in regional side-effects such as population growth, suburban sprawl, traffic congestion, and unaffordable housing. Additional jobs may actually lower standards of living if the employers pay low wages or attract new workers to the region. A study of 100 metropolitan regions by Paul D. Gottlieb for the Brookings Institution found that employment growth leads to population growth but not necessarily to per capita income growth.[41] In times of rapid growth local governments often have difficulty keeping up with a flood of development applications, and have little time to plan proactively for smarter development. Conversely, during the "bust" phase of economic cycles cities and towns are often desperate for development of any kind, and will approve virtually anything that is proposed. A more stable middle ground is needed, in which regions and cities do not depend on continuous economic expansion for their well-being. Gottlieb argues for a strategy of "growth without growth" through which regions would seek to increase income levels – through a focus on the types and quality of jobs – without the usual business-chasing that passes for economic development.

Regions such as the Bay Area and Atlanta offer perhaps the most extreme examples of the negative effects of regional economic growth. Atlanta expanded very rapidly during the late twentieth century, but became the nation's worst example of rapid suburban sprawl, surpassing even Phoenix and Los Angeles. Traffic congestion grew enormously, producing the nation's longest average commutes, and air quality declined, leading the US EPA to slap sanctions on the region. In the Bay Area, Silicon Valley generated extremely rapid growth in wealth and average incomes during the 1990s. However, this was accompanied by a dramatic increase in traffic jams, long-distance commuting, and housing prices. By the end of the decade the region had the most expensive housing in the nation, and many workers were being forced to commute 100 miles or more from new subdivisions in California's Central Valley. Yet growth was so deeply entrenched as a regional priority that the topic of reducing the extent to which cities zoned land for new jobs was rarely broached directly even among environmentalists.

Sustainability planning implies a sea change in economic development priorities. Instead of simply seeking to add jobs as fast as possible, regions will need to nurture industries that can provide quality jobs for existing residents. Instead of trying to become the next Silicon Valley, regions will need to prioritize growth in quality of life, community livability, ecological health, and human well-being. This more balanced set of goals is what sustainable regional development is all about.

* * *

At the regional scale, then, sustainability planning calls for much-improved coordination to strengthen relatively weak planning institutions, and for action in areas such as metropolitan growth management, planning of large-scale transportation systems, air quality, water quality, and equity. For many of these areas, such as air quality and transportation planning, the region is the natural scale at which action can be taken. For others, such as growth management and equity, local actions are insufficient since neighboring municipalities can so easily undercut each others' initiatives. Although many debates exist over how regional planning agencies are best structured – multi-issue or single-purpose, directly elected or appointed, with wide boundaries or relatively narrow – they can potentially play a powerful role in coordinating sustainable development regionally. Where new agencies cannot be developed, better coordination between existing institutions is needed instead.

11 Local government planning

It is at the local scale – consisting of city and county government plus a wide variety of special-purpose districts – that most urban development planning is done in the United States and many other nations. Cities and counties represent the front lines of planning in that they have primary control over land development, local streets and roads, bicycle and pedestrian facilities, recycling and waste collection, local parks and greenways, K-12 education, and many economic development, housing, and social welfare programs. Although higher levels of government establish the framework within which local planning takes place and may provide incentives, mandates, and funding, implementation of many important policies and programs occurs locally.

For decades local planners and government officials have thus played the lead role in approving suburban sprawl and other unsustainable development. These local leaders face pressure from landowners, developers, and businesses with interests oriented around traditional forms of urban growth and economic development. Information related to more sustainable paths of development is often lacking, and short-term pressures related to the next election cycle work against a long-term local planning perspective. NIMBY attitudes of local residents also enforce parochial attitudes, as neighborhoods seek to protect the status quo and local property values by keeping out much-needed affordable housing, community facilities, infill development, and even public transportation. Though neighborhoods do often have legitimate concerns, they frequently oppose any change even if it would have benefits for the city, the region, or the planet. In part an oppositional local politics has become a way for people to express their frustrations with modern life and concern about larger scale social and economic change is displaced onto local issues that residents feel they have somewhat more control over.

As a result of such forces there has been an enormous disconnect between local government decision-making and regional or global problems. The challenge is to put in place a different set of incentives and processes to change this situation and encourage more proactive local sustainability planning. Enlightened local officials and residents can and should be at the forefront of efforts to create more livable communities, preserve or restore elements of local ecosystems, and improve social equity and human well-being. There is already movement in this direction in many localities. Since the early 1990s local governments have played a much more active role within international sustainability conferences, as a part of the broader growth of civil society, and have often networked with each other and with NGOs directly, bypassing national governments.[1] Examples of local action have in turn inspired action in other cities and towns, regardless of national policy. In the more flexible, multi-level world of governance that lies ahead, there is certainly room for local officials to play a leading role in this way.

Who plans at the local level?

As mentioned previously the planning profession originated in large part through local government efforts to implement zoning and build infrastructure. Most urban planners are still employed by institutions at the local level. The agencies doing local planning have proliferated over the years and are far more numerous and varied than is commonly realized.

Local government in the United States is made up cities, counties, unincorporated towns or townships, and special districts created on an ad hoc basis to handle everything from schools to flood control. Cities are incorporated entities governing contiguous urbanized areas of land. City governments are typically formed when citizens in an urbanizing area (often led by local business or community groups) petition state legislatures or counties for authority to govern themselves. The resulting municipalities may be of several types. "Home-rule" cities operate with their own charters that have been approved by state governments, whereas small towns and all counties operate under state law with a more limited set of powers. "Cities" are not necessarily huge urban places; most suburban or rural towns are also incorporated in this fashion. A typical large metropolitan area now contains dozens or hundreds of city governments.

Over time, municipal boundaries often expand to include newly urbanizing land through a process of annexation. However, older cities may be hemmed in by other municipalities and cannot expand further. During the twentieth century most suburban communities in US metro regions incorporated as cities in part to prevent their takeover by older municipalities. The unfortunate result is a highly fragmented political landscape within each metropolitan area, with numerous suburban municipalities clustered around the original core, and each jurisdiction on the fringe rapidly gobbling up any remaining unincorporated county land. This fragmentation leads to competition between cities for economic development and tax base, and encourages each city or town to protect its own local interests rather than thinking regionally or globally.

Elected City Councils, usually chaired by mayors, run municipal government. These councils vary in size, often from around five to eleven members. Development and planning decisions are typically handled by appointed Planning Commissions, Zoning Boards, Design Review Commissions, and other city bodies, as well as by city planning staff who are permanent, paid municipal employees. Staff may in turn hire a variety of consultants to assist in planning processes, for example to examine environmental impacts of proposed development, to develop detailed designs for new public spaces, or to conduct public workshops. In many cities appointed City Managers coordinate staff and run day-to-day operations of city government. All these individuals and groups have input into decisions that affect community sustainability, and can initiate actions that promote sustainability goals at a local level.

Counties are larger jurisdictions governing land use in non-incorporated areas, often providing services such as police and fire protection, parks, schools, and sanitation outside city boundaries. With control over rural or sparsely populated land at the urban fringe, counties play an important role in determining the character and pace of new development. Typically county governments are more strongly pro-growth than incorporated cities, and have fewer planning and zoning mechanisms to limit or regulate development. They are therefore a key battleground in fights against suburban sprawl, for example in efforts to set up UGBs or to establish design standards for more compact and livable neighborhoods. Elected Boards of Supervisors, analogous to City Councils, run county government, although these typically have less power since counties operate more closely under the guidance of state governments and do not have their own charters. Counties also maintain their own planning staffs and planning commissions.

Special districts are single-purpose entities set up to handle particular functions ranging from public schools to street lighting to parks across a geographic area. Special districts often overlap cities and may bear no relation at all to other jurisdictional boundaries. Large metropolitan areas may have thousands of special districts that make the governmental picture very confusing. The San Francisco Bay Area, for example, has 9 counties, 101 cities, and at least 721 special districts, not counting school districts. The New York metropolitan region covers parts of 3 states and 31 counties, and is estimated to contain more than 2000 local governmental units.[2] This proliferation of special districts poses enormous problems of coordination, but also many opportunities, in that each can play a role in increasing urban sustainability. While local governments may be hemmed in by political or institutional limitations, autonomous or semi-autonomous special districts may be more free to take action. A park district may undertake an intensive ecosystem restoration program, a transit agency may work to improve bus service, or several cities together may establish a "joint powers authority" to purchase and protect open space.

The landscape of local institutions varies somewhat from state to state. The State of Massachusetts, for example, has no county governments – cities or towns have jurisdiction over all the state's land. Some of these local governments are also still governed by the traditional town meeting format, one of the few forms of direct (as opposed to representative) democracy. In many western states, by contrast, cities are few and far between. Most land is controlled by counties, which are dominated by rural political interests and are often strongly pro-development. Some states have township governments that are less formal than cities and have less "home rule" power under state law. Louisiana uses the term "parishes" instead of counties. And so on.

Outside the United States, variation in the nature and extent of local government powers is even greater. In Canada provincial governments – analogous to US states – have almost total power over cities, and can dissolve municipalities or rearrange their boundaries at will. Provinces often review municipal land development decisions as well. In much of Europe and Asia, national governments play a more active role in overseeing local planning, especially for capital cities that are of great national importance. National governments in Britain, France, Denmark, and Sweden for example have historically made major physical planning decisions for the capital cities of London, Paris, Copenhagen, and Stockholm. The same pattern holds true for large Asian cities such as Bangkok, Tokyo, and Seoul.

Local sustainability issues

As with other scales of planning, some issues are particularly salient at the level of local government. Following are some of these.

Zoning and development permitting

The basic mechanism through which cities in the United States regulate allowable types and densities of development, following land use guidelines established in planning documents, is known as zoning. (British planning, in contrast, places a smaller emphasis on zoning and instead relies on the discretion of local authorities in interpreting the public interest when deciding whether to permit development. National authorities may also review local development applications.)[3] In those countries that rely on it, zoning occupies a central place within the set of local planning tools that can promote sustainability.

However, zoning has been problematic since New York City adopted the first comprehensive citywide ordinance in 1916.

The case against zoning is strong. It has helped to institutionalize sprawl by mandating low residential densities and designating large areas within cities for strip development, regional malls, and office parks. Parking standards contained within zoning codes typically mandate that new development must provide several parking spaces per residential unit, encouraging a suburban style of development, raising housing costs, lowering densities, and promoting automobile use. Zoning typically prohibits apartments, duplexes, townhouses, and many other forms of higher-density housing from single-family-home neighborhoods, reducing the range and affordability of housing types and further reducing urban densities. Most troubling of all, zoning has been used historically as a device to keep poor people and minorities out of affluent areas,[4] usually by establishing requirements for large lots and setbacks that will ensure that only expensive housing is built ("exclusionary zoning"). For these and other reasons, rethinking zoning must be a key element of sustainability planning.

Extensive zoning codes are a relatively recent planning tool. Throughout the nineteenth century and the early decades of the twentieth, land development in North America was far less regulated than at present. Cities and counties did not restrict the allowed uses for given parcels of land or regulate the height, volume, density, or setbacks of buildings. During the Progressive Era the idea arose that government should regulate such criteria in order to protect the health, safety, and property values of homeowners. Citizens often felt a need to protect residential neighborhoods from smokestack industries or rapid encroachment of apartment buildings. The rise of scientific management approaches to city planning also contributed to the attractiveness of zoning – it provided a set of formalized regulatory tasks to rationalize the emergence of a new profession (planning) that would use scientific analysis to manage urban development.

Although pioneered in German cities such as Frankfurt in the 1890s, zoning laws first appeared in the US in the late 1910s and 1920s. These early versions were relatively simple, dividing cities into a few basic categories of allowed land uses. In the decades after the US Supreme Court upheld their constitutionality in the landmark 1926 decision *Euclid* v. *Ambler*, zoning schemes became increasingly detailed. Planners added requirements to cover building setbacks from lot-lines, maximum lot coverage ratios, minimum parking requirements, and many other aspects of site and building design. Today virtually every city and town has an extensive zoning code in place, with a few notable exceptions such as Houston. Zoning even extends to rural areas; counties typically zone agricultural land and open space outside cities for minimum parcel sizes of 5 to 100 acres or more. Zoning is the main tool for implementing the land use policy contained in a General Plan, and state law generally requires local zoning designations to be consistent with this General Plan land use vision. Unfortunately, zoning has proven to be primarily a negative planning power, in that its main function is to prevent certain types of unwanted development from happening. In the United States this restrictive tool was generally not balanced by positive and proactive planning to ensure that desired forms of development do in fact happen. Such an active public sector role in planning, land ownership, and development happened more frequently in European countries.[5]

Given its unfortunate impacts in the past, should zoning be eliminated or reduced in scope within sustainability planning? The answer is probably no. For one thing, eliminating zoning, although enticing as an idea, would be greeted by storms of political opposition in most communities, and may not be desirable given the need to respect existing neighborhoods and homeowners. The need is not so much to eliminate zoning altogether, or for that matter to add a host of new zoning requirements, but to use the zoning tools we have more wisely. In some cases that will mean loosening up the overly stringent and bureaucratic requirements that have been set up over the past century, for example by allowing a greater range of land uses or housing types in a given area. In other cases it will mean adopting new standards to nudge an often recalcitrant building industry towards more sustainable development, for example by adding minimum densities for residential development and requirements for water-conserving landscaping and setbacks from creeks. Maximum floorplate sizes for commercial buildings might also be set, to prohibit "big box" retail stores in many areas.

New Urbanist architects such as Andres Duany and Elizabeth Plater-Zyberk have developed a somewhat different approach of establishing a detailed urban design code for each new development that takes the place of zoning. Such "form-based" codes provide easy-to-understand diagrams showing what street, lot, and building designs are permissible. Form-based codes may be easier to understand and implement – and may better respond to needs for livability and to mesh with architectural traditions – than traditional lengthy, bureaucratic zoning codes.[6]

Since the 1990s many communities have been revising zoning codes to implement principles of the New Urbanism and smart growth, and further changes are likely in the years ahead to promote sustainability. Major areas of focus include:

- Producing a greater mix of land uses by reducing large areas of single-use zoning and allowing or requiring shops, workplaces, and community facilities within residential districts, typically at transit stops or as neighborhood centers (pros and cons of mixed-use development are discussed further in Chapter 12);
- Increasing residential densities and the diversity of housing types by allowing second units, duplexes, townhouses, and small apartment buildings to be mixed with single family homes;
- Establishing minimum densities and building heights in many locations (especially downtowns or infill sites) rather than focusing on maximums; and
- Adding provisions to require development to be set back from creeks and shorelines, and to preserve areas of important wildlife habitat.

Box 11.1 on pages 158–9 provides a summary of some key modifications to local zoning codes that may help create more sustainable communities.

The development approval process represents the "front lines" of much urban planning. How and where development occurs – the subdivision of land and construction of buildings – is a central concern of urban planning. Development decisions made now will determine the form and character of communities for centuries to come, as well as influencing how cities or towns relate to features of the natural environment, how they use energy and natural resources, and where and how different groups of people can live (especially those of different income groups).

City governments issue a variety of permits allowing property owners to legally develop their land and have a significant degree of control over what gets built. At minimum the development approval process typically in the US includes review of building plans by city staff at the zoning or "current planning" counter and issuance of permits for site grading, construction, and/or building use. But if a project requires any variances from zoning ordinances it will probably also be reviewed by a Zoning Adjustments Board, which can deny, approve, or require modifications to the application. Large development projects may be scrutinized by a Design Review Committee in some cities. If they require amendments to a city's General Plan these projects may be routed through the municipal Planning Commission, whose job is to oversee such plans. Decisions of the Zoning Board and Planning Commission can be appealed to the City Council or County Supervisors, resulting in yet another layer of review. So developers typically face a lengthy, multi-stage process to acquire permits for building.

Although the development approvals process often simply checks to ensure that projects meet the city's zoning and building codes, there is frequently leeway in the process for planners and public commissions to shape the character and impact of development. If a project requires zoning variances for certain things, planners and commissioners can use these approvals as leverage to negotiate with the developer about other changes to the project. They may end up allowing the zoning variances if an overall package of changes is made. There may be negotiation as well around environmental mitigations and various fees that the city charges for transportation, affordable housing, infrastructure, or other potential project impacts.

Development projects that require subdivision of land – the creation of separately owned parcels – trigger an additional approvals process governed by state law and typically overseen by Planning Commissions. This "subdivision mapping process" often involves the creation of new streets and blocks, and is vital in helping to establish a compact urban form, a connecting street pattern, a diversity of lot sizes and housing forms, adequate park space, and protection for creeks and other ecosystem elements. Through this procedure city officials establish the density of new development by approving certain lot sizes and building types (single family homes, duplexes, townhouses, apartment buildings, or other building forms).

Subdivision review can also influence how much land is set aside for parks, greenways, schools, and other public facilities, and how well the new development relates to existing urbanized areas. Setting a pattern of street and pedestrian connections to existing neighborhoods helps make a neighborhood pedestrian- and transit-friendly and creates a unified urban fabric. Typically during the past 50 years new subdivisions have been inwardly focused with little relation to other urban areas around them, helping to create a fragmented, unwalkable landscape. Much the reverse must happen if we are to create more sustainable communities – each subdivision must be planned with the larger city and region in mind.

Much new urban fringe development in recent years has taken place within districts zoned for "planned unit development," a situation in which there are few pre-set zoning requirements but planners and the developer must agree on an overall plan for the new neighborhood. Such a plan includes subdivision of land, creation of new streets, selection of lot sizes and building types, and addition of parks, schools, services, or other community facilities. In this case planners have much more flexibility to ask for changes that will improve community sustainability.

Environmental Review is a crucial part of the development approvals process. In California it is required of all projects under the California Environmental Quality Act; in other states it may only be required under the National Environmental Protection Act for projects that receive federal funding. City staff may issue a "negative declaration" (finding no significant environmental impact), require developers to implement a modest package of mitigations for environment impacts, or force them to prepare an Environmental Assessment or Environmental Impact Report that will guide more extensive mitigations and compare the proposed project with alternatives. These documents are often subject to litigation by outside parties who contend that impacts were not properly analyzed or alternatives not properly considered. NEPA and CEQA are often the only legal handles environmental groups possess to challenge development projects. However, these environmental review laws only require that potential impacts of the project and alternatives be *studied*; they cannot by themselves stop environmentally destructive development.

Box 11.1 Sample zoning changes to promote sustainable development

	Typical current practice	Smart growth alternative
Minimum lot sizes	6000 sq. ft. or more	1500–4000 sq. ft., if any
Maximum lot sizes	Rarely regulated	5000 sq ft. or less for single-family homes in many infill locations
Dwelling units allowed per lot	Most urban land zoned for single family detached housing	Allow second units on existing lots; allow multiple units on vacant lots in single family districts if building design conforms to neighborhood context
Allowable densities, downtown areas	Many suburban cities specify maximum residential densities of 20–40 dwelling units per acre even in high-density zoning districts	Eliminate maximum densities; rely on height, bulk, and/or design restrictions instead. Institute minimum densities of 20–30 dwelling units/acre
Allowable densities, residential areas	Many suburban cities have maximum residential densities of as little as 1–4 units per acre in low-density zoning districts	Establish minimum residential densities of 8–10 units per acre for new single family development and 20 units per acre for multifamily development; allow residential infill at this level
Height restrictions, downtown areas	Often 2–3 stories even in town centers; no minimum	At least 3–5 stories in downtowns and neighborhood centers; a 2–3 story minimum
Height restrictions, residential areas	2 1/2 stories or 30 feet	At least 3 1/2 stories or 40 feet
Lot coverage	Often less than 50 percent of the site	No maximum if parks and other public open spaces are nearby; encourage use of rooftops for open space
Floor area ratio	Often 0.50–0.80 maximum in downtown locations	At least 1.0–2.0 maximum, 0.5 minimum in downtowns, or height limits instead
Front setbacks	Often 20–40 feet minimum except in downtown areas; no maximum	No minimum necessary in many areas; consider adding maximum
Side setbacks	Often 5–15 feet	Allow zero-lot-line construction with appropriate design

Setbacks from creeks	Usually none	At least 30 feet from the centerline of the creek
Lot widths	Some cities require minimum widths of at least 50 feet for single family housing, 70 feet for duplexes	No minimum necessary
Mixture of land uses	Only homes, stores, or workplaces allowed across large areas of cities	Allow a finer mix of land uses to reduce driving and enhance community vitality; allow housing and shops to be added to office parks, offices and shops to housing districts
Mixed-use buildings	Not permitted in most places	Allow mixed-use buildings within neighborhood centers and along arterial strips; provide incentives for these
Secondary units	Prohibited or subject to conditional use permits	Allowed as of right in single-family residential districts
Parking for downtown or transit-oriented locations	1–2 spaces per unit minimum	1 space per unit maximum; car-free housing allowed in certain transit-oriented locations; car-sharing encouraged in large projects
Parking for residential neighborhood locations	2 off-street spaces per unit minimum	1 off-street space per unit minimum; 1 additional on-street space required for larger unit sizes; consider parking maximums
Parking charges	None mandated	Mandate a monthly fee per space for rental and condominium units
Parking for retail	3–4 spaces per 1000 square feet minimum	1 space per 1000 square feet minimum in downtown, transit-oriented, or neighborhood center locations; businesses allowed to contribute in-lieu free instead of providing parking on-site; 2–3 spaces per 1000 square feet in other locations
Parking for offices	3 spaces per 1000 square feet minimum	No minimum in downtown, transit-oriented, or neighborhood center locations; 1–2 spaces per 1000 square feet in other locations; employers required to charge for parking and provide incentives for alternate travel modes; local hiring policies encouraged

Redevelopment is a separate mechanism through which cities coordinate revitalization of areas designated as "blighted." A redevelopment agency within the city government then takes responsibility for carrying out extensive improvements within the targeted area. These improvements may include assembly of multiple parcels of land into a smaller number of easily developable sites, and provision of infrastructure such as roads, sewers, water mains, parks, schools, and other community facilities. In states such as California, cities are allowed to use the mechanism of "tax increment financing" to pay for new infrastructure or purchase of property in redevelopment areas. This allows them to issue bonds to fund initial improvements such as new streets, sewers, parks, and urban design changes. The bonds are then paid off from the increment of increased property taxes that will presumably come from the designated area as a result of new development. This financing mechanism has proven a powerful way for cities to leverage improvements in otherwise run-down neighborhoods, building urban sustainability by revitalizing downtowns, promoting transit-oriented development, developing more mixed-use urban neighborhoods, and so forth.

Historically, redevelopment – especially the mid-twentieth century form known as "urban renewal" – was often misused to bulldoze working-class neighborhoods containing racial and ethnic minorities and replace them with generic modernist development for middle- and upper-class residents. Racist motives have been present at times; during the 1950s and 1960s "urban renewal" was known colloquially as "Negro removal" because so many African-American neighborhoods were targeted. As Jane Jacobs pointed out in 1961 in *The Death and Life of Great American Cities*, planners failed to understand the benefits and potential of older urban neighborhoods, which often contained dense and vibrant social networks as well as historic buildings and a pedestrian-oriented, human-scaled building and block pattern. In one of the twentieth-century's great urban planning tragedies, residents displaced from these neighborhoods were often forced into far more difficult living conditions without their social support networks. Ironically, the fine-grained street, lot, and building fabric of bulldozed older urban neighborhoods is now an ideal that New Urbanist planners strive for.

Despite this checkered history, redevelopment powers can in principle be used by cities to bring about more sustainable forms of development. Redevelopment mechanisms may be particularly useful in coordinating transit-oriented development around new rail lines, in cleaning up and rebuild older industrial areas with brownfield sites, and in catalyzing investment in abandoned downtown districts. British planner Paul Winter argues that local authorities need to be more aggressive in using their powers for compulsory acquisition to ensure that sufficient housing is built,[7] and perhaps to save open space and bring about better-designed development as well. Redevelopment agencies could also prioritize creation of new urban parks, greenways, and ecological restoration sites as well as affordable housing (in some states they are already required to provide the latter within redevelopment projects).

The city of San Jose represents a good case study of both negative and positive forms of redevelopment. The city's downtown was extensively damaged by mid-twentieth-century urban renewal as neighborhoods were razed, parts of the fine-grained street grid turned into "superblocks," and vast parking lots and empty lots created in the belief that development would soon fill them. These sites languished in the 1950s, 1960s, and 1970s while suburban sprawl throughout the Santa Clara Valley sucked businesses and residents from the central city. With an influx of funds from Silicon Valley development and a new understanding of the need to create human-scale, pedestrian-oriented urban spaces, the city's Redevelopment Agency began actively rebuilding the downtown in the late 1980s and 1990s, constructing a host of new buildings, parks, greenways, public plazas, and streetscape improvements in conjunction with a new light rail system. In effect the agency was repairing the damage of earlier decades, seeking to recreate the vibrant urban downtown that had existed 50 years before.

Urban form

Given the need to rethink the landscape of sprawl, one of the most urgent areas for local sustainability planning is in urban form. Municipalities in the United States have historically been weak in terms of ensuring livable and sustainable urban form – adopting a profusion of zoning codes to be sure, but letting private developers take the lead in determining street patterns, street design, block and lot layout, land use mixtures, and open space configuration. Cities have typically responded to development projects as they come up, that is as developers walk in the door of the zoning office and present their plans. The more progressive jurisdictions may work extensively with the developer on design questions internal to the subdivision, for example requiring sidewalks and attractive landscaping, but usually don't focus on the bigger picture of how each subdivision will relate to existing development around it. The result is a landscape of fragmented land uses and inwardly focused communities that do not relate to one another or the city as a whole. (See Figures 11.1 to 11.3.)

In past centuries urban form has gone through a distinctive evolution in North America and much of the rest of the world. Medieval cities tend to be characterized by organic street patterns of tightly connected, winding streets that fit the needs of those traveling by foot or horse. Gridded street patterns with origins in Greek and Roman planning re-appeared during the Rennaissance in Europe and were repeated throughout the New World by the colonizing powers, being well-adapted to the quick establishment of new cities for military or speculative land development purposes. Somewhat looser, rectangular-block grids appeared in late nineteenth-century North America along the new streetcar lines, and are often referred to as "streetcar suburbs."[8] Early experiments with "garden suburbs" characterized by lower density and curving roadways also appeared at this time, beginning

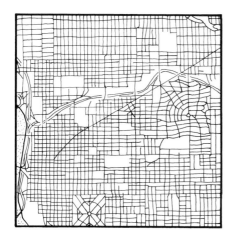

Figure 11.1 Nineteenth- vs. twentieth-century urban fabrics. These nine-square-mile areas compare an urban fabric created between 1880 and 1920 (eastside Portland, Oregon – left) with one created in the late twentieth century (suburban Washington County – right). Street patterns in the latter are far more disconnected. Many of the large open areas contain office parks.

Sacramento

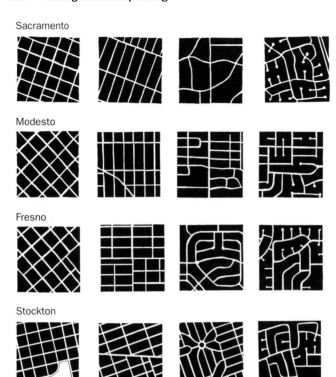

Modesto

Fresno

Stockton

Figure 11.2 The evolution of street patterns in four Californian cities. Most western US cities have followed a similar pattern of physical evolution toward more disconnected urban fabrics. The one-mile square areas here represent neighborhoods created in the mid-1800s, the late 1800s, the early 1900s, and the mid-to-late 1900s.

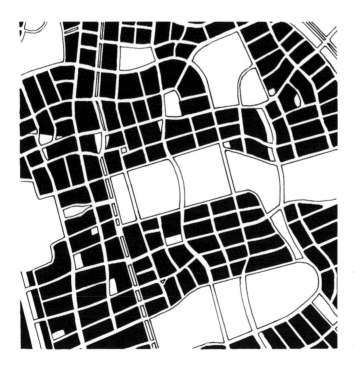

Figure 11.3 A New Urbanist street pattern. Cornell, a New Urbanist development in the Toronto suburb of Markham, shows a return to fine-grained street patterns, along with a generous amount of open space. Compare this urban fabric with those in Figure 11.2. This development suffers, however, from its suburban fringe location far from jobs and shopping.

with Frederick Law Olmsted's community of Riverside planned near Chicago in 1869. In the twentieth century, especially after the Second World War, the spread of the automobile encouraged increasingly disconnected subdivisions with cul-de-sacs and loop roads. Urban regions became vastly larger and more spread out, and new types of urban form appeared such as the office park, the shopping center, and the regional mall. Since these types of urban form accreted onto one another in each metropolitan area, the result in many cases was a pattern that might be labeled "Vienna surrounded by Phoenix," as one observer has characterized Toronto[9] – relatively dense, tightly connected grids of streets created in the nineteenth and early twentieth centuries surrounded by a much more loosely connected sprawl of suburbs built in the mid-to-late twentieth.

Looking ahead, certain urban form values are likely to be particularly essential in the future for sustainability goals such as preserving open space, reducing automobile use, enhancing equity, and improving community vitality. The work of urban designer and MIT professor Kevin Lynch, especially his landmark 1981 book *A Theory of Good City Form*, is an important precedent to such analysis, in that Lynch was among the first to systematically analyze the values and characteristics of different types of urban form. But, as mentioned earlier, Lynch did much of his writing before the influence of the modern environmental movement had been fully felt, and his work needs to be updated in light of current sustainability concerns. Expanding on his efforts, five urban form values now seem particularly important to the challenge of developing more sustainable cities and towns.

1 *Compact urban form* limits suburban sprawl and makes more efficient use of land than in conventional suburbia. The challenge is two-part: to preserve open space and to design a more efficient, compact, and livable urban form inside growth limit lines. Regions such as Portland, Oregon, which have sought to manage growth through an Urban Growth Boundary have learned this lesson the hard way. Much sprawl has occurred inside the Portland UGB because it was set too far out initially and the characteristics of new development were not a major focus until recently. Now the region must rethink how to create more compact development within its existing urban area.

2 *Contiguous urban form* implies that new expansion takes place next to existing urban areas. If new development projects are not contiguous, then inefficient, disjointed land use patterns are likely to result as the spaces between projects fill in haphazardly, and street connections between subdivisions are likely to be poor. The opposite of contiguous development is often referred to as "leapfrog" growth, in that development jumps from place to place across the landscape to wherever developers can find cheap available land.

3 *Connected urban form* features good street, path, and visual connections within the region, and is also relatively "legible" and easy for people to find their way around. Without these connections, a disjointed landscape is created in which walking, bicycling, using public transit, and even driving are difficult and involve circuitous routes. Arguably it then also becomes more difficult for residents in disconnected subdivisions to gain a sense of participation in the broader urban and regional environment. The nineteenth-century, square-block grids at the core of many older cities provide an extremely high degree of connectivity, promoting travel through the city by a variety of transportation modes. Not surprisingly, winding suburban street patterns feature very low connectivity.[10]

4 *Diverse urban form* contains a mixture of land uses, building and housing types, architectural styles, and prices or rents. If development is not diverse in these ways, then the result is a homogeneous built form, monotonous urban landscapes, segregation

of income groups, and increased driving, congestion, and air pollution. Nineteenth-century neighborhoods with diverse building types and land uses are today among the most vibrant, attractive, and popular districts in many North American cities. Twentieth-century, single-use zoning was a major force preventing diversity of urban form. In addition, the large scale of recent homebuilding and office park construction often prevents the creation of a diverse urban fabric, in that each builder is often unwilling to create more than a single type of land use.

5 *Ecological urban form* integrates features of the natural landscape into the form of the city in a way that protects and restores local ecosystems while providing recreational amenities for residents. In most urban areas little thought was given to this urban form value until the last third of the twentieth century. Developers simply bulldozed hills, culverted streams, and generally treated the landscape as a slate to be wiped clean for human use. Even garden suburb developers treated ecosystem elements primarily as aesthetic amenities for human benefit, not as valuable entities in their own right. In the last few decades, however, planners and citizen activists have begun seeking ways to protect or enhance ecosystem elements during the urbanization process. Regional and local planning agencies have designated park and greenway networks, placed some wetlands and stream corridors off-limits to development, and changed zoning codes to require park or open space dedication for most projects of any size. These agencies increasingly seek to identify key areas of ecological concern well in advance of development and integrate them into local or regional planning frameworks. Where simply zoning land off-limits to development is not an option, officials may choose to negotiate with developers, environmentalists, and other constituencies to develop "habitat conservation plans," under which some land is protected while other sites are developed.[11] This controversial approach has the advantage of leveraging protection for some areas without enormous expenditure of public funds, but the disadvantage of allowing much other development to go forth. In other cases local governments, park or open space districts, or NGOs such as The Nature Conservancy may purchase fragile habitat or conservation easements on the land to prevent development. Such emphasis on incorporating environmental concerns into the development of urban form, however, is still in its early stages and is far from universal.

These five forms are illustrated in Figure 11.4.

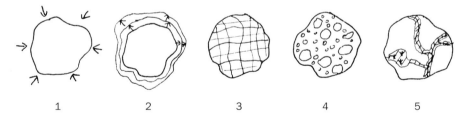

1   2   3   4   5

Figure 11.4 Five sustainable urban form values. Sustainable urban form is likely to be compact, contiguous, connected, diverse, and integrated with the natural landscape.

These urban form criteria vary independently. Development can be contiguous to existing urban areas without being connected to them. It can be compact without being contiguous. And it can be all three of these things without having the diversity, human scale, or environmental elements that help make places attractive, livable, and sustainable. Nineteenth-century urban fabrics tend to be fairly compact, connected, and diverse, though not always contiguous or ecological. Twentieth-century fabrics tend to score low on all criteria. Recent New Urbanist–inspired planning tends to rate more highly on each, though still facing huge obstacles because of the large-scale nature of many development projects, the need to accommodate the automobile, and the lack of effective ways to implement these design values on a regional scale.

Producing more sustainable urban form will require more proactive steps by local government to fulfill these urban form values by requiring connected road patterns, smaller block sizes, more pedestrian-friendly streets and intersections, park and green-way systems, and a mix of land uses that places jobs, housing, services, and community facilities in proximity and balance with one another across the urban landscape. City governments can use existing General Plan and Specific Area Plan processes to estab-lish a vision of such urban form, or can work more incrementally within particular areas of the city to achieve it as projects come up. But if cities do not establish a proactive vision in advance that developers can work within, they may have a harder time denying projects during the review process because they do not meet these sustainability considerations.

Taxing land at a higher rate than improvements (buildings) is another mechanism local government can use to encourage compact urban form and infill development. This step would increase pressure on landowners to make more productive use of vacant lots, parking lots, or leftover areas between existing developments, and could do a great deal to counter the current wasteful use of land within suburbia. Nineteenth century reformer Henry George advocated taxing only land, the "single tax," as a way to more fairly recover wealth created by the entire community (as reflected in rising property values) and to reduce the tendency of speculators to hold vacant land for years hoping for price increases.

Financial benefits can accrue to the municipality from meeting these urban form criteria. Compared with more compact forms of development, sprawl costs local govern-ments more to support and maintain – to build and service roads, sewers, water supply infrastructure, schools, and other services. Although there has been much debate about the exact costs of sprawl, they are substantial. In a 1997 review of more than 475 other studies of sprawl, Rutgers professor Robert Burchell and others found significant savings in compact development. The three studies bearing most directly on this topic found savings of 25 percent in roads, 20 percent in utilities, and 5 percent in schools compared with sprawl development.[12] Another recent study in the *Journal of the American Planning Association* found that residents of compact developments are in effect subsidizing those in more sprawling developments $17.24 a month in infrastructure costs, or $2567 over a 30-year period.[13] Development that is compact, contiguous, and offers connected street networks will also be easier and cheaper for public transit to service. Communities designed to create ecological urban form are likely to reduce the risks and costs of flooding, landslides, fires, and other natural disasters. Diverse communities with mixed-use development that reduce driving may lower regional infrastructure costs for roads and bridges. In these and many other ways, sustainable urban form can prove cheaper and more cost-effective than the alternatives.

Protection of open space

Very specific efforts are usually needed at the local level to protect open space within and around the community. These efforts ideally take place as part of an overall growth management program or park system or ecosystem planning, and follow from the city or county's General Plan. But occasionally cities may also have to act entrepreneurially to save parcels of land that come on the market unexpectedly within urban areas or in fringe areas where protection has not yet been adopted.

There are many different ways to protect open space within and around cities, each with its pros and cons. In the United States agricultural zoning is probably the most common method; areas around a community are simply zoned for agricultural use only, usually with a minimum parcel size that may be from 5 acres to 160 or more. The disadvantage of this method is that zoning can be easily changed by a vote of the city council, usually in response to developer pressure. Also, unless a large minimum parcel size is stipulated landowners can subdivide large farms into 5 or 10-acre "hobby farms" that amount to very low-density rural sprawl. Once agricultural land is subdivided in this way, traditional farming uses no longer become viable and political pressure may mount for further subdivision into low-density suburban development.

Urban growth boundaries have been adopted by many cities in Oregon, Washington (both under state mandate), and Northern California, as well as a few other jurisdictions in Tennessee, Pennsylvania, and Colorado. They are a more stable, long-term way to protect open space at the urban edge, usually being established for a period of up to 20 years with periodic reviews to allow modest expansions of the urban area if population pressure warrants. Often a vote of the electorate or a two-thirds majority of a City Council or county Board of Supervisors is required to change the boundary. Local jurisdictions agree not to allow land to be subdivided outside the boundary or to extend the infrastructure that would make development possible. Because of their relative permanence UGBs are a favored strategy of environmentalists. Opponents argue that UGBs lead to housing price increases within the boundary as the supply of greenfield land diminishes, and that development may simply leapfrog to the next county. A potential solution to the former problem is to establish strong municipal policies encouraging space-efficient infill development and the provision of affordable housing; a potential solution for the latter concern is to broaden the area covered by growth management protections or to move to a statewide framework for protection.

Municipal purchase of open space land for local parks or wildlife preserves is another option that has the advantage of being permanent but the disadvantage of requiring large amounts of public funds. Private groups such as the Nature Conservancy and the Trust for Public Land also make such purchases or provide loan funds to allow local organizations to do so. Municipal acquisition of parkland has been immensely important in most cities, but is often a difficult and slow process, as funds may need to be raised through a bond measure or other procedure. Not infrequently by the time a bond measure has passed and the funds are available, the price of the land in question has risen so much that the original intended purchase is no longer affordable.

Purchase of conservation easements is a less costly way to ensure that land remains undeveloped. A community group or city purchases the development rights to the land from the owner, and a deed restriction is then recorded, preventing development. This measure has been widely pursued in recent years, although there are some concerns about whether such easements will be enforced in perpetuity.

Transfer of development rights is a relatively rare method in which a local government essentially establishes a market in development rights, allowing rural landowners to sell their rights to developers, who use them to obtain higher densities than would be allowed by zoning for their projects in urban "receiving" areas. This method has been in place in Montgomery County, Maryland, since the 1970s. However, it is based on the assumption that urban zoning is less than optimal, raising the question of why the municipality in question does not permit denser development there in the first place.

Finally, land banking has been used extensively in Europe historically although not so much in the United States. Essentially the municipal government buys up land at the urban fringe and releases some of it for development in times and places that it thinks make sense. (Obviously, some of this land could be preserved as open space as well.) The local government must capitalize such a fund initially, but in theory it could then be self-sustaining or could even provide the city with a return on investment. The city of Stockholm, for example, at one point owned more than 60 percent of the land in its metropolitan area and managed it through land banking mechanisms. This method has the further advantage that the public recaptures the value of its investment in infrastructure such as rail or road systems, since it rather than speculators profits from the corresponding increase in land values.

Landscape planning

While much watershed, open space, and park planning is done at a regional level, cities and counties also play key roles in planning for parks and healthy landscapes. Historically, municipal parks and recreation departments have managed parkland accumulated through gifts from wealthy landowners, city purchases, or exactions from developers. Although the Parks Movement of the late nineteenth century encouraged cities to develop comprehensive park systems, these grand plans depended on political backing and the passage of funding measures and often languished. As a result most cities wound up with a relatively meager and motley assortment of open spaces. This was particularly true of Sunbelt cities such as Los Angeles that grew very rapidly during the twentieth century and allowed developers to build on vast areas of land very quickly. From the air over LA one sees an unrelenting sprawl of gray concrete and rooftops, with cemeteries and golf courses rather than parks forming the most obvious green spaces. Much of the region's parkland is in the Santa Monica Mountains, far from most people's homes. Even the LA River is hemmed in by development and encased in a channel of concrete. The failure to reserve parks and other open spaces is one of the great tragedies of development in that region, as it is in many others.

Increasingly, however, cities are taking a more proactive and systematic approach to creating and protecting greenspaces. One way to do so is by planning for an integrated open space system, preferably with parks in each neighborhood connected by greenways. Grants, bond measures, or other funding devices can then be developed to acquire those lands. Another key step is to systematically protect existing creeks and wetlands through municipal ordinances. A creek ordinance, for example, might prohibit development within 30 feet of any creek, forbid culverting or channelization of open waterways, establish a program for acquiring trail easements along creeks, and create guidelines for reestablishing native riparian vegetation. The cities of Berkeley and Oakland have established such ordinances. A wetland ordinance might go well beyond federal Section 404 regulation (which requires developers to obtain permits to fill wetlands) by actually prohibiting any fill, grading, or land clearance, and by regulating other structures such as docks, walkways, fences, and observation decks. An urban forestry ordinance might establish guidelines for street tree planting and park landscaping, and protect "heritage trees" above a certain diameter from being cut down. These and other steps can help local government plan for a healthier natural landscape within urban areas.

Clean-up and restoration

Local governments can and must take the lead in many activities cleaning up contaminated or previously used brownfields sites, and in restoring key features of local ecosystems. Private industry is often reluctant to undertake such efforts, and may find it cheaper and easier to relocate to greenfield sites elsewhere. Local community groups and neighborhood associations usually do not have the resources to pursue such clean-up activities themselves, although they can be an important source of volunteer labor and ongoing maintenance of restored ecosystem features. National initiatives such as the US Superfund program may assist with some of the most contaminated sites but do not cover others. So it may fall to local planners and elected leaders to initiate the reuse of much previously urbanized land.

As mentioned, many municipal efforts have focused on restoring watershed features such as creeks, shorelines, and wetlands. These are particularly important elements of local ecosystems, habitats for birds and many other forms of wildlife, and public amenities for many forms of recreation. Citizen groups often identify with watershed features and have taken the lead on many restoration activities. But incentive grants by local or state agencies, technical assistance by local Public Works departments, and land acquisition may be required as well.

Coordination between cities and local businesses or developers may be necessary to restore brownfields sites for future housing, economic development, or other urban uses. Cities can keep track of toxic chemicals and contaminated soils in an online database that will tell prospective developers the exact issues for particular sites, and can help ensure the reuse of contaminated parcels through technical assistance, grant or loan funding, or actual city purchase and clean-up. (See Figure 11.5 and Box 11.2.)

Figure 11.5 Vancouver's False Creek. On former industrial lands Vancouver has created highly livable new communities with attractive parks, shops, and plazas around an inlet.

Box 11.2 Emeryville, California's brownfields program

The city of Emeryville, California offers a dramatic example of how a municipality can transform itself through brownfield reclamation and infill development. Using pilot project grant funds from the US EPA, Emeryville has set up a "one-stop shop" to provide information on contamination at all sites within the city. Property owners and potential developers can go online to determine the exact condition of any parcel. The city also acts as an intermediary between developers and regulatory agencies, maintains an environmental Geographic Information Systems (GIS) database of the city, and handles some groundwater cleanup tasks itself.

Emeryville's transformation from a decaying industrial enclave into a leading high-tech community has been led by a very active Redevelopment Agency, using redevelopment powers – granted by the state to help cities rebuild blighted areas – to clean up and rebuild on urban land. Improvements undertaken by this agency include street redesign, toxic clean-up, land assembly, and development of parks and infrastructure. Virtually all of this small city has been declared a redevelopment district. Under redevelopment law, this agency has the power to buy small properties, assemble larger buildable lots, clean up sites, and build infrastructure, using the device of tax-increment financing to raise money. This tool allows redevelopment agencies to raise capital by issuing bonds based on the expected increase in property tax receipts, which for Emeryville is about $5.4 million per year.

The city's infill development has mixed new stores and office buildings with housing. Some 561 apartments, lofts, townhouses, and condos were built between 1995 and 2000 alone, of which 224 are affordable to those with low and moderate incomes. The city has adopted inclusionary zoning requiring that 20 percent of units in new projects of 30 units or more be affordable.

Although sometimes criticized for permitting "big-box" retail, Emeryville's entrepreneurial approach helped turn a city of decaying industrial buildings into one of the Bay Area's redevelopment success stories.

For more information or to visit the city's One-Stop Shop with online environmental data for city parcels, visit the city's website at www.ci.emeryville.ca.us/.

Resource use

Another area in which local governments have considerable potential to improve urban sustainability concerns the use of energy and materials. Cities and counties often run programs to help residents make their homes more energy efficient, improve recycling and materials reuse, encourage businesses to reduce waste, and lower water consumption.

One of the biggest resource policy challenges for cities and local agencies is what balance to strike among the "Three Rs." Traditionally the greatest emphasis has been on recycling – the collection of used materials (typically glass bottles, aluminum and metal cans, and newspaper) for remanufacturing. Recycling programs are the most public and visible type of ecological resource use activity, something that ordinary citizens can easily participate in. However, much greater sustainability benefits can potentially be derived from reuse (washing and reusing glass bottles uses a fraction of the energy of recycling the glass, for example) and from reduction of the solid waste stream in the first place.

Unfortunately these alternatives require more fundamental changes in consumer attitudes and how businesses use materials. Yet these changes are not impossible – bottles and other materials were routinely reused in the United States until the spread of plastic and aluminum containers in the 1970s, and such reuse still occurs in much of the rest of the world. The Swiss, for example, require beverages to be distributed in certain standard-sized reusable bottles. In France, Guatemala, and many other countries one can see trucks carrying empty drink bottles back to the distributor. Throw-away containers are gaining ground in many places, but it is certainly not too late to reverse the trend.

Increasingly the focus of action by local, state, and regional governments is on "integrated waste management" in which all of these strategies are employed in varying combinations to achieve the greatest possible improvements in resource use. Municipalities typically collect solid waste and are under the most direct pressure to reduce the waste stream because of rising landfill costs ("tipping fees"), a scarcity of landfill space, and/or state legislation requiring such a reduction. In 1989 the State of California, for example, required cities to reduce their solid waste stream by 50 percent by the year 2000 (the actual reduction achieved was 47 percent); New York State set a similar goal in 1988. In 2003 San Francisco planners set even more stringent goals of reducing the city's solid waste stream by 75 percent in 2010 and 100 percent in 2020, reaching a "zero waste" situation. The city aims to do this through a combination of increased monthly pickup fees for waste; extensive curbside recycling for many materials including electronic components, hazardous materials, yard waste and food scraps; and programs to reduce construction and demolition waste.[14] British cities are a bit further behind. Although the country's recycling rates for particular materials have been high, such as rates of 77 percent for paper and 89 percent for metals in 1999, overall only 9 percent of household waste and 22 percent of commercial waste was recycled.[15] The Blair administration subsequently announced a goal of increasing the household rate to 30 percent by 2010.

Percentages of major city solid waste budgets spent on waste diversion in the US range from 4 percent in San Diego and Dallas to 23 percent in Los Angeles. For their part, states such as Vermont, Connecticut, Delaware, Massachusetts, Oregon, New York, Iowa, Maine, California, and Michigan have passed bottle deposit legislation to promote recycling of beverage containers; this has been effective in increasing collection rates and reducing litter in participating states (and has spawned a cottage industry of "scavengers" who collect containers illicitly from household bins before municipal recycling trucks come around). However, deposit rates (typically $.05 per container) have not risen in many states since the 1970s, and the incentive may decline over time if this situation does not change.

Aluminum has traditionally been the most successfully recycled material, given high prices for collected cans, and recycling this metal produces huge energy savings over new production. The percentage of steel cans and glass bottles collected in recycling programs has been somewhat lower, around 40 percent and 24 percent respectively in the US. Newsprint collection has also averaged around 40 percent, with historic problems in maintaining stable prices for collected newspaper. Plastics recycling is highly controversial. Although plastics represent the fastest-growing part of the waste stream, recycling rates are low and falling, totaling only around 5 percent overall. The plastics industry has stepped back from a 1991 commitment to reach a rate of 25 percent recycling; and only two of seven main types of plastic resins (PET and HDPE) are recycled to any degree.

Local governments also frequently run programs to promote composting of kitchen scraps or to pick up and compost yard waste, since this organic material is a major element of the waste stream. Other programs collect used motor oil for recycling, promote recycling of construction waste, and even reuse surplus paint (Seattle has a program in which collected latex paint is blended and resold at a discount to local schools and hospitals as "Seattle beige"). As part of their economic development activities, cities can also promote eco-industrial parks in which industries use each other's byproducts as inputs. The most fully developed example of this remains the industrial ecosystem in Kalundborg, Denmark, in which a power plant, cement plant, oil refinery, pharmaceutical plant, wallboard factory, sulfuric acid producer, farmers, and residences use one another's waste steam, water, gas, fly ash, gypsum, and other materials. With time and active coordination by local authorities, other such systems may emerge.

Changing public sector procurement policies to require reused/recycled products is a further way in which local agencies can stimulate ecological resource use. Eco-labeling programs are also essential in establishing norms for recycled content and in ensuring consumer confidence that materials sold as recycled or reused are the real thing.

Liquid wastes pose a different set of recycling and reuse problems. Cities have great control over how wastes such as sewage and stormwater runoff are handled, even if these effluents are treated by a local or subregional utility and not the municipal government itself. Separating sewage and stormwater systems, so that heavy storms do not overwhelm sewage treatment facilities and flush contaminants into nearby waterways, is one basic step that many cities in North America are still working on. Stenciling storm drains (e.g. "No Dumping: Drains to Bay") to prevent people from dumping used motor oil and other materials into aquatic systems is another basic way to reduce improper waste disposal. Some cities, such as Arcata on the Northern California coast, have created wetlands for ecological sewage treatment, generally after primary treatment (settling of solids) and secondary treatment (aeration of the effluent) have been provided in a traditional facility.[16] A more radical method of ecological wastewater treatment has been undertaken by the Nova Scotia town of Bear River, which installed a Solar Aquatic "Living Machine" sewage treatment facility in 1995. This system handles sewage for 881 residents and consists of a series of large tanks inside a greenhouse to recycle liquids and sludge using plants such as water hyacinths, flower ginger, watercress, willows, mints, and grasses. The installation has proven a major tourist attraction in addition to performing its primary function.[17]

Energy is a different type of resource that local government has a great deal of control over. Unlike solid materials, energy cannot be recycled or reused; instead, the Second Law of Thermodynamics ("entropy increases") applies. Energy in the typical end-use forms of electricity, natural gas, or liquid fossil fuels is consumed and dissipates primarily as heat, although to some extent it can be captured for human purposes. So the focus must be on reduction of energy use through energy-conservation measures. Since the 1970s energy crises, utilities, states, local planners, and national governments have developed a wide array of "demand-side management" (DSM) programs to reduce energy use, including initiatives promoting compact fluorescent lighting, setting energy-use standards for appliances, helping residents insulate their homes, and providing energy audits to homes or businesses. As in the field of transportation planning, the emphasis is changing to managing demand from increasing supply. City governments in cold climates have been particularly active in helping low-income residents weatherize their homes, since these residents frequently live in the most poorly insulated dwellings and may not be able to afford to do so otherwise. Such programs can be seen as promoting equity (improving quality of life and saving money for the most needy). However, much more remains to be done on this front.

Greener production processes within industry are one of the most important avenues for improved resource use and lowered pollution. An extensive field of "environmental management" has emerged within business, concerned both with helping businesses meet environmental regulations and with making processes more resource-efficient. The ISO 14001 set of international standards represents a comprehensive approach to environmental management within industrial processes. These standards are now employed on more than 700 sites in the UK alone, and their use is growing rapidly worldwide.[18]

Economic development and growth

Local governments often feel compelled to plan and zone for rapid, unsustainable growth in jobs, population, and/or commercial development. This juggernaut of "progress" is taken for granted by many local officials.[19] Politicians and planners often feel that rapid growth is the only way they can ensure jobs for local residents and raise municipal tax revenues. They may seek growth as a way to upgrade deteriorating urban infrastructure and services. They may see taxes from new development as a way to pay off past municipal debt. Or they may simply seek growth because of a boosterish political climate in which the local economic and political leadership forms a "growth machine" for which civic and economic development objectives coincide.[20]

In many places the state tax structure provides powerful incentives for local governments to zone for big-box commercial development, suburban strips, office parks, automobile dealerships, and other forms of rapid, sprawl growth – the phenomenon known as "fiscalization of land use." In these places anti-tax politics or state tax limitation referenda have limited cities' and counties' ability to raise revenue through property taxes and bonds. Local governments are therefore forced to zone for land uses that will produce high sales tax revenues and require few public services. The fact that taxes for important services like schools are raised primarily at the local level – where land use decisions are also made – puts in place a structural incentive for bad planning. If taxes were shared across metropolitan regions, as mentioned earlier, or were collected primarily by state or federal government and then channeled back to local governments on the basis of population or need, then such incentives for bad local land use planning and rapid sprawl-style growth would be dramatically lessened.

The idea that rapid growth in the number of jobs or the physical development of land is the solution to local government problems must be questioned. Too often it simply produces short-term benefits while creating longer-term difficulties. This does not mean that local government should be in favor of no growth or even slow growth – such positions often camouflage exclusionary attitudes towards lower-income and minority populations. Rather, it means to seek "smart growth" which creates a better balance of economic, environmental, social, and fiscal well-being within the community. This may entail providing more affordable housing to correct an existing deficiency and improve equity, adding local shops and parks to improve quality of life, nurturing locally owned businesses that employ local residents and meet local needs, or creating a revitalized downtown and neighborhood centers that can provide stable centers of community and economic activity.

A sustainable economic strategy will take advantage of local resources, history, and skills to attract types of businesses that will provide decent-paying, long-term jobs for existing residents. If these businesses are relatively diverse and small-scale, the local economy is likely to be more immune to recession or economic disruption than if jobs are concentrated in one or two large firms that could leave overnight. If businesses are locally owned, they are also likely to be more active corporate citizens and more extensively involved in the community than if they are branches of multinational corporations.[21] And if businesses that don't pollute, don't produce toxic wastes, specialize in environmental clean-up, or produce products and services really needed by people, they will clearly bring other long-term benefits for the community as well.

Many cities must look in particular at the amount of land they have zoned for new, large-scale commercial or industrial development. Following the growth ideology, cities typically over-zone for potential new employers, designating vast tracts of land for new office parks or malls. But these businesses with their new jobs may simply attract new residents that will require additional services in the long term, drive up local housing costs, congest local roads, and lead to many other secondary growth problems. Instead, a focus on a stable, high-quality employment base for existing residents may be more appropriate. If new jobs are in fact seen as desirable, for example,. to achieve a better balance of employment and housing, then the type and location of these employers need to be carefully considered. The new jobs should match the skills of existing workers and pay decent wages. And companies should be encouraged to locate in existing urban areas rather than building sprawling office parks. Their workers will then be more likely to patronize existing local businesses and help create vibrant downtowns, and will have better access to public transportation and existing residential neighborhoods. (See Figures 11.6 and 11.7.)

Figure 11.6 A farmers' market. Farmers' markets are a way for cities to support local agricultural producers, ensure residents access to healthy food, and build community.

Figure 11.7 Cows on parade. Public art campaigns can add vitality and humor to cities. The "Cows on Parade" concept pioneered by Zurich has been emulated by Chicago and many other cities (often using local symbols). In Chicago, artists were commissioned to decorate 320 fiberglass cows that were placed around the city and later auctioned off for charity. The city's Department of Cultural Affairs estimates that the exhibit drew more than a million visitors over a single summer, with a $200 million economic impact.

Transportation planning

Local governments are directly responsible for most local road maintenance as well as bicycle and pedestrian planning and – if the city is large enough – at times transit planning. (This is may instead operated by a separate agency that may be regional in nature.)

For decades local transportation planning has typically focused on improving automobile circulation and providing parking. Roads have been widened, one-way streets added to speed automobiles through downtowns or residential areas, signals synchronized on main streets to smooth the flow of traffic, and new turn lanes or intersections created to serve new developments. Now, on the other hand, we are beginning to recognize that a focus on other forms of transportation is needed, and that environments designed for automobiles do not necessarily work well for pedestrians or local residents.

Planning for pedestrians requires a comprehensive reevaluation of the street environment. Sidewalks, which are deficient or completely missing in many newer communities, and some older ones, are a good place to start. Often developers of new neighborhoods construct only a four-foot sidewalk, if they include one at all; this is too narrow for two people to walk abreast. Six-foot sidewalks are much preferable. If they are separated from the street by a planting strip (typically three to six feet), then the pedestrian is further insulated from traffic on the street, and the grade of the sidewalk doesn't change as driveways cut through it (driveways often dip sharply to meet the street in their last few feet; this can create discontinuities in a sidewalk). Sidewalk curb cuts at intersections to allow wheelchair use are desirable as well.

   Intersections are another key area for pedestrian design. The person walking must be able to get across easily and safely. As intersections have become monstrously huge in many suburban areas – with five or more lanes of traffic on each street, complicated turning movements, and very long cycles for signals – this simple task has become increasingly complicated and stressful. One obvious solution is to keep streets small and distribute traffic across a grid of streets rather than channeling it all onto a few enormous arterials. But other things can be done as well. Pedestrian-activated crossing signals, count-down and/or audible signals, pedestrian islands half-way across each street, and bulbouts at corners to decrease the distance pedestrians need to travel and improve their sight lines are all strategies that can help.

   The pedestrian-friendliness of a street environment depends on many other factors as well. Having buildings lining the streets helps create a more intimate and human-scale environment than the current situation in many suburban areas where parking lots front the street, creating a vast landscape of pavement through which pedestrians must wander. Providing parking along the street buffers pedestrians from fast-moving vehicles. Short blocks, human-scale buildings, and a relatively fine-grained mix of land uses makes for a more interesting street environment and a greater number of route options. Prohibiting blank building walls along the street and encouraging street-front retail and sidewalk seating for restaurants and cafés helps greatly. Attractive landscaping, closely spaced street trees, pedestrian-scale lighting, and occasional benches add to the pedestrian environment.

   The overall street network must be examined to determine if pedestrians can travel easily from one point to another. In more recent suburban street patterns there is often no direct walking route through neighborhoods. Pedestrian paths may need to be created, occasionally by the city actually acquiring a key parcel of land or right-of-way and creating a new road connection or trail. When developing zoning and design guidelines for new developments, cities can require a certain number of street connections per mile (Portland, Oregon, requires five), maximum block sizes, or the addition of pedestrian and bicycle trails within new developments.

   Bicycle planning is also a rapidly growing field in most cities. The issues here are often similar to those for pedestrian planning; a street environment that is pedestrian-friendly will usually be bike-friendly as well. But other steps need to be taken in addition. Since bicycles generally travel on streets, a variety of measures may be needed to provide them space, to slow traffic, to improve visibility, and otherwise to accommodate their presence. Three main categories of bike routes can be created: designated bike routes on streets where the route is clearly marked and perhaps measures are taken to slow traffic and improve bike safety at intersections; painted or colored bike lanes on streets to create a perceived bike space at the edge of the road; and separated bike paths that are removed from motor vehicle traffic. These three types have uses in different contexts. While bike paths might be desirable in many cases, there is not always room to create them, particularly within existing cities and towns. The only solution then is to figure out better ways of accommodating bikes on streets through bike lanes or rates.

   Around the world we have seen a large resurgence in planning for cyclists, although unfortunately some nations such as China that traditionally have had high rates of bicycle use are now discouraging this in the name of progress. Rio de Janeiro, for example, has developed more than 84 kilometers of bicycle paths, including a trail from a large middle-class residential area in the south of the city through downtown. Inspired by Dutch bicycle planning, a municipal Working Group for Cycle Systems began in 1993 to determine priority routes and gather public support. Despite initial resistance from motorists and shopkeepers, the city succeeded in creating a number of grade-separated bicycle lanes, including a 3-meter wide cycle track between the Ipanema and Copacabana neighborhoods.[22]

A variety of traffic-calming measures can improve the street environment for pedestrians, cyclists, and neighbors who live nearby. An international traffic calming movement has in fact been underway for several decades.[23] Early examples included Copenhagen's pioneering pedestrian street, the Ströget, created in the 1960s, and the Dutch woonerf designs in which road space is shared in residential neighborhoods between very slow speed automobiles and residents. Downtown pedestrian districts were then created in many European cities in the 1970s and 1980s. Since traffic calming is often best undertaken on a neighborhood-wide scale, these and other approaches are discussed further in the next chapter.

A proliferation of new strategies for improving local public transport has appeared in recent years. One of the most popular is "bus rapid transit." Under this approach, high-tech buses travel on routes designed to speed up service. These routes may provide the buses separate lanes, queue-jumper lanes at lights, signal-preemption abilities at lights (the driver is able to trip the light by remote means), and quick-loading bus stops designed like light rail boarding platforms. The buses themselves may be low-floor models (to speed boarding) and may use alternative propulsion methods such as compressed natural gas or fuel cells. The idea is to provide service comparable to light rail systems at a fraction of the cost. One extremely successful bus rapid transit program has been implemented in Los Angeles, where passenger travel times have decreased up to 29 percent on BRT lines. Only three years after implementing the technology ridership had grown 40 percent.[24]

Light rail systems are also popular in many cities, and generally offer faster service and greater benefits in terms of encouraging more intensive land use development along transit routes (property owners have greater certainty that rail-based systems will be permanent). These systems vary from streetcars on roadways with very frequent stops, to more suburban-serving trams with their own rights of way and longer routes. Frequently cities have placed light rail lines in freeway medians to save money compared with acquiring rights-of-way. This may work well in some suburban contexts, but has major disadvantages in more urban areas in that the environment for transit-users is then loud and unpleasant. Transit-oriented development around stations is also more difficult, in that the presence of the freeway discourages many new potential land uses. Heavy-rail systems, with heavier-duty trains and more widely spaced stops are generally used to meet longer-distance commuter needs.

A wide variety of other transit modes holds promise as well.[25] One of the most useful in low-density suburban areas may be on-demand service, where a transit user requests a pickup by phone or on-line, and a van or small bus detours to pick him or her up on its next pass through. Vanpools and carpools to workplaces are being developed for many large employers, and many efforts are underway to make mode combination trips more feasible, for example, by providing bike racks on buses and shuttle service to commuter rail stations. In developing-world cities, a variety of privately run transit options often provides a better level of service than official agencies. Typically, vans or microbuses ply main corridors picking up passengers wherever they appear. The correos of Mexico City and the dolmush in Turkey are examples of such "jitney" service. Finally, taxis are an important travel option in many cities, particularly those that are relatively dense and have destinations close together, to keep fares down. (See Box 11.3.)

Box 11.3 Curitiba's transportation planning

The Brazilian city of Curitiba, a metropolis of 1.6 million in the southern part of the country, has emerged as one of the world's leading examples of creative urban development (see, for example, Rabinovich and Leitman, 1996). The city's success began in the 1960s when planners first laid out a concept of growth concentrated along structural axes. Since the city could not afford a rail-based metro transit system, it opted for a low-cost but highly innovative bus network. Double-articulated buses speed along on their own rights-of-way, with feeder networks funneling passengers into the main routes. Raised "bus tubes," where passengers pay in advance, speed up boarding at each stop. The system cost $200,000 per kilometer, as opposed to estimates of $60 to $70 million per kilometer for a subway.

Despite having one of Brazil's highest rates of automobile ownership, three-quarters of all commuters in Curitiba now take the bus. The city's transport system is an impressive illustration that intensive development along transit corridors, coupled with a highly efficient transit service, can dramatically reduce driving.

Equity

Although equity issues are rarely top on the agenda of local government, usually taking a back seat to economic development, for example, much can be done to promote equity at this scale. Living wage ordinances are one strategy, aimed towards at least somewhat improving conditions for those at the bottom of the economic hierarchy. Affordable housing programs are another, usually accomplished through direct public subsidy of nonprofit affordable housing builders, "inclusionary zoning" requiring that for-profit builders include a certain percentage of affordable units within their projects, or (usually outside the US) direct public construction of affordable housing units. Economic development strategies aimed at promoting decent-wage jobs, locally owned businesses, and training for low-skill workers can also help. Local government decisions about where to site locally unwanted land uses ("LULUs") – such as landfills, incinerators, bus depots, corporation yards, and waste transfer stations – have profound equity impacts. As environmental justice advocates have argued, such siting decisions have exposed lower-income or minority communities to noise, pollution, toxic chemicals, and other unsavory side effects. Lastly, local tax policy has strong equity implications. The rising reliance on sales taxes to fund local services hits the poor hardest, since purchases take a higher percentage of their income; property taxes or even local income taxes tend to be more progressive, taking a larger share of municipal revenue from wealthier taxpayers.

Housing

Housing is one of the most basic human needs, a commodity that despite large programs to provide social housing during much of the twentieth century (called public housing in the US) is now provided primarily by the market. However, leaving the provision of housing entirely to the market has its problems. For-profit builders tend not to provide sufficient quantities of affordable housing for lower-income residents, and their developments may not meet other criteria that local governments would like to see satisfied (for example, in protecting the local environment, providing parks and public space, and linking appropriately to existing urban areas). So a strong public sector role is required in regulating housing development and/or supplementing the efforts of the market, and this task generally falls to local government.

Providing sufficient quantities of housing is one main task important for sustainable communities. Otherwise, as is happening currently in many urban areas around the world, housing prices escalate rapidly, desirable areas become enclaves of wealth, and the poor are displaced through processes of gentrification and must either pay a large portion of their income for housing or travel long distances to jobs from distant communities with more affordable housing. With development becoming more difficult in many areas because of decreasing supplies of centrally located greenfield land, increasing public resistance, environmental requirements, and other factors, many urban areas are simply not producing enough housing to meet population needs. This situation highlights the need to stabilize population, but it also highlights the need to increase production in the short-term and to do it by creating attractive, well-located neighborhoods.

Local governments can address this need through a variety of strategies. They can directly construct housing themselves, a strategy that has fallen into disfavor in the US and many other countries after decades of cheaply built, poorly designed public housing. They can provide loans, grants, and other assistance to nonprofit builders. They can require for-profit developers to include a certain percentage (usually 10 or 20 percent) of "affordable" units within market-rate developments through inclusionary zoning. And they can seek to ease the process of new housing development by speeding up permitting processes, preparing plans that indicate sufficient locations for new development, making sure that zoning codes do not prevent local development of appropriate intensity, and acting as intermediaries with concerned neighborhood groups.

These activities all require heightened attention to good urban design and planning, especially if housing is to take place in infill locations. Finding ways to placate NIMBY opposition, and to ensure that new infill development provides benefits for existing neighbors, is especially important, as is changing the makeup of the housing development community. In recent decades much home-building has been taken over by large-scale, mass-production builders (in North America, companies like KB Home, Pulte, Trammel Crow, and Shea Homes). In many places these companies have displaced the small- and medium-sized builders who once built the majority of new housing units, and who potentially have greater flexibility in responding to local contexts. Active local government support of small and nonprofit builders may be needed to counter the centralization trend within the housing industry.

Public health

Last but by no means least, a variety of public health issues, many of which have already been described, are best tackled at the local government level. Public health–related topics range from the identification and clean-up of brownfield sites to the construction of parks and recreational facilities to the provision of pleasant and walkable street environments. Many cities and counties also run health clinics, homelessness prevention programs, and other social service agencies. Still other local public health issues, such as air and water quality, will require regional, state, or national action since they go beyond the boundaries of any single local government.

Comprehensive "healthy city" programs have been adopted by a number of jurisdictions attempting to think holistically about how they can create healthy living environments for their residents. For example, the city of Glasgow, Scotland, has developed a Healthy City Partnership that is a member of the European Network of Healthy Cities. Begun in 1988, the Partnership's members include community, academic, housing, and business organizations as well as the City Council. It has undertaken a range of actions including programs for minority communities, for women's health, for parenting and children's health, and for healthier food and eating habits.[26]

Comprehensive local sustainability plans

Since the early 1990s a growing number of cities in North America and elsewhere have explicitly oriented their General Plans or other planning documents around the concept of sustainability. Other communities internationally have adopted Local Agenda 21 plans or other broad sustainability planning frameworks. However, such efforts often represent more window-dressing than actual change. In the study mentioned previously, Berke and Conroy (2000) found that US General Plans with specific language about sustainability were not significantly different from other high-quality General Plans without such language. Similar research in 2000 by Caroline Brown and Stefanie Dohr assessing national, regional, and local level plans in the United Kingdom found that although sustainability terminology has become widespread, these concepts haven't always made it into policy documents.[27] In another assessment of UK planning, Daniel Mittler found that, although more than 80 percent of local authorities were expected to produce Local Agenda 21 plans and this had substantially increased public involvement in many places, these efforts hadn't changed the basic power of planners to bring about more sustainable development. Finally, researcher Kent E. Portnoy (2003) developed an "Index of Taking Sustainability Seriously" and has applied this to 24 large US cities.[28] He found these communities generally lacking in systematic implementation of sustainability initiatives in areas such as transportation, brownfields revitalization, and biodiversity. The top-scoring cities on Portnoy's ratings were all on or near the West Coast: Seattle, San Jose, Scottsdale, Boulder, Santa Monica, Portland, and San Francisco.

The London Plan under development in the early 2000s also featured a sustainability theme; Mayor Ken Livingstone called for London to become "an exemplary sustainable world city."[29] However, at least in the draft of the plan it was far from clear how the sustainability goal would be realized, and the assumption of rapid economic and population growth made this objective problematic.[30]

Such findings may simply indicate that sustainability planning is in its early stages, and that consensus or political backing has not yet emerged for the most meaningful changes. There may also be a tendency by mainstream planners and politicians to co-opt terms such as "sustainable development," using them to justify efforts that are essentially business-as-usual, even though some more committed jurisdictions may be embarking on dramatically new planning directions using the same concepts.

In any case, there is a large amount of variation in the nature and success of comprehensive municipal sustainability frameworks. Some very successful and progressive local governments, such as that of Portland, Oregon, use the term little if at all, yet pursue planning that meets a wide range of goals that others might consider to be sustainable. Other cities such as San Francisco have adopted broad sustainability platforms, but have made little systematic progress on implementation owing to political disinterest or opposition. What matters in the end is not the terminology but the results.

Still, focusing directly on sustainability objectives within a General Plan or other comprehensive planning document does provide a way to think through systematically a long-term approach to local planning, and to reconcile objectives of economy, environment, and equity. It may also be an opportunity to develop specific sustainability-related goals, and to use tools such as sustainability indicators to measure progress towards these goals.

If cities do develop comprehensive sustainability frameworks, it seems essential to structure conditions to ensure that they have a decent chance of succeeding in the long run. A general sustainability vision must be connected in a meaningful way to specific policy and program changes, for example zoning revisions, changes in energy and recycling policy, and programs to redevelop certain areas or to ensure sufficient affordable housing. These steps must be monitored and evaluated over time, preferably by city or county staff rather than by nonprofit organizations who may not always have time, money, or political backing to do the job. Policy changes must have political buy-in from key interest groups, politicians, and the general public. And they must be institutionalized so that implementing staff and watchdogs will exist long term to ensure that changes come about. All these conditions will come about slowly, through consistent effort by planners, politicians, and local residents.

*   *   *

With or without a comprehensive sustainability plan, planning at the local government scale is crucial to sustainable development, since this is the scale at which most day-to-day land use and economic development decisions are made. Efforts to balance the Three Es, to establish sustainability-oriented criteria for new development, and to monitor sustainability indicators are especially vital here. Planning for sustainability within local government is challenging because broader perspectives are easily lost at this scale and local politics may be very pro-development. However, opportunities for change exist in every local jurisdiction, even those that seem most conventional. If we can find ways to show how these changes will improve local quality of life as well as meet regional and global needs, much progress can be made.

12 Neighborhood planning

Neighborhoods are one of the basic building blocks of cities, modest-size physical units that make up the residential portion of the urban area and form the environment that we all inhabit every day. Planning and design at the neighborhood scale affect our daily lives, determining what facilities are available locally, how far we need to travel, and much about our opportunities for interacting with our neighbors. This is also the scale at which individual development projects occur – whether they are large new subdivisions or smaller, more incremental changes to the urban fabric. Relatively small design and planning decisions such as the width and design of streets, the size of blocks, the mix of land uses, and the location and nature of parks and public spaces can have huge implications in urban livability and sustainability.

"Neighborhood" is a subjective term. For some, it refers only to a few blocks around the home, or even just a few neighboring houses or buildings. For others it may include a square mile or more, a large area containing hundreds of blocks and tens of thousands of residents or workers. The "hood" may also denote a particular cultural or social grouping of people living in proximity to one another, with little relation to any physical attributes of blocks or streets. All these definitions can be useful at different times. But typically the term has been applied to an area that a resident can easily traverse on foot – that is, with dimensions of one mile or less. This area should also possess some unifying social, architectural, economic, historical, or physical characteristics so that it can be distinguished from surrounding neighborhoods.

In practice, neighborhoods often have their own historical self-definition, and neighborhood associations that have defined their own boundaries based on local tradition may exist. The original land developers may have also established defining features such as village centers, gateways, and unifying street design or architecture, and may have given neighborhoods distinctive names as a part of marketing schemes. Thus, appellations such as "Forest Hills," "Glen Ellen," and "Elmhurst" are common. A frequent joke is that newly built neighborhoods are named for the natural elements the developer has destroyed in the process of construction, for example "Emerald Glen" or "Woodland Estates."

Main transportation routes such as freeways, railroad tracks, or major arterial streets often serve as de facto neighborhood boundaries. Many neighborhoods have also traditionally been built around schools and parks, particularly by developers following the classic "Neighborhood Unit" model first proposed by Clarence Perry in the 1920s (see Figure 12.1).[1] This influential model established the principal of an inwardly focused residential community centered on a school and park, with most traffic and stores relegated to large peripheral roads with shopping centers at their intersections. Partly in an attempt to insulate neighborhoods from automobile traffic many twentieth-century developers followed this model, although often without creating the interconnecting street network, parks, shopping, and community facilities that Perry envisioned.

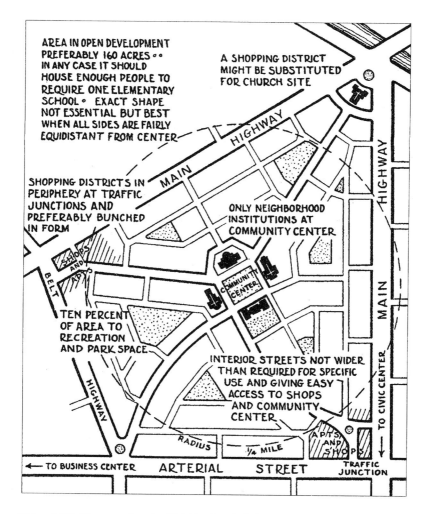

Figure 12.1 Clarency Perry's Neighborhood Unit concept. Clarence Perry's ideal of the "Neighborhood Unit," first presented in the 1929 Regional Plan for New York and New Jersey, developed the notion of an inwardly oriented neighborhood with traffic relegated to large arterial streets on the edge. In practice the development industry has taken this model to an extreme, insulating the neighborhood from surrounding areas. Parks, shops, and civic buildings are often missing.

Each neighborhood tends to be characterized by a particular type of urban form. It typically possesses a certain sort of street fabric – a particularly type of grid, or else looping streets or cul-de-sacs – used by the developers who first platted the area. This street fabric determines to a large extent how pedestrian-, bicycle-, and transit-friendly the neighborhood will be, and how well the neighborhood is connected to the surrounding urban area. Neighborhoods may also have consistent street design (street widths, sidewalks, street trees, and the like), a particular type or era of housing, a particular style of park or greenway system, and typical building forms and setbacks. In addition, demographics, income levels, culture, and housing prices may be similar through the neighborhood.

The neighborhood scale is particularly important because of a widely perceived need to reinvigorate a sense of community in postindustrial society. For at least 50 years sociologists have lamented the weakening of ties between people and the growth of individualistic attitudes instead. David Riesman's 1953 classic *The Lonely Crowd* described this process within conformity-oriented postwar society.[2] More than three decades later Robert Bellah and his colleagues chronicled the often-desperate search for community within 1980s America in *Habits of the Heart*,[3] and Harvard political scientist Robert Putnam documented the long-term decline of individual ties to social organizations in his 2001 book *Bowling Alone*.[4] The reasons for the decline of community ties are varied, according to Putnam, but include the physical design of neighborhoods and cities in addition to other factors such as the growth of television viewing, economic pressures, the loosening of traditional family structure, increased mobility, and generational change. Neighborhood design is also responsible for a large number of environmental problems, including loss of open space, destruction of wildlife habitat, and excessive resource consumption.

Because they have historically been in charge of regulating land development and subdivision, planners have a great deal of control over the character and form of neighborhoods, and can potentially help bring about more sustainable types of neighborhood design. Legal and institutional mechanisms, including zoning ordinances, subdivision controls, design review standards, and the processes of development approval, are relatively well developed for action on this scale. Neighborhood-scale planning is therefore one of the more promising areas for sustainable urban development.

Who does neighborhood planning?

Historically, private developers have played the most active role in planning neighborhoods, since they actually plot and/or construct them. Within the large-scale development model that has become the norm for new residential communities, these entrepreneurs obtain large chunks of land, lay out streets, subdivide property into small parcels, sometimes add parks and other amenities, and either construct homes and infrastructure themselves or sell parcels to other builders who do so. Even if buildings are not constructed immediately but are added decades later, the street and subdivision plan decided on initially determine much about the eventual character of the neighborhood. Although most large-scale developers are for-profit corporations, nonprofit builders also have opportunities to engage in neighborhood planning, usually on a more incremental scale around specific building projects.

Having developers as the primary designers of new neighborhoods, typically working within a framework of relatively weak city or county regulation at the urban fringe, has been a major problem. These builders are not necessarily motivated to produce sustainable communities in the long run, just buildings and lots that will sell for a substantial profit within a few years. Developers also have little incentive to examine how their own designs relate to other neighborhoods or the city as a whole. The result is a fragmented, chaotic urban landscape in which different neighborhood-scale developments – including residential subdivisions, office parks, and shopping malls – do not connect well to one another or meet broader urban or regional objectives.[5] Streets may not meet one another, sidewalks may not be present, buildings and site design may be unattractive and unecological, and no public places may exist that are comfortable for pedestrians or children. Leftover land between large construction projects develops haphazardly and adds to the confusion.

City planners and zoning boards exert substantial control over designs for new neighborhoods and improvements to existing ones, in that they approve or deny project applications and grant requested zoning variances. However, while planning staff may negotiate with developers on project details they rarely exercise the authority they might to revise subdivision layout. The current development approval process often simply codifies conventional wisdom, creating low-density, single-family-home subdivisions with wide streets, no shops or neighborhood centers, little variety of housing types, and little connection to surrounding areas. The public sector role in bringing about more sustainable neighborhood form remains weak.

To be sure, some cities are revising zoning codes and adopting design guidelines to produce more livable places, often under the influence of the New Urbanism. Changes to allow mixed land uses (homes, shops, and jobs near one another), mixed housing types, more pedestrian-friendly streets, and appropriate forms of infill development are particularly common. Some municipalities are also undertaking preparation of area plans for new or existing neighborhoods. But such actions are still in the early stages and often lack the political backing to truly meet regional and municipal needs.

A third main set of groups doing neighborhood planning – and often leading the charge for more sustainable communities – are advocacy organizations, community development corporations (CDCs), business groups, and neighborhood associations. These groups have greater freedom to develop a strong vision for a neighborhood than city planners hemmed in by political concerns. However, advocacy organizations must work with city planners or developers to get their visions implemented; they have little power to undertake development or fund improvements themselves. Consequently these plans usually work best if done in consultation with other actors who can provide resources and implement agreed recommendations. (See Box 12.1 and Figures 12.2 and 12.3.)

Box 12.1 The East Clayton Neighborhood Concept Plan

The rapidly growing city of Surrey, British Columbia is pursuing a sustainable development process known as the Headwaters Project, in conjunction with experts at the University of British Columbia. The first phase of this project is the East Clayton Neighborhood Concept Plan, a new neighborhood for 13,000 people on 560 acres of land. The product of extensive community design charettes, this project first developed consensus on seven sustainability planning principles, and then came up with a physical design for the site.

The new neighborhood will be compact, relatively high density (about 13,000 people on 550 acres), highly walkable, mixed-use, and in a traditional grid-like form which disperses traffic and facilitates public transit. Narrow streets will be well-shaded by trees, garages will be placed behind houses, all homes will be oriented to take advantage of passive solar heating, and extensive greenways and natural drainage systems will be created. Homes will cost about 20–30 percent less than comparable suburban houses because of more efficient use of land and infrastructure. The number of miles that residents drive is estimated to be 25 percent less than in comparable developments, resulting in a significant reduction in greenhouse gas emissions.

For more information, see Gilliard (2003, pp. 13–14); www.sustainable-communities.agsci.ubc.ca/projects/Headwaters.html; and www.smartgrowth.bc.ca. Also see Figures 12.2 and 12.3.

Figure 12.2 A new, ecological neighborhood: East Clayton. Billed as "the largest sustainable community in British Columbia," the East Clayton neighborhood in the city of Surrey features compact, walkable neighborhoods, natural filtration of stormwater runoff, a balance of jobs and housing, an average 2-minute walk to a park or greenspace, and units costing 20–30 percent less than standard homes in the area. Credit: James Taylor Chair in Landscape and Liveable Environments.

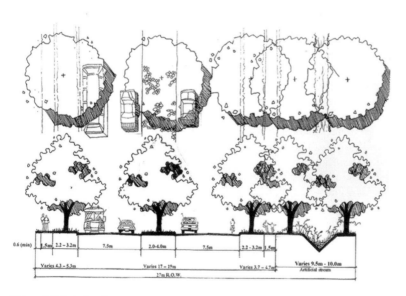

Figure 12.3 A riparian greenway corridor at East Clayton. This "Riparian Parkway Greenway" features a constructed stream channel to provide habitat and biofiltration of surface water. Credit: James Taylor Chair in Landscape and Liveable Environments.

Neighborhood sustainability issues

Certain sustainability planning issues are particularly important at this scale. Achieving compact and relatively mixed-use neighborhoods is often seen as a goal, since this reduces suburban sprawl and the distances people need to travel to go to shops, schools, workplaces, and recreational facilities. The mix of jobs and housing can be balanced at a neighborhood scale, as well as at city and regional scales. Integration of high-quality public transit into neighborhoods is vital. Traffic calming is often best planned on a neighborhood-wide basis. Preservation and restoration of creeks, wetlands, slopes, wildlife habitat, and other ecological features is a frequent concern, along with creation of local parks, community gardens, greenways, and other public spaces. Other urban design and planning initiatives at the neighborhood level can help create a "sense of place" by preserving historical structures, celebrating local culture and tradition, and linking the neighborhood to the natural landscape.

Neighborhood design

Many urban form elements contribute to the feel and function of a neighborhood. The nature of the street fabric is one of the most basic and determines many other elements of neighborhood form. Grids, curvilinear streets, cul-de-sacs, and other street forms all produce radically different neighborhood characteristics, even with the same overall density and building form. Some street patterns connect neighborhoods to the surrounding city much better than others; some are best adapted for exclusive, insulated communities that want little to do with the outside world. The size of blocks and lots, the design of streets, the arrangement and design of parks, and the presence of shops or community facilities also greatly affect neighborhood livability and sustainability. As Jane Jacobs argued in 1962, small blocks make for a more pedestrian-friendly environment, giving those traveling by foot or bicycle far more potential routes between two points and avoiding long, monotonous streetscapes with large-scale buildings. Sidewalks and safe intersection crossings are two other neighborhood design features that greatly improve pedestrian friendliness; both are featured in the "Pedestrian Environment Index" developed by planners in Portland. Park and greenway design, discussed further later, are also key factors in neighborhood livability.

The evolution of neighborhood street fabrics has gone through a number of stages, influenced by evolving transportation technologies, urban design philosophies, construction techniques, planning regulation, and development economics. Most North American development in the early to mid-nineteenth century used a grid with relatively small, square blocks. A look at the map of most cities or towns will show this square-block grid still existing at the historic urban center, though now perhaps streets are lined by office buildings or stores instead of homes. In the late nineteenth century as streetcar lines proliferated (the first electric streetcar was introduced in Richmond in 1879), a new grid form emerged with longer, rectangular blocks. Streetcar company owners frequently bought land and platted large new neighborhoods along their routes. This "streetcar suburb" grid can also be identified on maps of most North American cities. Within these nineteenth-century grids, blocks filled in slowly over decades with buildings built individually by hand, typically using post-and-beam construction as opposed to the more rapid "balloon frame" method that came later. The gridded streets of each new neighborhood did not always link up to the existing grid perfectly but generally formed a relatively well-connected overall street network.

In the late nineteenth and early twentieth centuries a new neighborhood design philosophy arose – the "garden suburb" based on ideals of English country living. This vision of green, leafy, picturesque suburbs was propagated in the United States through the writings and drawings of Andrew Downing, a popular arbiter of taste in the 1840s and early 1850s as editor of the *Ladies' Home Journal*.[6] Ironically, picturesque design principles were initially applied in the United States to cemetery design, beginning with Mount Auburn cemetery in Cambridge in 1831.[7] These facilities pioneered naturalistic scenery and curvilinear streets to provide a semi-rural atmosphere in contrast to the gridded regularity of the nineteenth-century city. A few pioneering mid-nineteenth-century developments, such as Frederick Law Olmsted's 1869 design for the Chicago suburb of Riverside, then applied this philosophy to residential development in North America. With its curving roads, wide lots, and large setbacks, this subdivision on a commuter rail stop helped establish a new ideal of the low-density suburban environment. But except for a few experiments such as Riverside, the garden suburb model would wait for more widespread implementation until the advent of the automobile in the early twentieth century made such decentralized development more feasible.

In his turn-of-the-century Garden City concept Ebenezer Howard presented a different and more radical suburban vision – a constellation of relatively self-sufficient new towns circling a large older city, connected by rail and separated by greenbelts with agricultural land and public institutions. In Howard's vision a grid of streets, albeit molded into a circle, would still form the framework for each community, but other qualities would change, especially the regional form, the placement of employment and shopping along central avenues, and ownership patterns, which would become essentially cooperative rather than private. But with the partial exception of the British and Swedish New Towns built after the Second World War and the three American Greenbelt communities built in the 1930s, this vision of community form remained unexplored. Garden suburbs, with their winding streets, lower densities, and exclusively residential land use, triumphed over garden cities.

Modernist philosophies of neighborhood design, promoted by the Congrès International des Artes Modernes (CIAM) in the 1930s, advanced another alternative. This neighborhood model emphasized sleek, functional buildings, large public spaces, the creation of much larger block sizes or "superblocks" with internal pedestrian circulation, and (as implemented in Europe) good public transit and extensive public services. "Towers in a park" described many modernists' vision of the neighborhood. This model was mainly embraced by public authorities in socialist or social democratic countries and in North American public housing projects. But unfortunately these developers gave too little thought to how these environments would actually fit the needs of residents or natural ecosystems. The result was frequently disastrous. Bleak, sterile public spaces and buildings dwarfed residents while frequently proving just as automobile-dependent and wasteful of land and materials as suburban tracts.

Meanwhile, the private market embraced the garden suburb approach to neighborhood construction and expanded on this during the twentieth century. As automobiles proliferated designers realized the advantages of residential streets with no traffic, and began to add loop roads and dead-ends to their new neighborhoods. In 1916 American developer Edward Henry Bouton praised this model of the non-connecting residential street although the French term "cul-de-sac" had yet to be widely applied to it:

> In many places where topographic or other conditions make it difficult or undesirable to extend a street to its intersection with another, such streets may be designated with "dead-ends," or returned upon themselves, forming "places,"

to which great charm is attached by the sense of privacy and seclusion which they impart.[8]

Developments such as Roland Park in Baltimore, built in the 1910s, added a bulbed end to short dead-end streets, producing the distinctive cul-de-sac form that would dominate later suburban neighborhoods. Cul-de-sacs proliferated steadily throughout the twentieth century, their numbers reaching a peak in 1980s and 1990s development. In Canada developers took a somewhat different approach, favoring loop roads over cul-de-sacs, but the end result was the same: to create neighborhood environments completely insulated from the rest of the metropolis.

Towards the middle and end of the twentieth century a growing chorus protested both modernist and garden suburb philosophies of neighborhood design. Mumford had been a strong critic of mainstream development throughout his career,[9] but was joined by writers such as William H. Whyte, who argued for clustering suburban development to save open space,[10] and Jacobs, who described the advantages of traditional urban neighborhoods with highly connected street patterns.[11] The discipline of environmental design arose in the 1970s in part to look at how people actually use neighborhood environments, employing means such as post-occupancy evaluations to see how residents liked the buildings and exterior spaces that had been created for them. The results were often shocking, and pointed to the need for more human-scaled public spaces, "eyes on the street" to reduce crime, better places for children to play, and a variety of semi-private outdoor spaces that apartment residents could personalize and have control over.[12]

In the 1990s the New Urbanism proposed to rethink neighborhood design by returning to many features of the old streetcar suburbs – in particular highly connected street patterns, pedestrian-oriented street design, mid-block alleys, and architectural features such as front porches and garages behind the house. Referred to initially by labels such as "Traditional Neighborhood Development," this movement began with a few pioneering projects such as Seaside, Florida (designed by the husband-and-wife team of Andres Duany and Elizabeth Plater-Zyberk in the early 1980s), and Laguna West, California (designed by Peter Calthorpe in the late 1980s). The movement was institutionalized through annual Congress for the New Urbanism conferences beginning in 1993, and quickly spread into the mainstream of planning thought if not into actual neighborhood development. Movements for smart growth, livable communities, and sustainable development dovetailed with many aspects of the New Urbanism. The result of this re-evaluation of urban form was a much more sophisticated understanding of how neighborhoods could be better designed to meet human and environmental needs.[13] In fact, the New Urbanism does not simply repeat the streetcar suburb model of neighborhood design, but creates a range of highly connected street forms that include many small parks and public spaces. Rather than rigid grids, New Urbanist communities often employ more organic, slightly curving street forms, but the end result is the same: a highly walkable urban fabric. (See Figures 12.4 to 12.7.)

Figure 12.5 (opposite) Riverside, an early garden suburb. Frederick Law Olmsted's 1869 development of Riverside, outside Chicago, provided an influential model of the garden suburb.

Figure 12.4 The beginnings of the garden suburb. British designers pioneered picturesque garden suburb design in the early 1800s. In a late 1820s perspective showing homes similar to today's "McMansions," Decimus Burton envisions villa residences in Hove.

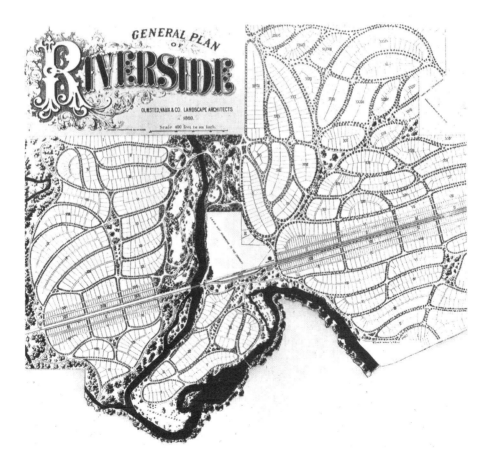

Figure 12.6 Radburn develops the "superblock" model. At Radburn, built in suburban New Jersey in the late 1920s and early 1930s, Clarence Stein and Henry Wright created a new version of the garden suburb: the superblock with houses located on cul-de-sacs and an internal network of green spaces. Only a portion was built due to the arrival of the Great Depression. Mainstream developers copied the cul-de-sac and superblock ideas without the green spaces.

Density

For many North Americans, "density" is a four-letter word. They associate the term with large, impersonal apartment buildings, public housing projects, or physical environments like Manhattan. Yet adding residents, jobs, and businesses to a neighborhood provides many advantages, including improving safety; increasing the viability of local businesses, cafés, and restaurants; providing sufficient ridership for transit; enhancing community interaction; and saving open space.

Rather than use the "d-word," planners and elected officials often talk instead about "compact development," "smart growth," or "walkable neighborhoods." Oakland Mayor Jerry Brown has used the phrase "elegant density." One approach pioneered by Rutgers professor Anton Nelessen to help reduce local objections to denser development has been to conduct a "visual preference survey" of local residents. Researchers show people images of typical lowdensity suburban development and other types of higher-density communities such as turn-of-the-century streetcar suburbs and well-designed urban infill projects. Most residents find they prefer the higher-density alternatives because these include more attractive streetscapes, local shops and restaurants, and a greater diversity of housing choices. Over 25 years Nelessen and his colleagues have administered this survey to approximately 50,000 people nationwide, with fairly unanimous results in all geographic regions.[14] Public workshops and design charettes are also useful tools to help citizens see that increasing neighborhood densities can be desirable. Again, when asked

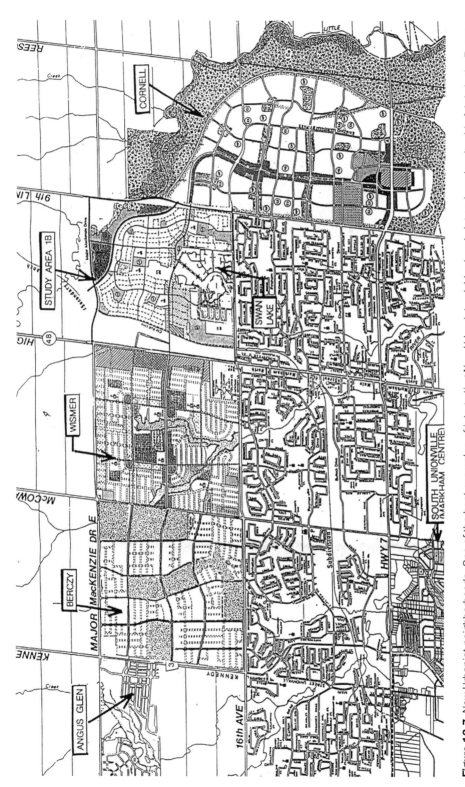

Figure 12.7 New Urbanist neighborhoods. One of the best examples of integrating New Urbanist neighborhoods into a broader citywide fabric is the Toronto suburb of Markham. The Cornell New Urbanist community designed by the firm of Duany Plater-Zyberk and others is at lower right. 1970s and 1980s subdivisions are at the bottom. In principle, cities can connect such neighborhoods to form a more coherent urban form. Credit: Town of Markham.

to choose among many housing and land use patterns, residents often select traditional town forms with higher densities and mixtures of land uses rather than typical suburban sprawl.

Recent US suburban densities have been relatively low, often four to eight dwelling units per net acre before local roads and public facilities are factored in (gross densities are even lower). Average densities were lowest after the Second World War, when residential lots of a quarter-acre or more were common, and have risen in recent decades as land prices have escalated, though much development is still below eight units per acre. By contrast, densities in many older neighborhoods built around the turn of the century – including the streetcar suburbs that often score best in visual preference surveys – are often 10 to 16 units per acre. Densities for apartment buildings in downtown locations can range above 200 units per acre for attractive five-story buildings that fit well along existing streets. (A five-story, 50-unit apartment building on a quarter-acre, 100 by 100 foot lot represents a density of 200 units per acre, and still can have an attractive courtyard, entry plaza, and rooftop deck.)

Traditional British suburban densities are somewhat higher than in the US. The garden suburb designs pioneered by Raymond Unwin and Barry Parker in the early twentieth century were around 12 units per net acre, while areas of London such as Bloomsbury, Regent's Park, and Bedford Park achieve densities of up to 40 units per acre (100 units per hectare) with many single family homes and substantial amounts of private open space. Friends of the Earth in the UK proposes 28 units per acre (69 units per hectare) as a sustainable urban density;[15] some North American smart growth advocates call for similar intensities of land use. The minimum density usually seen as necessary to support frequent public transit service is 12 units per acre, so this average density, if combined with a highly connecting street fabric and good street design, should make for a walkable and transit-oriented neighborhood environment.

As sociologist Amos Rapoport pointed out in 1975, perceived density can be wildly different from actual density in a neighborhood, and is a function of traffic, noise, safety, greenery, reduced open space and many other factors rather than the number of dwelling units or people per acre.[16] An urban neighborhood that is green, quiet, attractive, and composed of modest-sized buildings will strike observers as far less dense than it actually is. Conversely, an environment that is dirty, noisy, and full of traffic is likely to be perceived as more urban and dense, even if it takes a low-density suburban form. Perceptions of density are also highly related to culture; in many parts of the world a large number of people in a neighborhood, such as in the fashionable arrondissements of Paris with their five-story buildings, is considered quite civilized and desirable. Planners and developers can make density livable – and lessen perceived density – by combining relatively intensive residential development with attractive streetscapes, public spaces, and amenities such as parks, shops, restaurants, and child care centers. (See Figure 12.8.)

Infill development

According to British writers David Rudlin and Nicholas Falk, "the most fundamental feature of the Sustainable Urban Neighborhood is its location – the fact that it is located within existing towns and cities."[17] Such development is the main alternative to continued suburban sprawl on greenfield sites. If done well infill can create not just attractive new buildings and housing units in existing urban areas, but entire neighborhoods that are more pedestrian-oriented, vibrant, diverse, and ecological than our present communities.

HOUSING TYPES AND DENSITIES

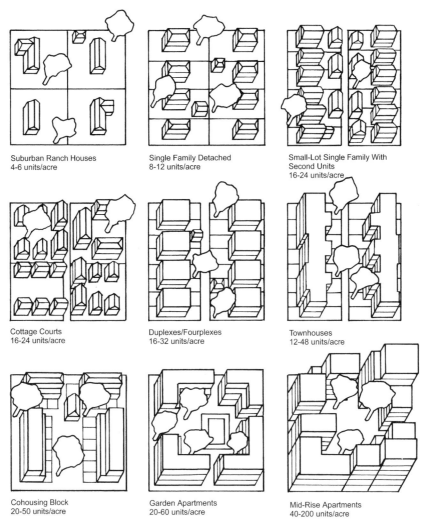

Suburban Ranch Houses
4-6 units/acre

Single Family Detached
8-12 units/acre

Small-Lot Single Family With
Second Units
16-24 units/acre

Cottage Courts
16-24 units/acre

Duplexes/Fourplexes
16-32 units/acre

Townhouses
12-48 units/acre

Cohousing Block
20-50 units/acre

Garden Apartments
20-60 units/acre

Mid-Rise Apartments
40-200 units/acre

Stephen M. Wheeler

Figure 12.8 Housing types and densities.

Four main types of infill locations exist within our cities – neglected downtowns, underutilized arterial strips, existing single-family-home districts where property owners might add second units, and large sites formerly occupied by factories, malls, office parks, or military bases. If contaminated the latter are often known as "brownfield" sites.

Large-scale infill development opportunities are available in many North American downtowns that deteriorated enormously during the second half of the twentieth century. These centers of small towns as well as large cities are now characterized by vacant lots, surface parking lots, shabby older buildings in need of rehabilitation or replacement, and one-story fast food restaurants and other low-intensity land uses. All these properties could be redeveloped to create new urban neighborhoods. Three-to-five story apartment buildings – or even taller buildings – with ground floor shops and restaurants can help make older downtowns into exciting 24-hour communities rather than depopulated wastelands after office workers go home for the day. In such locations high-rise apartments may be appropriate. Americans have an inordinant fear of high-rise buildings, perhaps associating them with failed public housing projects of the 1950s and 1960s, but cities such as Vancouver, British Columbia, show that high-rises can help create attractive urban neighborhoods. The key to Vancouver's use of high-rises is to keep the buildings slender and to set them back from the street, so that they don't block light and so that they allow the street to be a green, human-scaled environment. Two-to-three story townhouses – attached homes with private front and back yards – are also a traditional, highly livable urban housing type that could be reintroduced in many urban neighborhoods. Washington, DC, Baltimore, Boston, and many other older American cities are full of attractive townhouse neighborhoods built almost 100 years ago. Such downtown infill may need to be coordinated by city redevelopment agencies in many cases, since these entities have the power to assemble parcels of land, improve infrastructure, and add parks and streetscape improvements to support infill. Care should be taken to respect historic buildings and existing residents. Rather than repeating the bulldozerdriven urban renewal projects of the twentieth century, a much more sophisticated, contextual, and incremental process of rebuilding urban neighborhoods is needed. Generally existing housing should not be redeveloped – sites such as parking lots, failed shopping centers, and old industrial areas offer infill potential without risk of destroying historic properties or dislocating residents.

Since they are such a ubiquitous feature of the North American landscape, arterial strips represent enormous opportunities to create new infill neighborhoods. However, steps must be taken to make these wide, often heavily trafficked streets more attractive and pedestrian-oriented, for example, by adding sidewalks and street trees, by reducing or narrowing lanes when traffic volumes are not too high, or by creating plazas, mini-parks, or courtyards off the main street. (See Figures 12.9 and 12.10.)

Infill can take place very unobtrusively in existing neighborhoods through adding second units to existing single-family homes. In many older neighborhoods these have already been added, legally or illegally, often by converting basements, attics, or garages into small apartments. These units can house students, elderly parents, and a variety of other residents who do not need large amounts of space, while providing supplementary income to homeowners. But city zoning codes often prohibit such accessory units or require extensive procedures for use permits. Making second units legal as-of-right would be a great step towards adding additional housing and raising neighborhood densities to levels that better support public transportation and local shops. Allowing duplexes or townhouses on vacant lots within single family home districts would be a further step towards accommodating additional housing without significantly changing the form or character of existing low-density neighborhoods.

Large reuse sites – once occupied by factories, railyards, shopping malls, office parks, airports, or military bases – represent a final type of infill challenge. These locations offer the possibility of creating entire new neighborhoods from scratch. But often these brownfield sites must be cleaned up, in the worst case by removing many feet of soil and transporting it to a landfill. Local governments may need to work with property owners to develop a Specific Area Plan governing redevelopment of these sites, and political obstacles may need to be overcome, but the rewards can be enormous. Old railyards in

Portland, Oregon, and San Francisco have become vibrant new urban neighborhoods, the former Stapleton Airport in Denver has been redeveloped as a New Urbanist infill community, and former port facilities in Baltimore, New York, and London have become significant new additions to those cities.

Infill development is often vigorously resisted by residents and local politicians, and NIMBY opponents have killed many promising projects. This resistance may be due to fears that property values will decline (many studies have shown that in fact they do not), desires not to have less affluent residents or people of color living nearby, or a generalized fear of change. Yet infill can provide enormous advantages for existing residents of a community, for example by providing new restaurants, cafés, parks, transportation options, and public spaces, and can increase rather than diminish property values. Much of the challenge will be to win over opponents through communication, collaboration, education, and specific responses to their concerns. Cities can help overcome neighborhood resistance by conducting a neighborhood visioning process in conjunction with the preparation of Specific Plans. This task is not easy, but skilled facilitators can help develop participation processes that are constructive rather than oppositional.

As discussed previously, local zoning codes often work against new development in existing urban areas. It is literally impossible in most places to re-create the thriving downtowns of a century ago because they would be illegal under current zoning. These codes often limit building heights, prevent mixed-use buildings, require setbacks from the street, and mandate large amounts of parking. The solution is for local governments to

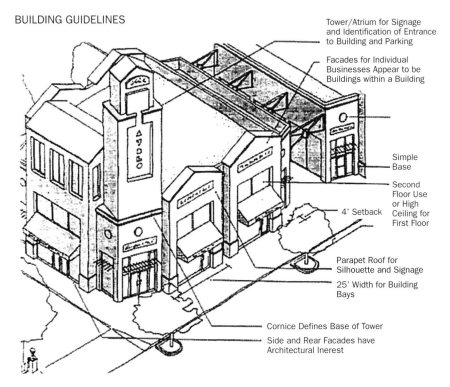

BUILDING GUIDELINES

Tower/Atrium for Signage and Identification of Entrance to Building and Parking

Facades for Individual Businesses Appear to be Buildings within a Building

Simple Base

Second Floor Use or High Ceiling for First Floor

4' Setback

Parapet Roof for Silhouette and Signage

25' Width for Building Bays

Cornice Defines Base of Tower

Side and Rear Facades have Architectural Inerest

Figure 12.9 Design guidelines can guide neighborhood infill. By adopting urban design guidelines for particular neighborhoods, cities can speed up the development review process, improve results, and create greater certainty about what is expected. Credit: City of Albany, California.

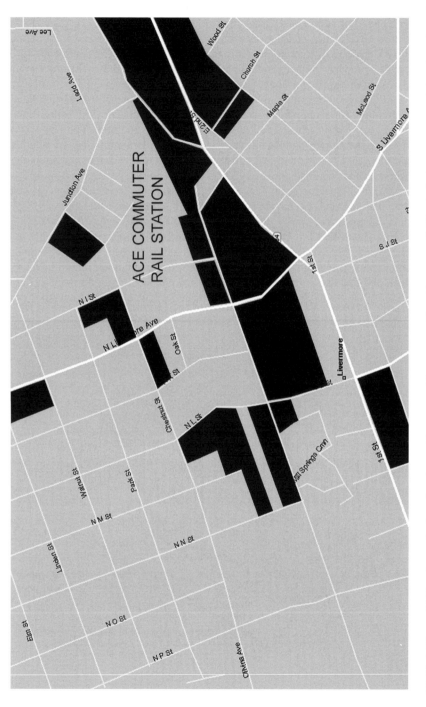

Figure 12.10 Infill possibilities in a suburban downtown. Like many older communities, Livermore, California features a compact downtown based on a walkable nineteenth century street grid. Although this center boasts a new ACE commuter train station, it is full of vacant or near-vacant lots that could be redeveloped into housing and shops.

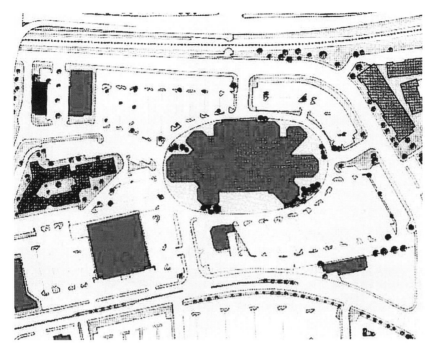

Figure 12.11 Redeveloping an old shopping center as a new neighborhood (before).

Figure 12.12 Redeveloping an old shopping center as a new neighborhood (after).

Figure 12.13 The Crossings. Figures 12.11, 12.12, 12.13 show The Crossings in Mountain View, California, which redeveloped a defunct 1960s shopping center with 359 homes on 18 acres next to a new CalTrain station. The new neighborhood includes townhouses, apartments, cottages, and single family detached homes. The city has helped make such transit-oriented infill development happen by preparing "Precise Plans" for the sites. Credit: Calthorpe Associates.

Box 12.2 Infill development can decrease driving

A 2000 study by the Natural Reseources Defense Council and the US Environmental Protection Agency of an infill subdivision in Sacramento vs. a greenfield counterpart found that the infill neighborhood substantially reduced driving and travel distances (see National Resources Defense Council and US Environmental Protection Agency, 2000).

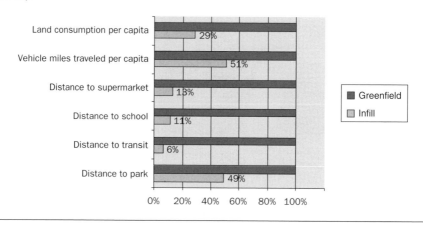

review zoning codes line by line, and make sure that they allow desirable forms of infill. For example, a city might raise its downtown height limit to five stories to accommodate housing above stores, while establishing a minimum height of at least two or three stories. Municipalities might also require buildings along main streets to establish a solid street frontage, and to include retail or restaurant spaces along the sidewalk.[18] (See Figures 12.11 to 12.13 and Box 12.2 above.)

Compact, mixed-use development

Before the advent of single-use zoning for broad areas of cities, neighborhoods typically contained offices, small shops, grocery stores, and restaurants in addition to housing. Zoning for most post-Second World War subdivisions prohibits these non-residential land uses. One of the main tenets of the New Urbanism and sustainability-oriented design in general is to include this variety of land uses within communities once again, typically within neighborhood centers or along main streets. If jobs, housing, shops, and recreational facilities are closer together, the theory goes, then people will need to drive less and neighborhoods will be more vibrant and livable.

Although mixed-use development does provide these benefits, it has proven somewhat difficult to bring about in practice. In most actual New Urbanist developments, such as Seaside, Kentlands and Laguna West, the "mix" includes workplaces and only a very small number of retail storefronts in a village center.[19] Developers have had to subsidize many of these shops or allow the storefronts to remain empty, since particularly in the early stages of a new development there is not sufficient population density to support them. Local shops also face competition from a wide range of big-box retailers, chain stores, and shopping centers in the surrounding area.

Although a number of good mixed-use developments have been built in North American cities, this more integrated mix of land uses remains more an ideal than a reality, especially in suburban locations. Token amounts of retail within large residential projects, New Urbanist or otherwise, on the edges of metropolitan regions are unlikely to change existing patterns of automobile use or long-distance commuting. Much more fundamental changes in land use are needed instead.

What sustainability planning might envision is a far more radical mix of land uses throughout urban and suburban neighborhoods, coupled with strong restrictions on single-use and large-scale developments. Real neighborhoods need to include grocery stores, hardware stores, drug stores, cleaners, child care centers, places of worship, medical or dental offices, fitness centers, and much more – all within close proximity. Office buildings and much light industry should also be located near where people live. This, after all, is the model of the traditional town before the age of the automobile.

To help truly mixed-use neighborhoods come about, cities will need to end current zoning for large expanses of homogeneous land uses such as office parks, shopping centers, and residential tracts. Cities may well also need to amend their zoning codes to prohibit stores of greater than 30,000–50,000 square feet, since these big box stores tend to kill smaller, neighborhood-oriented retail businesses and cause a great deal of long-distance travel. (Certain types of big box stores are in fact known as "category killers," in that they drive every other merchant in the same category out of business.) Exact size limits will depend on the type of retail; grocery stores may need the higher amount to offer a wide range of products; hardware stores and most other types of retail business do not. Once an environment is created that does not actively undercut neighborhood-scale, mixed-use development, then more locally-oriented businesses can emerge.

Particularly valuable will be placing a range of jobs near residential neighborhoods. In an age in which most forms of economic activity take place in office buildings or other non-polluting forms of workplace, there is no reason for businesses to be widely separated from homes. Instead, offices can be located along arterial streets and transit lines, in transit villages at commuter rail stops, and within city, town, and neighborhood centers (where they can be joined by new housing). Planners and community development groups should encourage types of workplaces that are appropriate to the income level of residents in a given neighborhood. Jobs and housing should be balanced, in other words, not just regionally or at a county scale (as progressive planners often seek to do currently), but within each city, town, and neighborhood. The result will be greatly reduced pressures for long-distance commuting and improved quality of life for residents who can walk or bike to work.

A growing number of new neighborhoods illustrate principles of relatively compact, mixed-use design. In the United States, new communities such as Orenco Station outside Portland, Oregon, and Playa Vista in Los Angeles (though controversial for environmental reasons) provide such examples. In Britain, Poundbury, a 400-acre extension added to Dorchester, is a good illustration.[20] This new neighborhood, designed by Leon Krier under the aegis of the Prince's Foundation, is almost two-thirds mixed-use buildings, and provides 20 percent affordable housing along with 21 commercial and seven retail businesses. To be inhabited by 5000 people, Poundbury replicates local architectural styles and uses local and recycled building materials.

Streetscape design

One of the biggest challenges at a neighborhood level is making arterial streets more pedestrian-friendly and livable. These streets are often congested with traffic and lined by strip businesses – fast-food joints, gas stations, auto repair shops, and other one-story, drive-in businesses. Writers such as J.B. Jackson have suggested that they are an essential and underappreciated part of the American cultural landscape, and perhaps in some ways they are, yet to many of us these arterial strips are unpleasant and stressful places to be whether on foot or in a car.

These arterial corridors within almost any city or town offer extensive opportunities for infill development. Luckily, there are well-established traditions of large streets in many countries that both carry substantial volumes of vehicle traffic and are green, pedestrian-friendly places to be. In particular the "multiple roadway boulevard" model places fast-moving vehicles on center lanes and separates this traffic from slow-moving side streets using landscaped medians.[21] Designers such as Frederick Law Olmsted originally developed this strategy in the nineteenth century. Examples are provided by the Champs-Élysées in Paris, the Paseo in Barcelona, Atlantic Parkway in Queens (designed by Olmsted), and the Esplanade in Chico, California. Similar designs can be used today to retrofit some of today's least attractive arterials.

Converting existing arterial streets into pedestrian-oriented boulevards will take proactive planning by local government. Planners can rezone land along these streets for denser, mixed-use development, redesign streets to include wider sidewalks and landscaped medians, and, where possible, create full-fledged boulevards with separate fast and slow travel lanes. The traffic volume on these routes can be reduced by taking steps to move jobs and housing closer together, improve public transit, and provide economic incentives for people to use other modes of transportation. Gradually such actions can help humanize one of the most unpleasant and problematic elements of many neighborhoods currently. (See Figures 12.14 and 12.15.)

Figure 12.14 Redesigning a suburban arterial as a walkable boulevard (before).

Figure 12.15 Redesigning a suburban arterial as a walkable boulevard (after). Figures 12.14 and 12.15 are a visual simulation showing how a typical suburban arterial in Pleasanton, California might be transformed into a pedestrian-oriented street. Credit: Steve Price/Urban Advantage (www.urbanadvant age.com).

Traffic-calming

As automobiles multiplied rapidly in industrialized nations in the early and mid-twentieth century, many observers realized that they were degrading neighborhood quality. Proliferating motor vehicles were a central element of the inhuman "technopolis" that Mumford warned against. Jane Jacobs, although not necessarily a fan of pedestrian-only districts in American cities that couldn't support them, believed that too much traffic leads to an "erosion of cities" and that there should be an "attrition of automobiles" through widening sidewalks, narrowing streets, bottlenecking traffic lanes, and other steps that would make driving less convenient.[22] Later in the 1960s establishing "streets for people" was part of the vision of humanist Bernard Rudofsky, who provided an erudite, illustrated history of American streets and examples worldwide of how streets have historically served a glorious profusion of public uses other than conveying motor vehicles.[23] This line of thinking has expanded in more recent decades through the work of scholars such as Donald Appleyard, whose 1981 book *Livable Streets* was a milestone of research into neighborhood traffic management,[24] and activists such as Australian David Engwicht, who successfully rallied opposition to a local expressway by chalking out its route through a neighborhood park.[25]

The modern traffic-calming movement began in Europe in the 1960s in response to accidents in which children were killed or injured by speeding drivers. Cities in Germany and the Netherlands adopted particularly active programs to reduce vehicle speeds. The word "traffic-calming" in fact is a translation of the German term *Verkehrsberuhigung*. In cities such as Cologne citizens began organizing in the early 1970s to close or calm residential streets where speeding drivers had endangered children. In the Netherlands, advocates developed *woonerf* ("living yard") designs through which local streets were made into pedestrian-priority areas. *Woonerfs* generally include measures such as trees or planters within the road area, lack of differentiation between pedestrian and automobile space, and special paving treatments to designate certain street areas as part of the living area associated with houses. The traffic-calming concept has spread steadily in North America since the late 1970s. Traffic-calming mechanisms fall into two main categories: those that seek primarily to reduce vehicle speeds, and those that focus on lowering traffic volumes. Speed humps – relatively broad "vertical deflection devices" six feet or more across – are perhaps the most common form of traffic-calming and have appeared in countless cities throughout the world. They are very effective at reducing vehicle speeds to around 15 miles per hour without placing any restrictions on traffic volume, and have the additional advantage of being relatively cheap. Speed humps have been probably the most widespread method of traffic calming in the United States. In contrast, sharper speed bumps are used extensively on private access roads such as around shopping mall parking lots. Cities in Mexico and other parts of the world use them on public streets as well, where they are very effective at reducing speeds to around 5 mph, but are seen as a nuisance by many drivers. Both speed humps and bumps are viewed as a health hazard by some disabled activists who dislike being jarred by paratransit vans traveling over them.

Other devices to reduce traffic speeds include speed tables (raised sections of pavement much broader than speed humps), chicanes (offset planters or sidewalk extensions forcing vehicles to travel a zig-zag path down local streets), bulbouts (curb extensions into the street at intersections to facilitate pedestrian crossing and slow cars entering the street), chokers (parallel intrusions into the street mid-block), traffic circles at neighborhood intersections, and various forms of colored and/or textured pavement (which can remind drivers they are in a pedestrian priority zone).

The other main approach to traffic-calming – focusing on traffic volume rather than speed – uses means such as diverters (preventing traffic from entering streets), semi-diverters, diagonal diverters, and other forms of street closure. Berkeley was one of the US pioneers with such devices in the late 1970s, using rows of concrete barriers called bollards to divert traffic from neighborhood streets south and west of the University of California campus. This method greatly improved quality of life on the calmed streets. But as Berkeley and other cities have discovered, these devices have the disadvantage of diverting traffic onto other streets, which may then experience even worse problems. In effect they make a gridded urban neighborhood into a system of cul-de-sacs, with through-traffic concentrated on arterial streets. Such cities now approach diverters with great caution.

The current wave of traffic-calming devices in North America appears to encompass circles, humps, and bulbouts, which slow traffic without such diversion. Circles in particular are used extensively by Portland, Seattle, and Vancouver. Planting trees along streets, adding mid-block chokers, and narrowing street or lane widths can also help reduce vehicle speeds.

Many European cities have gone much further by creating pedestrian-only districts in central cities. Almost every major city on the continent has such an area where motor vehicles are only allowed in the early morning for deliveries, if at all. Frequently these districts are the most dynamic, interesting, and economically vibrant parts of the city. Copenhagen's Ströget, a 2-mile-long pedestrian shopping street closed to traffic in the 1960s, was the first major example. There have been several waves of efforts to create pedestrian streets or malls in the United States since the 1960s, but these have not worked well in situations in which a critical mass of residents or tourists did not exist to support local shops. On the other hand, examples such as Santa Monica's Third Street Promenade have been successful (in its post-1990 incarnation). Closure of certain streets in urban parks such as Rock Creek Park in Washington, DC and Golden Gate Park in San Francisco, especially on weekends, is also on the increase.

As far as motorists are concerned, the ideal (given that unlimited high-speed travel is not possible) may be residential streets that have relatively slow speeds (20–30 mph) but smooth and continuous travel. This approach minimizes the aggravation of stopping and starting, while creating a street environment that is relatively safe and pleasant for other users. Practical implications of this approach include minimizing use of stop signs in residential neighborhoods (relying on traffic circles and other methods to keep traffic slow and intersections safe) and synchronizing lights on arterial streets to serve traffic moving at relatively slow speeds. This approach also has significant benefits in terms of fuel economy, pollution, vehicle wear, and noise. (See Figures 12.16 to 12.22.)

Parks, gardens, and open space

Since a better connection between human and natural environments is a central challenge of sustainable development, neighborhood planning should seek to create a variety of open spaces and natural areas. To date many neighborhoods have been built with little regard for natural or public space. Schoolyards, with their asphalt playgrounds and standardized athletic fields, form the most common type of open space in many neighborhoods. However, school facilities do not serve the recreational needs of many residents and with their bare expanses of hardtop or grassy playing fields hardly represent a local connection with nature. New forms of neighborhood open space are essential – neighborhood parks, trails, gardens, greenways, and outdoor recreational facilities – as urban areas become more intensively developed.

Whereas nineteenth-century park designers often sought to create picturesque urban parks as picnic grounds for working-class families, today's urban green spaces need to serve a far greater range of functions. Different types of spaces are needed for gardeners, young children, older children, teens, elderly residents, team sports players runners, bicyclists, hikers, and many other groups. The traditional model of the one-square-block neighborhood park doesn't meet the needs of many of these groups. As previously discussed, neighborhood open spaces should also be thought of as part of a regional green infrastructure network, providing habitat and wildlife corridors throughout the urban region. Restored creek corridors, wetlands, shorelines, and riverbanks are particularly important in this regard, and can also provide long, interconnected jogging or biking trails for active recreational needs.

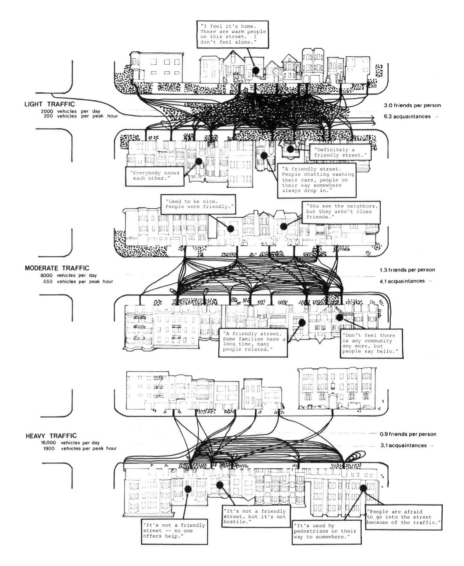

Figure 12.16 Traffic volume vs. neighboring. In this classic study by Donald Appleyard of three streets in San Francisco, the amount that people knew their neighbors varied in inverse proportion to traffic volume.

Urban gardening is on the increase in many urban areas, although gardeners have often had to fight for scraps of land to serve as community gardens. New York City urban gardeners had to struggle especially hard in the 1990s against Mayor Rudolph Guiliani, who wanted to sell off city parcels that were being used as community gardens for development. Luckily, most of these areas were saved through a public campaign featuring celebrities such as Bette Midler. Rather than fitting community gardens in haphazardly on leftover scraps of land, municipalities should look on their gardens as an essential strategy for making dense neighborhoods livable. Planners can identify surplus

Figure 12.17 Traffic-calming devices: a speed table. Speed tables, as here on a residential street in Utrecht, create a broad, raised area of paving.

Figure 12.18 Traffic-calming devices: chicanes. Chicanes, as here on a street in Basel, alternate parking and landscaping from one side of a street to the other to slow traffic.

Figure 12.19 Traffic-calming devices: bulbouts. Bulbouts, as here in Vancouver, push out the curb at an intersection to slow traffic.

Figure 12.20 Traffic-calming devices: circles. Traffic circles, as here in Arcata, California, force traffic to move more slowly at intersections.

Figure 12.21 Sign for a pedestrian zone in Heidelberg. Most European cities and towns have pedestrianized their downtowns. Deliveries are often allowed in the early morning. These pedestrian areas succeed because of their relatively high density, mix of land uses, and excellent access by public transit.

Figure 12.22 A pedestrian street in Santa Monica. Third Street Promenade in Santa Monica offers an example of a US city creating a new public space by closing a street to motor vehicle traffic.

city land or acquire strategically located parcels for this purpose or can require developers to add garden space within large development projects.

With an increasingly elderly population in North America, Europe, and Japan as baby boomers age, cities also need to focus on providing neighborhood open spaces for seniors. The elderly (especially those who can no longer drive) benefit from attractive public spaces that are within walking distance of homes and that are accessible to individuals with wheelchairs or other disabilities. Vast playing fields are not particularly useful for this population compared with smaller, attractively landscaped parks or "healing gardens" with multiple walkways, benches, and a variety of plantings. Community centers with programmed daily activities are also desirable for senior citizens.

For children, the standard playground with its swings, slide, and climbing equipment offers a very limited range of outdoor activities. Less structured play areas, including fields, woods, wetland areas, and areas with movable objects that can be actively re-arranged, are also valuable. Children often benefit from having some "magical spaces" or "areas of mystery" that are not neatly created for them by adults. Vacant lots, patches of woods, and areas of overgrown landscaping have often served this purpose historically. Of course such areas must be safe. Ensuring their safety can be accomplished both through site design and by having an entire city or town that is safer. Means to achieve this include building design that puts "eyes on the street," relatively compact and walkable neighborhoods that put more people on the street, and, at a much broader level, steps to increase equity, reduce disparities between the rich and poor, and provide decent services, education, and employment for the less well-off.

Creating a range of attractive open spaces needs to become a much more integral part of neighborhood planning. For municipalities this may mean adding or strengthening open space requirements within zoning codes and subdivision regulations, developing specific design guidelines for such spaces, and actively acquiring land or easements for open space. State and federal governments can assist them through grant funding and modifications to subdivision law requiring additional open space. In many cases city governments will also need to actively schedule and manage events in these public spaces, as is already done by cities such as Portland in its Pioneer Courthouse Square and Tom McCall Waterfront Park. It is not enough simply to create great public spaces – a range of scheduled activities in them ranging from farmers' markets to concerts and cultural fairs can ensure that they are extensively used.

Ecological restoration

Ecological restoration can add greatly to neighborhood livability and sustainability. Restored ecological features can improve habitat for wildlife, create recreational amenities, and help bring urban residents closer to the natural landscape. Natural areas within a neighborhood are particularly important to kids who may explore them in far more intimate detail than adults, learning a great deal about the non-human world by doing so.

Virtually every neighborhood of any size has a creek, waterway, or wetland associated with it, although some of these may have been destroyed or covered up by development. Traditionally developers or other landowners have channelized, culverted, straightened, or otherwise extensively altered creeks. Riparian vegetation (species adapted to growing along waterways) has been completely or partially removed. Invasive species have frequently taken over from native plants that are usually best-suited to provide habitat for birds, amphibians, reptiles, and mammals. The water flow itself has often been radically altered by large amounts of pavement in the neighborhood or upstream, typically creating strong, "flashy" runoff after storms that can quickly erode creek banks and cause floods. Impermeable paving in the neighborhood – roads, parking lots, driveways, school grounds, and even some backyards – also allows less rainfall to enter the groundwater table, lessening creek flows and water for vegetation during dry periods.

Creek restoration offers a prime opportunity to remedy these problems at a neighborhood scale. It can help bring neighborhoods together and create valuable scenic and recreational amenities for nearby residents and property owners. Relatively easy creek restoration activities include removing trash and debris from stream channels, improving signs and trails, and digging out invasive, non-native plants and replanting native species. More difficult or long-range activities include removing concrete rip-rap, stabilizing banks with log lattices and willow plantings instead, and unearthing and reconstructing culverted creek channels. The banks and flood plains of larger rivers can be similarly restored, in some cases by removing existing development that was inappropriately allowed to encroach on the waterway and its floodplain. Beginning in the 1980s a citizen movement in North America has sought to carry out creek restoration. Early Californian, examples include Wildcat Creek in Richmond, Strawberry Creek in Berkeley, the Guadalupe River in San Jose, and St Luis Creek in the center of downtown San Luis Obispo.[26] In the eastern United States, community groups such as the Anacostia Watershed Council in Washington, DC have taken on responsibility for cleaning up and restoring their local waterways, in this case Washington's "other river," the Anacostia.

Habitat restoration activities – often in association with creek restoration – gained momentum at about the same time. Within such programs, a particular strategy is worked out to restore or create wildlife habitat on a given site, taking into account the history of the land, the local microclimate, soils, and hydrology, and any problems resulting from urban use or contamination. Invasive plant species are typically removed and replaced with natives, and native fish or animal species are restocked. Attention must be paid to

Figure 12.23 Creek restoration. Volunteers plant seedlings of native shrubs along a re-created stream channel for Sausal Creek in Oakland.

Box 12.3 Arcata, California ecological wastewater treatment marsh

For more than 20 years the city of Arcata, California has processed its sewage through a series of constructed wetlands that double as a wildlife sanctuary, educational facility, and popular recreational site. The idea for the project first came about in the 1970s, when the city was faced with the prospect of spending millions of dollars to help construct a traditional wastewater treatment plant. The proposed plant was energy-intensive, required much new infrastructure, and raised problems related to ocean disposal of effluent. Instead, citizens worked with the city to explore environmentally sound alternatives. The current system was approved by the state in 1983 and began operation in 1986.

Wastes first pass through a settling tank from which solids are removed and composted on-site. Next the effluent passes through oxidation ponds heavily planted with bulrushes, where microbes help break down wastes. A final stage of treatment consists of larger wetlands in the Wildlife Sanctuary, which are accessible to the public and ringed by pedestrian trails. The treated water eventually passes into Humbolt Bay.

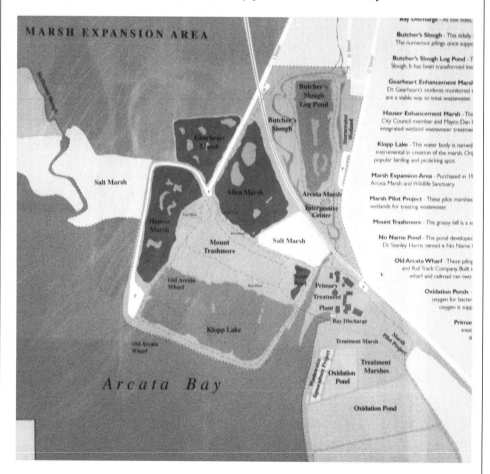

Figure 12.24 Plan of the Arcata ecological sewage treatment marsh.

The 100-acre system of freshwater and saltwater marshes now attracts more than 150,000 visitors annually, and is one of the best birding sites on the West Coast, with more than 200 species sighted. The project has also proven reliable and extremely cost-effective.

Figure 12.25 Arcata treatment ponds.

creating both "edge" and "interior" habitat spaces, since different species may inhabit each area. The size of the patch of restored habitat determines much about what species can be introduced and how, and corridors linking to other habitat patches in the vicinity must be considered. The ideal is to create a series of habitat patches – including some backyards – throughout urban neighborhoods, and to link many of these into a broader mosaic using greenway corridors, often along creeks or rivers. (See Figure 12.23, Box 12.3 and Figures 12.24 and 12.25 above.)

Improving neighborhood equity

Equity concerns are present at the neighborhood level, as at others. Potential equity-related questions include: neighborhood accessible to people of all ages, races, and economic groups, both as residents and visitors? Does it contain a variety of affordable housing options and transportation options for different groups? Does it expose certain groups of residents to toxic contamination problems in soil or groundwater, or to air pollution? Does it saddle some groups with the externalities of other urban activity, such as the noise and pollution of freeways or contamination from oil refineries? Do concentrations of poverty exist because of the reluctance of other neighborhoods to plan for or accept affordable housing and social services?

Historically, the neighborhood scale is one at which segregation has been most vividly expressed in the United States, especially through exclusion of African Americans. Sustainability at the neighborhood level implies making every neighborhood accessible to all, particularly less affluent individuals and families and persons of color. The objective of sustainable development is not green enclaves in upper-middle-class areas, but well-rounded neighborhoods that are diverse and equitable as well as ecological and livable.

Taking steps to ensure a range of housing types and sizes and a sizable number of affordable housing units in each neighborhood is essential. Some cities seek the latter through inclusionary zoning requirements. But this housing does not reach the very poor. Cities need to target very low income households earning 50 percent or less of area median incomes as well, typically by providing rental housing or "supportive housing" which offers services such as employment counciling, substance abuse programs, and childcare for working parents. Whereas currently local governments often only require that low income housing remain affordable for 20 or 30 years, the affordability of these units should be locked in permanently through deed restrictions and binding language in city permits. The objective of a diverse range of housing types should also be sought through revisions to zoning codes requiring a mix of unit sizes in all new developments. Other strategies historically have included limited-equity cooperative housing and public housing (cities building this now tend to opt for scattered-site housing that is identical in appearance to market-rate housing, rather than the large-scale public housing projects of the past). Strict state, federal, and local enforcement of regulations prohibiting discriminatory lending and real estate practices is also essential.

Along with affordable housing, neighborhoods should feature a range of jobs and services available to all residents, the less affluent and less educated as well as the wealthy and well-trained. Municipal governments can seek to ensure this through economic development strategies emphasizing a diverse employment base, and through requirements that businesses give preference to applicants living locally. Living wage ordinances, requiring any company doing business with the city to pay its workers a wage that a family can live on (usually about double the federal minimum wage), are also a way to ensure that all residents will have adequate income.

Beyond these steps, equity can be improved through efforts to ensure that decent education, recreational facilities, and other public services are available to all neighborhood residents. Neighborhoods that have suffered historically from entrenched poverty and social problems will need special attention not just to improve the housing stock and physical form of the neighborhood, but to address the historical factors that have led to its marginalization, including poor education, economic isolation within the region, and a lack of protections for renters and families. Cities may also want to map and identify particular hazardous materials threats, mitigating these where necessary to ensure that exposure to toxic pollution does not affect some segments of the population more than others.

Addressing equity at any scale is a difficult challenge, but it is especially so in neighborhoods where residents these days are so likely to oppose any change at all, let alone initiatives that might introduce poor people or other groups into an existing residential environment. There is no easy way to get around this problem. Nevertheless, a number of incremental changes, including mandates for fair lending and real estate practices, and a great deal of public education and political leadership can help more inclusive, equitable neighborhoods come about.

If done through inclusive and participatory processes, preparation of neighborhood or community plans can be an important mechanism to enhance local democracy and equitable decision-making. It can give residents a greater degree of control over their everyday environments, and provide them with an opportunity to get to know one another and to work together around common concerns. It can even be a path towards fuller involvement in citywide politics – many neighborhood activists have traditionally gone on to seek elected roles within city government. However, public participation at the neighborhood scale is often problematic. Although local opposition to heavy-handed planning schemes is needed at times, too frequently public involvement becomes channeled in negative directions, essentially opposing any change, even if it is to include affordable housing, calm traffic, add community facilities, or improve public transit. The challenge is to promote decision-making that involves every constituency in the community, that is constructive, proactive, and broad-minded, and that keeps regional and global needs in mind. There is no magic way to do this, but early public involvement, full notification of interested groups, careful advance work with community leaders, and context-sensitive proposals and project designs can help.

Economic issues

Neighborhood-scale economic development planning frequently dovetails with equity goals. The question is how to ensure an appropriate mix of jobs and services for local residents, preferably building on local culture, skills, and resources. This goal is often best met through a diverse range of locally-oriented businesses – a situation that is radically undercut by the large-scale manufacturing and big-box retail economy that is ascendant in most industrial countries. Limiting the invasion of big-box stores and chain retailers is not easy. As previously mentioned, setting maximum building footprint limits within zoning codes can help. Zoning to prohibit particular types of businesses has a long history as well, and can be used to restrict fast-food outlets, drive-through businesses, or other types of unwanted enterprises. Zoning restrictions or urban design standards prohibiting large parking lots or one-story buildings – frequently desired by strip-type developers – can also be useful.

Meanwhile, a variety of proactive economic development strategies can help a more local economy emerge. Business incubators are one frequently used technique. Basically, a municipality or a local nonprofit corporation working with local officials sets up a building in which a number of entrepreneurs can rent cheap office or manufacturing space and take advantage of shared office equipment and support services. The city's economic development staff may offer these newly started businesses technical support and training in terms of analyzing markets, developing business plans, accessing capital, or establishing accounting procedures. Micro-enterprise loan funds are another technique used to support new businesses. Essentially these funds bring together a number of local entrepreneurs who take turns borrowing small sums for business start-up and support one another in these efforts. The Grameen Bank in Sri Lanka is the best-known example in a developing world context.

The globalization of manufacturing – and the consequent flight of local manufacturing jobs from communities within industrialized countries to low-wage labor locations in the developing world – is a fundamental force undercutting local employment. Although local economic development activities can help nurture other forms of business less likely to be affected by globalization, attention to larger-scale policy is necessary as well. Action at national and international scales to moderate or reverse global free trade policies is likely to be essential in the long run to address this problem.

Another economic issue relevant to neighborhood development concerns the scale of local development. In the nineteenth and early twentieth centuries most homes were built individually or in groups of two or three by local builders. During the second half of the twentieth century that situation changed, so that increasingly large-scale "production homebuilders" became responsible for creating vast tracts of housing at once. This scale of development fuels suburban sprawl – such builders vastly prefer greenfield land at the urban fringe where it is possible to build large numbers of units at once – and works against the emergence of a more stable local construction industry in the long term. The challenge here may be to promote a smaller-scale economy of local development, in which incentives favor small builders who hire locally and can take the local context into account when developing projects.

Public health

Many public health issues are particularly pressing at the neighborhood level. One of the foremost, increasingly recognized these days, is the degree to which the neighborhood supports walking and other outdoor activity. Such activity can help reduce obesity, improve cardiovascular health, promote community interaction, and enhance a sense of personal connection to the local landscape and place. For elderly people particularly, a daily walk is one of the best ways to maintain health and well-being. Many specific steps to improve neighborhood walkability have been mentioned previously. Particularly vital are the development of a highly connected street fabric; good sidewalks and pedestrian-friendly intersections; the addition of local destinations such as stores, parks, and community centers; and traffic-calming to make streets safer and more pleasant for pedestrians.

Provision of parks and open space at the neighborhood scale is important to public health for similar reasons. In many places this will require municipal governments instituting or strengthening requirements that developers include parks and greenways within new neighborhoods. Municipal investment may also be required to add parks to existing neighborhoods. Making small grants available to neighborhood-based groups for ecological restoration projects or the creation of new mini-parks can be a particularly effective way for cities to improve parks and public spaces within neighborhoods.

Reducing toxic hazards and moving towards use of safer products and techniques for local businesses, landscaping, lawns, gardens, and other neighborhood spaces are further public-health-related neighborhood goals. A variety of persistent organic pollutants and other toxins may be present in neighborhoods from past dumping practices, local gas stations with leaking underground storage tanks, past businesses, and utility transformers and facilities. The use of pesticides, herbicides, and fungicides on residential landscaping, road medians, or nearby farms is another potential threat. A good starting place for concerned neighborhood associations is simply to identify the potential threats and establish a dialogue with local officials on how best to address them.

Other local air quality and water quality problems may be further threats to public health. Differences in topography, hydrology, and prevailing wind patterns may mean that pollution problems are substantially greater in one neighborhood than in another a relatively short distance away. Ideally cities will collect and analyze information on pollution problems, warn neighbors, and take steps to reduce the extent of these problems in advance. But needless to say this does not always happen, and proactive investigations by neighborhood groups may be required to bring matters to public attention.

* * *

Undertaking sustainability planning at the neighborhood scale, then, is an opportunity to address very specific, on-the-ground issues related to the integration of the built and natural environments, the creation of walkable communities, and the existence of equitable housing options, employment, and educational opportunities. A main problem is that the basic components and physical structure of neighborhoods these days are frequently left up to the private sector builders who first develop communities. To ensure more sustainable neighborhood form and character, more proactive public planning will be necessary. City and county officials will need to establish stronger guidelines for new neighborhood and for infill of existing neighborhoods, enforceable through zoning codes and development approvals processes. Nonprofit community development corporations, affordable housing builders, environmental watchdog groups, and other neighborhood advocacy organizations can help enormously in this process of making neighborhoods both more livable and more sustainable.

13 Site planning and architecture

The buildings we inhabit and the lots these structures sit on represent the smallest scale of development affected by urban planning, a domain that overlaps with the professions of architecture and landscape architecture. Issues at this level include how development fits into and affects the natural landscape, how it uses energy and materials, how it influences the neighborhood and social interaction, how accessible and useful it is for different groups of people, and how it affects the health and daily lives of residents or workers.

We experience the results of site and building design decisions every day. Are the structures we live and work in comfortable, safe, and adaptable to our needs? Do they relate well to the street and sidewalk, creating a pedestrian-friendly neighborhood? Is the landscaping attractive, climate-appropriate, and respectful of local ecosystems? Is the flow of energy and materials into the site or building kept as low as possible, using renewable sources wherever possible? Are waste materials reused or recycled? In general, does this building or site represent a model that could be repeated widely in a sustainable society?

These questions are often overlooked in architectural education and practice, where the emphasis instead may be on the aesthetic merits of different building forms or the desire to establish a distinctive look. Many current buildings are also built only for short-term utility, and their nature is dictated largely by economic rather than environmental or equity concerns. Even if well-built, structures are often bland and colorless, using generic architecture that makes every place look like every other place and acculturates us into a homogenized global culture run by multinational corporations. The result is a world of generic office buildings, chainstores, and tract housing that looks much the same in Jakarta as Tuscon. We can do much better. Site and building design that enhances local ecologies, cultures, and communities is one of the core challenges of sustainability planning.

Who plans at the site and building scale?

The number of individuals and institutions who influence site and building development is much greater than one might think. Key actors include individual property owners, for-profit developers of varying sizes, nonprofit developers, architects, landscape architects, urban planners, engineers, bankers, insurance companies, construction contractors, neighborhood residents, and city, county, and state government.[1]

Developers – the individuals or companies that purchase land and coordinate development of sites and buildings for resale – are the best-known and most-often-vilified players in site- and building-scale activity. However, many different types of developers exist, some doing much more sustainable types of development than others. The mix varies considerably from time to time and place to place. One of the big stories of the twentieth century in terms of urban development was the rise of production homebuilding corporations that built subdivisions on a grand scale, using mass production techniques first pioneered at developments such as Levittown, on Long Island, in the late 1940s.[2] (The earlier advent of "balloon-frame" construction, using relatively cheap, small-dimension lumber, also helped pave the way for the mass production of housing.) Though their efforts provided cheap housing for millions in the decades after the Second World War, these companies institutionalized the automobile-dependent, single-use, suburban tract style of development that is responsible for many of today's sustainability problems. Their ability to provide cheap, mass-produced housing also drove many small- and medium-sized development companies out of business, fundamentally reshaping the development industry.

Other types of builders have specialized in office or commercial development, and likewise grew in scale and standardization during the last century. But some small-scale developers persisted, and new niches have appeared in recent decades. With people rediscovering the virtues of city living, some developers now specialize in urban lofts, townhouses, infill sites, or rehabilitation of existing buildings. Others focus on multifamily construction such as apartments or condominiums in suburban locations. As affordable housing has vanished in many regions, some individuals have founded nonprofit development companies to meet the needs of the lower end of the market neglected by the for-profit development sector. Some of these nonprofit housing organizations have become very large – San Francisco-based BRIDGE Housing, for example, builds as many as 2000 units of affordable housing each year. Habitat for Humanity, founded in Atlanta in 1976 by Millard and Linda Fuller, has built more than 100,000 homes around the world, largely through volunteer labor. States and municipalities often assist the efforts of nonprofit builders by making available loans, tax credits, and other forms of financial or technical assistance. Nonprofit builders are typically more willing to work on small, infill sites and to consider the surrounding neighborhood context.[3]

Further diversification of the development industry is essential if it is to become better at meeting the needs of sustainable development. Since many large development companies are wedded to particular "products" that are not particularly sustainable, smaller and more flexible developers may need to take the lead. Planners may want to ensure that a variety of incentives is in place to encourage the growth of alternative developers who are interested in creating green buildings, affordable housing, rehabilitated historic structures, and infill neighborhoods. In the long run mainstream developers can perhaps be enticed towards new methods of construction and design if these are supported in the marketplace and promoted or mandated by public officials.

City planners, engineers, and inspectors employed by local government are another main influence on site and building development. As should be clear by now, government action establishes the framework within which development takes place. Zoning codes, subdivision regulations, and parking requirements often drastically limit development options. Municipal inspectors also enforce building codes that regulate certain design elements, building materials, and construction methods. These codes are promulgated initially by organizations such as the International Code Council (and previously the Council of American Building Officials and regional associations of building officials). These professional associations produce a standard code intended to promote human health and safety (and recently energy conservation as well). State governments then adopt versions of this model building code, and cities in turn apply versions of the state code with relatively minor modifications to meet local conditions. In the past overly rigid building codes have often prevented use of alternative building materials such as rammed earth and straw bale, and have prevented exploration of alternative building techniques in general. As interest in alternative materials has growth, some municipalities have added special sections of building code to accommodate them. Some cities have also changed their codes to allow or require graywater systems, passive solar design, and energy efficiency.[4]

Architects have the challenge of designing buildings that will conform to these codes as well as meet client needs and incorporate sustainability principles. This task is not easy. Building designers may be constrained by the client's budget, choice of site, and lack of interest in alternative methods. In some cases there is leeway in working with the client to suggest ways to save energy, minimize nonrenewable resources, avoid potentially toxic materials, and relate the building to neighborhood, city, regional, and global contexts. However, in other cases clients are inflexible. Some ecological architects turn down potential projects rather than work with a site and building program that they do not feel is appropriate – for example, oversized houses on the urban fringe. In addition, many large developers these days have dispensed with architects altogether, relying on generic designs within a mass-production framework. As a result, many architects are in the often-frustrating situation of wanting to design better projects but feeling they have little leverage with developers or individual landowners.

Landscape architects, often crucial to the livability and sustainability of particular development projects but not necessarily employed at all by developers trying to cut corners, represent an even more underappreciated profession. Key areas of focus that would add to the sustainability of a particular development project can include preserving or restoring natural drainage on the site, specifying native or climate-appropriate plantings, using trees and shrubs to maximize natural heating and cooling for buildings, planting to provide habitat for local species, and creating attractive outdoor spaces to meet the needs of all building users. Traditionally many builders and their landscape consultants have simply stuck in a lawn and a minimum of shrubbery to hide a building's foundation, or have created a more elaborate landscape planting plan exclusively for aesthetic purposes. Ecological considerations and the needs of building residents were often not considered. But this situation is slowly changing. In particular, more detailed knowledge is being developed of how to design sites to meet the needs of children, the elderly, the disabled, and particular cultural groups. Since 1968 the Environmental Design Research Association has held conferences and published studies in part to share information on such strategies.

Like architects, landscape architects serve at the will of their clients and face a challenge of educating developers about green building and landscaping practices. However, the number of clients willing to experiment with native species, drought-tolerant landscaping, permeable paving, and other alternative techniques is growing these days, in part as overall public consciousness about the natural environment rises. Interest in creek restoration, innovative park design, and habitat protection, all areas in which landscape architects may be involved, has also grown.

Lenders are a hidden player influencing the character of many development projects, since developers (especially nonprofits) typically must borrow large sums for site acquisition and construction. Traditionally banks and other financial institutions have favored conservatively designed projects similar to others that have proven track records of repaying loans. Financiers have been skeptical about alternative building materials and mixed-use buildings in either greenfield or infill locations, often ignoring the retail component of a mixed-use project entirely in their calculations. A major problem for developers wanting to build infill projects has been lack of the "comparables" required by lenders – past projects of the same sort of development in the same neighborhood. What results is a chicken-and-egg problem: a desirable project cannot be financed if comparables don't already exist, but those comparables can't be created if no one will lend for them. This situation is changing as more infill projects get built, some with government assistance, and more progressive lenders emerge who are willing to commit to the revitalization of older neighborhoods. However, this obstacle is still substantial.

Other groups also influence the character of particular development projects. Neighbors and/or future residents may enter into extended negotiations with developers, planners, and architects to express their particular needs and desires. Engineers and building inspectors frequently determine what technologies can be used for the building or site landscaping. Hydrologists may provide information about groundwater, soils, and waterways, determining, for example, what kind of drainage or sewage system is needed on the site. On large development projects, other environmental consultants may be called in as well. Each of these participants can have a role to play in improving the sustainability of a given development. One major problem in the past has been that various parties involved in putting together a development project have often worked in isolation from one another, each with a different set of assumptions and expectations. In particular, ecological design consultants often did not have a chance to talk to all parties and were often not involved early on in the process. Getting them involved at the beginning to discuss opportunities and share knowledge with other team members can yield creative solutions and avoid unnecessary roadblocks later on. This is one of the most essential elements of sustainable design – to begin with environmental and social goals at heart.

Sustainable site design issues

At the site level how does one "design with nature," to use Ian McHarg's phrase?[5] More properly, from a sustainability point of view, how does one "design with nature *and* community *and* equity *and* a sustainable economy?" Many authors have developed lists of ecological design principles as a way of addressing this question.[6] The discussion that follows will apply such themes to the development of particular sites.

Lot location

In many ways the most significant sustainability consideration is the location of the lot. Is it contiguous with existing urban development? Is it near transit, shops, and existing neighborhood centers? If the site is to be used for housing, is it near jobs, schools, shops, and parks? Would development on this site avoid disrupting open space or habitat? If the answer is "no" on one or more of these counts, then perhaps this parcel should not be developed at all. Problems arise particularly with sites at the urban fringe or in rural areas, where available cheap land may not be appropriate to build on. If local governments have done their job they will have protected such areas by zoning them for agricultural uses, purchasing them for parks, placing them outside urban growth boundaries, or protecting them through other means. But even if the land is not protected, development on it should still be avoided. A basic mandate in sustainable site planning, then, is to develop in the right places.

This guideline raises difficult questions for design professionals, developers, and land-owners. Obviously, one solution is not to acquire such sites or accept work related to them in the first place. Alternatively, planners and designers can sometimes convince clients to preserve much of the site as open space, or can arrange for open space agencies to buy the land or purchase conservation easements preventing development while compensating the landowners. Questions of professional ethics may be involved if planners and designers go forward with conventional projects on these sites. The American Institute of Certified Planners' *Code of Ethics* prohibits planners from working on projects that may cause significant environmental harm (though there is little enforcement of this provision). It can also be considered unethical to work on projects that will impose large costs on society in terms of traffic, resource consumption, or social inequality. Consequently, it is best both for planners and developers to avoid questionable projects from the start and to focus instead on sites and forms of development with clear benefits.

Site planning

A related consideration is whether the proposed development is making appropriate use of the particular site. Is the proposed use (single family homes, apartments, offices, stores, or other types of enterprises) already over-represented in the surrounding area? Should other uses be sought instead? Should the site contain mixed uses? Are the proposed densities of housing or office buildings appropriate? Is the development creating housing types that meet the needs of the local community, especially less wealthy individuals and families? Again, planners and designers may be at odds with landowners about the appropriate use of the site and may need to try to persuade clients to consider better alternatives.

If location and use reflect sustainable development needs, then the focus can be turned on to the site itself. "Site planning" organizes the development of a single parcel of land by locating building(s) or other facilities in particular places, arranging for roads, sewers, water, electricity, and other infrastructure, and developing plans for grading, drainage, landscaping, lighting, and other site improvements. Developers must do site planning at the outset of a development project, and typically pay architects, landscape architects, or planning consultants to assist with this task. Site planning may be a relatively simple proposition for a single house on a small lot, essentially treated as part of the building design. Or it may be a very complex process for a large development covering many acres, done by a design and planning team entirely separate from the building architects.

One main site planning question is whether particular ecological features should be preserved or restored. Creeks, wetlands, and wildlife habitat on the site are of special importance and may be regulated or protected by local ordinances or state or federal law. Section 404 of the federal Clean Water Act in the US, for example, requires permits for the discharge of dredged or fill material into wetlands or waterways. Planners or developers wanting to fill such areas on a site must therefore receive approval from the Army Corps of Engineers. But the far better course of action is usually to preserve waterways and to establish buffers around them, for example, setting development back from a creek to leave a riparian corridor with both wildlife and recreational value. Hillsides and wetlands should also be left alone for the practical reasons of avoiding flooding, landslides, or other forms of ground movement. The presence of such landscape features on the site may lead to a site plan that clusters development at one side to avoid these areas, or that maintains greenways and open spaces throughout the site.

When it comes to positioning buildings on the site, additional questions come into play. How should the building(s) be positioned to create an attractive, walkable street frontage? To maximize yard or garden space for inhabitants? To maximize solar access? To respect neighbors? Such considerations must be addressed both within the building siting and within architectural design. They require a detailed study of the site and an ability to understand the many different issues and needs that converge on any particular development project. (See Figure 13.1.)

Figure 13.1 Fitting buildings to the landscape. Buildings at Sea Ranch, on the Northern California coast, are designed to blend into the landscape. Fences and decorative landscaping are not allowed.

Landscape design

Sustainable landscaping strategies may emphasize drought-tolerant species in arid or semi-arid areas, trees that provide shade and evaporative cooling in hot climates, and bushes or trees that furnish protection from the elements on windy, cold, or exposed sites. Planting a variety of native species can also often provide habitat and food for local bird and animal species. Edible landscaping – bearing fruit or nuts that can be consumed by humans – can provide food and interesting educational opportunities for adults and children alike. Including garden space within the site design provides opportunities for urban residents to grow some of their own food and develop their connections with the earth. Appropriate forms of irrigation are important for sites where additional water is necessary, for example, drip irrigation systems that minimize water loss in dry climates.

Good site planning can help restore ecological health on most sites. Invasive, non-native plant species can be removed, and native shrubs, trees, flowers, and groundcovers planted instead. Soil contamination from past uses of the site can be cleaned up. Impermeable paving can be removed and replaced with gravel, porous asphalt, or stone, concrete, or brick pavers that let water enter the ground to recharge the aquifer. Natural drainage swales planted with reeds, rushes, sedges, grasses, and other riparian species can also help stormwater runoff recharge the local water table (see Figure 13.2). Existing fences can be removed to re-create wildlife corridors and increase habitat. Additional buildings can be constructed at underused sites within existing urban areas, making more efficient use of them, adding needed housing, stores, or office space, and preserving open space elsewhere.

The nonprofit organization TreePeople and the US EPA held an excellent set of design charettes in the late 1990s to develop ideas for retrofitting existing urban sites in the southwestern US. Participants looked at a five examples throughout Los Angeles: a single family home site, a multi-family housing site, a public high school, a light industrial site, and a Jiffy Lube and mini-mall. Recommended sustainability strategies for these sites included adding systems to collect roof water and parking lot runoff in cisterns for later landscaping irrigation, shade trees to reduce air conditioning needs, "parking orchards" with trees and edible landscaping to improve surface parking lots, vegetated swales in

Figure 13.2 Natural drainage swales. Plantings in this parking lot in Portland, Oregon help filter runoff and allow it to drain into the soil.

parking lots to filter runoff, porous paving for some surfaces, graywater systems to reuse shower water for landscaping, and general reductions in paved surfaces.[7] Although the focus of this exercise was on saving water in a hot climate, a similar approach could be taken with other sustainable site planning goals such as saving energy and restoring elements of the natural landscape.

Site planning is typically also concerned with connecting a given parcel of land to road, water, sewer, gas, and electrical distribution systems. Sustainability questions arise in each of these areas. Road connections are important to connect a large site to existing neighborhoods all around, and site designers can seek to integrate their development into existing grids or street networks, thus improving the ease of biking or walking through the whole neighborhood. It is also important to minimize paved surfaces, especially if impermeable paving is used. A sustainable development project is likely to have relatively narrow roads or driveways and as few surface parking areas as possible, both to use the land efficiently and to avoid runoff and excessive use of asphalt.

For most urban and suburban projects there will be little alternative to connecting to a municipal sewer system. Municipal ordinances usually require it, and sufficient land may not be available in any case for either septic systems or more innovative ecological sewage treatment methods. But on large sites and in more rural areas alternative systems may be possible, such as constructed wetlands or "living machines" in which sewage is filtered through a series of tanks, usually in a greenhouse, that contain water hyacinths or other plants to purify the water. Other ecological wastewater infrastructure may include gray-water systems in buildings to use shower and sink water for toilet flushing or landscape irrigation. The graywater is routed to cisterns where it can be stored and then applied to secondary uses. Rainwater can be harvested in a similar way.

Most sites and buildings have little alternative but to hook up to the regional electric grid (although a few pioneering homeowners, mainly in rural areas, seek to live "off the grid"). But generating electricity on-site through solar or wind technology is often a possibility, and in many locations such current can be fed back into the grid when not used, literally running the meter backwards. Photovoltaic panels have traditionally been more costly than other forms of electricity, but are finally reaching the point where they make economic sense for homeowners, especially if utilities or state governments provide rebates or tax incentives for installing them. Solar hot water panels also help reduce use of gas or electricity for water heating, and are the most common type of solar installation on residential building roofs.

Other more philosophical principles can also be applied to site design. One frequent theme is that site planning should take into account natural flows of materials and energy in a particular place. Historically, many cultures have sought to do this through disciplines such as the Chinese practice of feng-shui, which considers the many ways in which chi flows through the landscape. Some feng-shui principles directly reflect ecological factors, for example, by taking into account the flow of air down a valley or the relation of a house to neighboring watercourses. In Western culture taking natural air, water, and energy flows into account may involve a more scientific or experimental process of studying the characteristics of each site and determining how best to mesh site design with sun, wind, slope, hydrology, soils, and vegetation. Landscape architects might, for example, plant windrows to shield buildings from cold winds or deciduous shade trees to reduce temperatures during the summer while allowing sunshine to enter south-facing windows in the winter. They might also map flows of groundwater and site buildings well away from them. Architects might analyze the energy used to make building materials ("embodied energy"), and minimize the total energy brought to the site during the process of development by choosing appropriate materials, such as locally produced brick, stone, rammed earth, straw, or remilled lumber. (See Figure 13.3.)

Figure 13.3 Native landscaping. Use of native species in residential landscaping can provide habitat for birds and other species and reduce the need for watering, fertilizer, or pesticides.

Use of natural forms, flows, and materials

"Making nature visible" is another ecological design principle that ecological architects such as Sim Van der Ryn argue can help the public understand the unique characteristics of each site or place.[8] There are many ways to do this within site planning. Creeks, drainage swales, rock outcroppings, or small patches of native vegetation can be made centerpieces of landscape design. Flowing water in any form is a pleasing and intriguing natural element on a site. Runoff from roofs can be channeled into "rain fountains" that provide a delightful design element. Edible landscaping helps residents and visitors understand how food can be grown locally. Composting facilities and gardens also help demonstrate connection to the earth. These and other techniques can make visitors to the site aware of the relation between natural and built landscapes.

Preserving the traditional forms and characteristics of the local landscape is a further common theme in ecological site design. Careful study of the natural landscape and historical building patterns is required to develop strategies for doing this locally. One example is Sea Ranch, a vacation community on the Northern California coast near Mendocino that was designed in the 1960s by architects Charles Moore and Donlyn Lyndon working with landscape architect Lawrence Halperin. At Sea Ranch both the buildings and the site plan have been designed to reflect the pre-existing landscape of wind-swept meadows along coastal bluffs, cypress windbreaks, and forested hills. Houses are made of weathered wood, roof lines are pitched to emulate the line of the wind-sculpted cypress, and parked cars and driveways are hidden behind weathered wooden fences. Owners are required to maintain naturalistic landscaping and forbidden from erecting fences, so that the overall effect is of homes sprinkled here and there among meadows and forests, rather than a patchwork of individual yards. A network of trails next to the sea and throughout the development provides public access. Sea Ranch does not necessarily represent a sustainable community overall – it is otherwise a low-density, automobile-dependent vacation development for the well-to-do – but its design illustrates how the form and character of a natural landscape may be preserved within development.

Ecological architecture

Although usually thought of as the exclusive terrain of architecture, building design has enormous implications for sustainability planning. Building construction consumes 25 percent of the world's wood harvest, heating and cooling buildings uses a similar proportion of fossil fuel production, and building demolition accounts for 44 percent of landfill waste.[9] Building construction, heating, and cooling account for about half of carbon dioxide emissions in the United Kingdom;[10] the figure is probably similar in North America. Green building design can employ many of the same principles as ecological site design, in particular letting natural water, air, and energy flows influence the architecture, rooting architectural design in the characteristics of specific places and landscapes, and "making nature visible" within built environments.[11] Other considerations include the needs of building users and potential future users, and the long-term environmental and social costs of building materials. Although 1970s-era solar architecture was often associated with a modernist design aesthetic, green building practices can be incorporated into virtually any architectural style and building form.

Vernacular architecture – the local building traditions that have evolved over centuries in response to a particular place and climate – is often a rich source of ideas for ecological buildings. Vernacular design elements help root buildings within particular natural and cultural landscapes and reflect local history and tradition. Such architecture typically uses local materials and skills, and is often more energy-efficient and flexible than generic modern designs that insulate buildings and their inhabitants from the local environment. Following design critic Kenneth Frampton, architect Douglas Kelbaugh argues for a "critical regionalism" in which ecological design incorporates the vernacular architecture of a particular region.[12] For example, a regional design ethic for the Pacific Northwest might emphasize use of wood (dense forests dominate the landscape), wide roof over-hangs to keep off rain, and generously sized windows to let in the soft, slanting light of this high latitude. Vernacular building design in Mediterranean, Latin American, and Southwestern US locales, in contrast, is quite different. Construction materials are often stone or adobe (bricks made from local clay or mud), colors reflect the warm earth tones of these regions, and buildings are often designed to protect against heat, with thick walls, small windows, and shaded interior courtyards.

One main concern of ecological architects is to meet building heating and cooling needs in ways that minimize energy use. Vernacular architecture can provide many clues, though other green design techniques are useful as well. On sites exposed to cold or chill-ing winds, such as along seacoasts or in mountainous areas, building design can reduce the number of windows on the windward side of the building and improve insulation, while maximizing solar heating through large south-facing windows. In hot climates, on the other hand, architects can design buildings to capture cooling breezes or to allow hot air inside to escape out of rooftop vents during evening or night hours. Large, south-facing windows can be avoided or shielded to capture only winter sunlight. Traditional vernacular building practices frequently make use of such principles, since before the mid twentieth century technologies such as air conditioning were not available. For example, Arabic architecture in the Middle East frequently includes chimney-like towers to capture cooling breezes and/or release heated interior air. Also, traditional architecture in the southeastern United States makes extensive uses of wide porches or verandas to provide cool, shaded seating areas and to screen interior rooms from the sun's heat.

The set of strategies known as passive solar architecture seeks to minimize artificial heating in the winter and cooling during hot summer months. Employing passive solar design means first being aware of which direction is south (or north in the southern hemisphere), and orienting buildings to maximize solar exposure during cold months. Winter sunshine entering the building through south-facing windows heats rooms; thick walls with high thermal mass can retain this warmth at night. Fans or other air circulation devices can then distribute heat throughout the building. Conversely, in the summer months these walls will retain night-time coolness during the day. Sunshades, overhangs, or deciduous trees can be used to shade windows during hot times of the year. Some architects have even experimented with cooling buildings by pumping water deep into the ground, where it is chilled by the constant earth temperature, and then recirculating it throughout the building in a form of natural air conditioning. One ambitious proposal called for many buildings in downtown Toronto to be cooled by pumping cold water from deep in Lake Ontario.

Reducing artificial lighting is another main concern for ecological architects. Clerestory windows (small secondary windows high up on a wall) can increase the amount of light allowed into buildings and reduce artificial heating and illumination needs. Adding skylights or light wells can also help reduce needs for artificial lighting, especially within office buildings that will be occupied primarily during daylight hours. Large, open rooms help natural light diffuse into the interior of buildings, as opposed to small spaces that will each require their own artificial lighting. Ecological building design may also include use of active solar energy systems (photovoltaic electricity or hot water) or wind energy (usually produced at some distance from the building).

Energy conservation mechanisms can be applied throughout the building and have been increasingly required by local and state building codes. Highly energy-efficient windows, wall insulation, appliances, water fixtures, and lighting are now widely available. Using double-glazed or triple-glazed windows (with two or three layers of glass) is one of the most effective ways to insulate a building as well as to reduce noise from traffic or neighbors outside. The introduction of compact fluorescent lightbulbs in the 1980s and 1990s has represented one of the biggest energy savings breakthroughs of all, since these bulbs typically use one-fifth the electricity of a conventional incandescent bulb. However, compact fluorescents and energy-efficient versions of standard fluorescents have yet to achieve nearly the widespread usage that they deserve.

Inside buildings, paints and finishes that contain low quantities of volatile organic compounds (VOCs) can greatly improve indoor air quality. Carpeting that is recycled and/or recyclable – not to mention made of natural materials such as wool or hemp – is desirable. Products that outgas dangerous chemicals such as formaldehyde can be avoided. Such measures, along with natural ventilation and openable windows, can help prevent the "sick building" syndrome that has struck many modern office buildings, in which workers suffer from a variety of mysterious and debilitating ailments. (See Figure 13.4.)

Figure 13.4 Green building strategies can be simple. Ceiling fans promote natural ventilation in Berkeley's renovated Civic Center building, while high ceilings and large windows let in daylight, reducing artificial lighting needs.

Green building materials

Often people associate green architecture with seemingly exotic construction materials such as straw bale and rammed earth. While these represent useful strategies in certain circumstances, a wide variety of other alternative building materials can be used in a much broader range of cases. Essentially, any material that reduces life-cycle resource consumption or long-term social and environmental impacts is worth exploring.

Sustainably harvested lumber is perhaps the resource with the widest applicability since much construction in North America will continue to be wood-based. (Buildings in Europe and elsewhere typically use higher proportions of non-wood materials such as pre-cast concrete, steel, stone, and brick.) Certified sustainably harvested wood is grown within forests that are managed under carefully developed rules, for example, prohibiting clear-cutting, requiring buffer zones around creeks, and mandating replanting and a rate of harvesting that preserves the viability of the forest. Several associations certify wood as "sustainably harvested"; these groups include the Rainforest Alliance (with the "Smartwood" label) and the Certified Forest Products Council. Since 1993 the Forest Stewardship Council, based in Oaxaca, Mexico, has sought to coordinate forestry certification programs worldwide and accredit those certification organizations that comply with basic standards. Certified wood products are now becoming widely available through mainstream outlets such as some Home Depot stores. Although somewhat more expensive than conventional lumber, certified wood products have also become more price-competitive in recent years, and the market for them could be expanded further by tax incentives or municipal requirements for their use in publicly funded projects.

Reused lumber from older buildings is another sustainable source of wood products. Frequently, demolished structures contain large beams and supports that can be salvaged, and such wood is often of higher quality than anything available now since it came from old-growth forests that have since vanished. To reuse such lumber, nails and other hardware are removed and the wood is re-planed in a mill to produce finished beams, studs, and boards. Alternatively, new structural wood and fiberboard can be produced from smaller pieces of recycled wood bonded together or laminated with resins or other adhesives. Old and weathered boards can be reused directly to achieve certain architectural effects. Many unusual sources of reused lumber exist; one firm, for example, now specializes in buying thousands of old railroad ties from Thailand – originally cut from beautiful tropical hardwoods – and remilling these as high-end flooring for North American homes.[13]

Steel framing may at times be an ecologically desirable alternative to wood, despite the fact that steel often costs more and may contain higher embodied energy. Metal studs and other structural building components reduce needs for logging, are relatively reusable and recyclable, and support jobs within existing domestic industries. The replacement of wood two-by-fours with steel framing has in fact been happening anyway within parts of the US construction industry owing to rising lumber prices.

Recycled concrete and asphalt is another alternative material that offers environmental benefits over new versions of these substances. Any rubble from past construction can be crushed on-site to produce aggregate for new foundations or paving. Materials can also be delivered to off-site recycling facilities. New paving materials are emerging that do not have the high energy and/or petrochemical inputs of concrete or asphalt (the latter is a mix of gravel and heavy hydrocarbons left over from oil refining). One such alternative uses pine pitch to bind a paving surface that is harder than concrete but less costly.[14]

A wide variety of other materials can be recycled into green construction products. Durable, attractive, "plastic lumber" made from recycled consumer plastics is ideal for decking, park benches, or other uses where wood might otherwise be subject to rot. Recycled glass collected through municipal recycling programs is now frequently mixed into asphalt used for roads and parking surfaces. Recycled wood from bowling alley lanes, as well as other recycled glass and plastics, can be used to create hard, durable, aesthetically pleasing kitchen counter-tops. Such reused or recycled building materials represent one of the most promising avenues to greener buildings.

Rammed earth represents a new variation on one of the oldest building materials – local soil, used for thousands of years within adobe buildings in many parts of the world. In the simplest versions of this technique, earth is simply compacted within wooden forms that are removed after the wall is created. Usually a small amount of cement is mixed in with the earth to provide increased strength. In the approach known as PISE ("Pneumatically

Impacted Stablized Earth") construction – based on earth-building techniques traditionally called "pise" in France – a mixture of soil and concrete is sprayed into forms to create a stabilized earth wall.[15] Steel reinforcing or sections of reinforced concrete can be added for earthquake safety. Use of rammed earth typically creates building walls of 12 to 18 inches in thickness whose thermal mass is extremely good at retaining heat to warm the building at night and coldness to cool it during the day. The result can be a comfortable and energy-efficient dwelling whose walls are also fireproof and resistant to rot or insect damage.

Straw bales have been widely used in recent years in a variety of buildings and provide exceptional thermal insulation as well as utilizing local agricultural waste. Typically bales are used to fill in a wall that is supported by post-and-beam construction; the wall is then plastered and painted. But load-bearing straw bale walls can also be created that are "pinned" with iron rebar or wood. If properly dried and insulated, straw bales are rot- and fire-resistant. Building codes in many western US states have now been changed to allow use of this material. In Northern California surplus rice straw from the Central Valley, often burned as an agricultural byproduct, has been used in a number of homes, helping to solve a major air pollution problem.

Bamboo is an alternative construction material traditionally used in Asia and Latin America that has great potential applications within North America as well. Incredibly strong and light, it can be used for framing small buildings or as a material for flooring, furniture, or fences. With a tensile strength greater than steel, it can be used to replace iron rebar in reinforced concrete. This extremely fast-growing resource could be cultivated sustainably in tropical or subtropical countries as a substitute for wood from temperate forests.

"Cob" construction, which mixes earth with straw or other fibrous materials, is also enjoying a comeback in England, where it has been used historically in 500-year-old cottages that are still standing. Other traditional practices such as thatched roofs are being explored again as well. Roofs made from a variety of natural materials – clay tiles, wood shingles, thatch, or sod – represent a green alternative to the use of asphalt shingles that tends to dominate current construction. Some of these other materials may also be far longer-lasting than flimsy asphalt-based roofing, which is typically intended to be replaced every 20 years.

Green building programs and standards

In recent decades many levels of government have adopted requirements for energy- and water-efficient construction that have greatly improved the efficiency of new buildings. California, for example, first adopted its Title 24 Energy Efficiency Standards for Residential and Nonresidential Buildings in 1978, covering everything from insulation and lighting to water heating, air conditioning, fireplace construction, clothes dryers, and pool heaters. Residents of the state had saved an estimated $20 billion in electricity and natural gas costs by 2001 according to the California Energy Commission. A 2001 update to the standards was expected to result in an additional $57 billion in savings by 2011.

Since the early 1990s a number of cities have started Green Builder programs to promote ecological building practices. The city of Austin, Texas, for example, began small with an energy rating system for new buildings during the 1980s. In 1991 the city expanded this into a more comprehensive Green Builder program. A central component of this system is a four-star rating scale assessing whether buildings meet criteria in areas of energy efficiency, water efficiency, materials efficiency, health and safety, and community. Municipal guidelines provide developers and architects with details about how these criteria can be met, and the city offers a variety of technical-support services and cash incentives for green buildings. As of 2001, 16 other cities nationwide had developed similar programs. Denver's Built Green program, for example, certified 3000 homes in 1999.[16] Although such programs currently affect only a small percentage of the building stock, their influence is growing.

Beyond such initiatives, many other cities have been reviewing their building codes to ensure that they allow and encourage practices such as graywater recycling, passive solar architecture, and straw bale construction.[17] San Francisco, for example, adopted an ordinance requiring dual plumbing systems, including one for graywater, on new commercial buildings in the late 1990s, with the water to be used to irrigate public landscaping. Municipalities can also issue nonbinding green building guidelines as another simple and basic step to promote more sustainable building practices. Such guidelines might spell out recommended practices for energy efficient design, use of sustainable construction materials, recycling of construction debris, and so forth. Planning commissions, zoning boards, and design review committees might be instructed to look favorably on projects meeting these guidelines, or even to give them density bonuses or other incentives.

At a national level, the US Green Building Council developed the LEED (Leadership in Energy and Environmental Design) Green Building Rating System in the late 1990s to establish uniform, comprehensive green building guidelines that could be applied across the country. Applied initially to industrial and commercial buildings as well as to residential buildings higher than four stories, the LEED system requires developers to complete a detailed checklist and awards 64 possible points for green building practices. Basic certification requires 26 points, while silver, gold, and platinum LEED certificates are awarded to buildings meeting higher standards. Main categories of the checklist include sustainable sites (site selection, alternative transportation, stormwater management, etc.), water efficiency, energy and atmosphere, materials and resources, indoor environmental quality, and innovation in the design process. Although open to criticism on some fronts, in particular for its "design by checklist" approach, which reduces the incentive for developers to greatly exceed the standard on any particular item, LEED nevertheless provides a valuable tool that city governments and public agencies in particular can use to specify particular levels of environmental performance. (See Box 13.1.)

Meeting social and equity needs at the building scale

Although sustainable architecture is often seen primarily in terms of ecological factors, many social factors and equity issues are equally vital in the long run. Particularly important are building and site designs that meet the needs of a wide range of users, that will be flexible and adaptable over time, and that will promote community and social interaction.

In a ground-breaking 1986 book of design guidelines entitled *Housing as if People Mattered*, Clare Cooper Marcus and Wendy Sarkissian pointed out how traditional design of medium-density housing has ignored the actual needs of users, and how simple steps such as adding porches, walkways, a range of private and semi-private outdoor spaces, and context-appropriate building design could greatly increase perceived livability of housing. They also championed the method of "post-occupancy evaluation" in which the actual experience of building users – rather than abstract architectural theory – is used to help design more user-friendly buildings in the future. The growing discipline of

Box 13.1 The LEED standards

Project Checklist

Sustainable Sites

14 Possible Points

Y			Prereq 1	**Erosion & Sedimentation Control**	Required
Y	?	N	Credit 1	**Site Selection**	1
Y	?	N	Credit 2	**Urban Redevelopment**	1
Y	?	N	Credit 3	**Brownfield Redevelopment**	1
Y	?	N	Credit 4.1	**Alternative Transportation**, Public Transportation Access	1
Y	?	N	Credit 4.2	**Alternative Transportation**, Bicycle Storage & Changing Rooms	1
Y	?	N	Credit 4.3	**Alternative Transportation**, Alternative Fuel Vechicles	1
Y	?	N	Credit 4.4	**Alternative Transportation**, Parking Capacity	1
Y	?	N	Credit 5.1	**Reduced Site Disturbance**, Protect or Restore Open Space	1
Y	?	N	Credit 5.2	**Reduced Site Disturbance**, Development Footprint	1
Y	?	N	Credit 6.1	**Stormwater Management**, Rate and Quantity	1
Y	?	N	Credit 6.2	**Stormwater Management**, Treatment	1
Y	?	N	Credit 7.1	**Heat Island Effect**, Non-Roof	1
Y	?	N	Credit 7.2	**Heat Island Effect**, Roof	1
Y	?	N	Credit 8	**Light Pollution Reduction**	1

Water Efficiency

5 Possible Points

Y	?	N	Credit 1.1	**Water Efficient Landscaping**, Reduce by 50%	1
Y	?	N	Credit 1.2	**Water Efficient Landscaping**, No Potable Use or No Irrigation	1
Y	?	N	Credit 2	**Innovative Wastewater Technologies**	1
Y	?	N	Credit 3.1	**Water Use Reduction**, 20% Reduction	1
Y	?	N	Credit 3.2	**Water Use Reduction**, 30% Reduction	1

Energy & Atmosphere

17 Possible Points

Y			Prereq 1	**Fundamental Building Systems Commissioning**	Required
Y			Prereq 2	**Minimum Energy Performance**	Required
Y			Prereq 3	**CFC Reduction in HVAC&R Equipment**	Required
Y	?	N	Credit 1	**Optimize Energy Performance**	1–10
Y	?	N	Credit 2.1	**Renewable Energy**, 5%	1
Y	?	N	Credit 2.2	**Renewable Energy**, 10%	1
Y	?	N	Credit 2.3	**Renewable Energy**, 20%	1
Y	?	N	Credit 3	**Additional Commissioning**	1
Y	?	N	Credit 4	**Ozone Depletion**	1
Y	?	N	Credit 5	**Measurement & Verification**	1
Y	?	N	Credit 6	**Green Power**	1

LEED™ Rating System Version 2.1

Materials & Resources   13 Possible Points

Y			Prereq 1	**Storage & Collection of Recyclables**	Required
Y	?	N	Credit 1.1	**Building Reuse**, Maintain 75% of Existing Shell	1
Y	?	N	Credit 1.2	**Building Reuse**, Maintain 100% of Shell	1
Y	?	N	Credit 1.3	**Building Reuse**, Maintain 100% Shell & 50% Non-Shell	1
Y	?	N	Credit 2.1	**Construction Waste Management**, Divert 50%	1
Y	?	N	Credit 2.2	**Construction Waste Management**, Divert 75%	1
Y	?	N	Credit 3.1	**Resource Reuse**, Specify 5%	1
Y	?	N	Credit 3.2	**Resource Reuse**, Specify 10%	1
Y	?	N	Credit 4.1	**Recycled Content**, Specify 5% (p.c. + [1/2] p.i.)	1
Y	?	N	Credit 4.2	**Recycled Content**, Specify 10% (p.c. + [1/2] p.i.)	1
Y	?	N	Credit 5.1	**Local/Regional Materials**, 20% Manufactured Locally	1
Y	?	N	Credit 5.2	**Local/Regional Materials**, of 20% in MRc5.1, 50% Harvested Locally	1
Y	?	N	Credit 6	**Rapidly Renewable Materials**	1
Y	?	N	Credit 7	**Certified Wood**	1

Indoor Environmental Quality   15 Possible Points

Y			Prereq 1	**Minimum IAQ Performance**	Required
Y			Prereq 2	**Environmental Tobacco Smoke** (ETS) **Control**	Required
Y	?	N	Credit 1	**Carbon Dioxide** (CO_2) **Monitoring**	1
Y	?	N	Credit 2	**Ventilation Effectiveness**	1
Y	?	N	Credit 3.1	**Construction IAQ Management Plan**, During Construction	1
Y	?	N	Credit 3.2	**Construction IAQ Management Plan**, Before Occupancy	1
Y	?	N	Credit 4.1	**Low-Emitting Materials**, Adhesives & Sealants	1
Y	?	N	Credit 4.2	**Low-Emitting Materials**, Paints	1
Y	?	N	Credit 4.3	**Low-Emitting Materials**, Carpet	1
Y	?	N	Credit 4.4	**Low-Emitting Materials**, Composite Wood	1
Y	?	N	Credit 5	**Indoor Chemical & Pollutant Source Control**	1
Y	?	N	Credit 6.1	**Controllability of Systems**, Perimeter	1
Y	?	N	Credit 6.2	**Controllability of Systems**, Non-Perimeter	1
Y	?	N	Credit 7.1	**Thermal Comfort**, Comply with ASHRAE 55–1992	1
Y	?	N	Credit 7.2	**Thermal Comfort**, Permanent Monitoring System	1
Y	?	N	Credit 8.1	**Daylight & Views**, Daylight 75% of Spaces	1
Y	?	N	Credit 8.2	**Daylight & Views**, Views for 90% of Spaces	1

Innovation & Design Process   5 Possible Points

Y	?	N	Credit 1.1	**Innovation in Design**	1
Y	?	N	Credit 1.2	**Innovation in Design**	1
Y	?	N	Credit 1.3	**Innovation in Design**	1
Y	?	N	Credit 1.4	**Innovation in Design**	1
Y	?	N	Credit 2	**LEED™ Accredited Professional**	1

Project Totals   69 Possible Points

☐ ☐ ☐ **Certified** 26–32 points **Silver** 33–38 points **Gold** 39–51 points **Platinum** 52–69 points

U.S. Green Building Council

environmental design has adopted such methods to explore the physical and psychological impacts of a wide range of urban spaces.

New forms of housing with shared facilities, gardens, courtyards, or other spaces may help provide residents with a wide variety of amenities. Apartment buildings or condominium complexes can feature interior courtyards, pools, gardens, laundries, games rooms, exercise facilities, and other amenities. Developers can add shared cars that can be signed out on an hourly basis, meaning that residents who don't need to drive every day do not need to own a car, or can get rid of a second car. Such amenities are extremely helpful in making urban living an attractive alternative to the large-lot suburban house.

There has, of course, been a long history of intentionally designed cooperative living communities that have sought to provide more community-oriented and equitable living environments. These include utopian communities of the nineteenth-century, cooperative housing projects in the America of the 1930s, the communes of the 1960s and later, and various more recent eco-villages. Intentional communities such as these offer one way to provide more supportive living situations for people – especially families with children, the elderly, or the disabled. Cohousing is one recent movement that has sought to create alternative living environments suited to promoting human community. Beginning in Denmark in the 1970s, the movement spread to the United States around 1990 and has resulted in hundreds of new developments in which a few dozen residents usually have their own private units (with their own kitchens) while sharing kitchen and dining facilities in a "common house."[18] Cohousing residents typically design their own living environment from scratch through several years of meetings. However, a wide variety of much less formally organized shared-living situations has been present in cities historically, and is being expanded today.[19] Cooperative houses, blocks in which neighbors take down fences and share backyards, and informal sharing of laundry and other facilities represent ways for residents to minimize resource consumption and improve urban livability.

One essential consideration in sustainability-oriented architecture is how buildings can evolve and be adapted over time to the needs of various sorts of users, often with different ages, family configurations, or cultural backgrounds than the initial residents. In particular, much of the housing built in recent decades has been in the form of single-family detached units appropriate for a family with young children. Typically this housing features large unit sizes, open floor plans, and suburban locations that may be less useful, for example, for elderly or single individuals or for shared homes of unrelated adults. Retirees might wish they could close off part of the house and rent it as a separate apartment for additional retirement income. Self-employed residents might prefer to have much of the floor space devoted to an enclosed home office or workshop with a separate entrance for clients. Specific details of floorplan and design can make units more or less adaptable to these different uses over time.

Adaptable housing and office space can in turn help promote neighborhood diversity and vitality by accommodating different family sizes and configurations, by providing a diversity of unit sizes and prices for varying income levels, by allowing home businesses and small-scale entrepreneurial activity to help unemployed or lower-income individuals increase their income, and by allowing unemployed, elderly, or lower-income households to easily rent rooms or apartments for supplementary income. Such flexibility helps buildings serve human needs well in the long term. By increasing the variety of residents and activities in a neighborhood, such building design can also avoid the sterile, homogeneous character of many current subdivisions and meet the needs of diverse households.

One further area that designers of green buildings must consider is indoor air quality. Paradoxically, highly energy efficient buildings which allow very little leakage of heated or cooled interior air to the outside are also at greatest risk of air quality problems, as any outgassing from construction materials, carpeting, paints, and furniture will remain within the building. But other, more conventional buildings are also at risk of "sick building syndrome," especially if they are entirely climate-controlled and lack openable windows, if they use carpets or particleboard rich in chemicals like formaldehyde, or if they have vents, ducts, insulation, or inadequately drained foundations that contain molds or other irritants. As buildings have grown more hermetically sealed and full of synthetic materials, this risk has grown. The solution is to design buildings with great care to provide sufficient ventilation, user-adjustable climate control mechanisms, and healthy materials.

Economics and sustainable building

Part of the reason green building isn't more widespread is that a whole set of economic incentives favors unsustainable building practices. Not only are social and environmental externalities not factored into the cost of building materials, but many architects, engineers, and contractors have little familiarity with alternative techniques, and the building industry itself has proven surprisingly resistant to new practices. This may be in part because the industry has become dominated by large companies that have become entrenched in particular ways of doing business. National production homebuilders construct tens of thousands of units each year, and their profits depend in part on cheap, mass-production techniques.

This situation is the opposite of the ways that homes were built prior to the twentieth century. As late as the 1940s most homes were built individually by relatively small-scale, locally owned companies, or in small groups of a few units at once. In older urban downtowns and many nineteenth- or early twentieth century neighborhoods, the buildings that appear distinctive and interesting today were added gradually over many years. This slower, more incremental style of building, epitomized by Europe's medieval and Renaissance cities, is at the core of the process proposed by Christopher Alexander and his coauthors (1977) in their classic urban design manifesto *A Pattern Language*. To create a beautiful and livable city, Alexander argued, it is necessary to add buildings one at a time and to design them according to time-tested principles that have been used in cultures the world over. Each building is designed in such a way as to add to the urban context around it and to help "heal the whole."[20] Development in this slower, more thoughtful manner may also provide greater opportunities to consider green building techniques than today's mass-production construction practices.

We certainly don't need to return to nineteenth-century styles of homebuilding to have greener building or more sustainable development. But a move towards building practices that are smaller-scale, more locally-oriented, and more sensitive to particular contexts does seem likely to produce significant social and environmental benefits. Ways of producing such housing so that it is still relatively affordable – or to increase the incomes of lower-wage workers so that they can afford housing produced in this way – must also be found.

Further, true-cost pricing of building materials could greatly increase use of more sustainable components. The price of wood products, for example, has traditionally not reflected the costs of deforestation and the subsidies provided to the logging industry by the US government in the form of underpriced timber and road construction on federal lands. Life-cycle pricing of materials can likewise help "level the playing field" economically between green and conventional buildings, and ensure that building construction takes place with the lowest level of impacts. State or national governments can begin to establish true-cost pricing of building components by establishing green taxes on materials that are nonrenewable or have major environmental costs. They might also consider extra charges for development that doesn't recycle construction debris (burdening municipal landfills), that exacerbates urban heat island effects (through excessive paved surfaces and poor landscaping), or that creates excessive amounts of runoff from surface parking lots (burdening municipal storm drain systems, exacerbating flooding, and reducing natural recharge of the water table). A related step to ease and lower the cost of green building would be to revise building codes to allow and encourage use of alternative materials, meaning that developers would no longer face the added uncertainty of getting new methods approved and the expense of hiring engineers to justify "alternative methods requests" under current building codes. Concerted action of these sorts by public officials can begin to make green building practices more economically viable.

Sustainability planning at the site and building scale therefore dovetails with architecture and landscape architecture to offer opportunities to apply sustainability principles to our everyday environments. The design and construction of buildings and landscaping affect us very personally, since these processes create the settings in which we live. As with any other scale, there is a great deal of momentum behind existing, unsustainable ways of doing things. Yet at this most immediate level we have the opportunity to do many things ourselves, even if these simply involve choosing locations and buildings to purchase, taking simple energy conservation steps, or installing native or climate-appropriate landscaping. These daily decisions can make a big difference, and set an example of personal action for a more sustainable world.

14 How do we get there from here?

Although the challenges of sustainable development seem overwhelming at times, the point of this book is that it is indeed possible to plan for a better future. Much positive change is happening every day, and through our actions we can bring about even more. Taking a long-term perspective on social evolution helps support this claim. Our species has managed to learn many things, such as to respect basic human freedoms and rights, in principle if not always in fact, and to develop institutions protecting these. It has learned much about human psychology and communication, although this knowledge has yet to be systematically employed. It has developed great spiritual wisdom in many different traditions. And it is learning to understand and respect the earth's ecosystems, though this too is very much a work in progress.

Nevertheless the situation is often daunting. Even where sustainability-oriented planning initiatives have been adopted, we often face what Owens and Cowell call an "implementation deficit."[1] Too often promising efforts succumb to entrenched social values, gridlocked institutions, or economic and political forces that promote self-interest, short-sightedness, and the phenomenon that historian Barbara Tuchman used to call "wooden-headedness" – the persistent following of strategies that are obviously counter-productive even when clear evidence exists that these are not working.[2] Americans particularly, isolated in our nation that epitomizes many of the excesses of capitalism and materialism, often seem unable to "connect the dots" – to recognize the hypocrisy of many of our leaders, to acknowledge the ways that power and money corrupt our ideals, and to see that we are interdependent with others in the world. Other nations share many of these problems. Many structural forces promote this blindness, including television, advertising, social norms, the power of economic institutions, and the physical environment we have created around us. At least some of these structural conditions will need to change for social and political values to change. Or perhaps, if we are lucky, both inner and outer changes will happen at the same time, in response to stimuli that we can as yet only dimly see.

Developing an awareness of opportunities

The foundation for change at either a personal or a collective level is awareness of problems and the ability to see opportunities for new ways of doing things. Not to be overly simplistic, but as a society are in an enormous amount of denial about many of the problems around us – the ways that we have degraded the landscape and impoverished certain groups, for example, and the ways that our lifestyles in affluent nations are related to suffering elsewhere in the world. The flurry of daily activities, economic pressures, our myopic media, advertising, and many other cultural forces constantly distract us from this awareness.

In terms of urban planning, we can all become more aware of the built landscape and social environments around us, and better at recognizing and understanding urban problems. In particular we can cultivate skills of "looking at cities," as former San Francisco planning director Allan Jacobs recommends in his book of that name.[3] We can observe urban environments carefully as we go about our daily lives, note how people use urban spaces, experience what different places feel like, and practice identifying opportunities for ecological restoration, neighborhood revitalization, a more vibrant local economy, and improved equity.

This sort of experiential learning often goes by the academic title of "phenomenology," and has been the focus of a small but growing movement since the late 1980s.[4] One of its proponents, David Seamon, defines phenomenology as "an interdisciplinary field that explores and describes the ways that living things, living forms, people, events, situations and worlds come together environmentally."[5] Whatever the label, these skills of seeing are particularly needed right now. Mainstream culture and even graduate programs in urban planning often discourage such direct observation, asking us instead to experience the world through computers, television, other electronic media, or overly abstract forms of analysis. Granted, theoretical analysis is very important. But gaining better understandings of the on-the-ground nature, history, and potential of particular places is an essential starting point for sustainability planning. From that awareness, opportunities for action emerge.

Understanding opportunities for change also depends on an appreciation of the forces and dynamics that affect how societies evolve. This is where theory and analysis come in. Gradually we all need to become more aware of these forces and ways to change them. Ecological economist Richard Norgaard (1994) has proposed one useful model of "co-evolutionary" process through which society evolves. In Norgaard's view, human values, knowledge, and organization evolve in conjunction with the natural environment and technology. Each of these factors influences the others, and all are important. Figure 14.1 shows how he diagrams his model.

The exact labels here are not as important as understanding how such forces interact to bring about social evolution. For "organization" the word "institutions" might also be used, referring to the whole range of government agencies, NGOs, private corporations, public–private partnerships, economic systems, and laws and regulations that structure the environment within which any form of action occurs. Changes in any of these institutions, such as the continued rise of NGOs promoting an environmentally and socially oriented agenda, can help society evolve towards greater sustainability. Along with the word "knowledge," other terms such as "cognition" might be associated, referring not just to the body of information available to people, but to the ways they process experience and

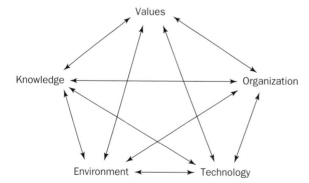

Figure 14.1 Coevolutionary forces. Understanding how societies evolve in response to various forces can shed light on how a context promoting sustainability planning can be structured.

the mental frameworks or paradigms they use for evaluating what goes on in the world. For "values" we might substitute terms such as "culture" or "society," that is, any set of social constructs that helps us decide what is important and what isn't. These priorities can evolve substantially over time. Political scientist Ronald Inglehart, for example, has tracked values within industrialized countries over several decades, and finds a pronounced shift towards valuing quality of life and the natural environment over the past several decades.[6] This is a hopeful trend for those of us interested in creating more sustainable communities.

A theoretical framework such as Norgaard's could go by many names – "coevolutionary," "ecological," "process-oriented," or "a systems perspective." The names are not important; what matters is our growing understanding of how different factors interact to bring about social change, and of how by influencing one element or another we can help society evolve. Helping people see the world differently (changing cognition), through teaching, writing, art, or even architecture, can thus have a very great effect in the long run. Changing institutional structures by the creation of new agencies and laws or the reform of existing ones can also have a large impact. Helping values change, through spiritual practice, teaching, personal example, work with children, or other means, can likewise help lay the groundwork for social evolution. Each of these strategies begins at one point on Norgaard's diagram and eventually affects all the rest.

One implication of such a coevolutionary or ecological perspective on the world is that contexts can be structured to maximize the chances that good outcomes will happen. In the case of sustainability planning, the structuring of institutions is particularly important. For example, the existence of federal and state Environmental Protection Agencies in the United States since the 1970s has made possible far more systematic approaches to environmental regulation and protection than before. At a local level, the establishment of design review commissions and zoning reforms can help set up an environment in which better-designed and more appropriate forms of development will occur. Giddens and others have written extensively in recent decades on "structuration" – the processes through which social systems are structured by rules and institutions – and other institutionalist researchers focus on this as well.[7] By structuring institutions well, we can create many opportunities for positive social change in the long run.

All the tools mentioned in previous chapters can be used to nudge urban planning in the direction of sustainability, structuring a context in which the ecological or social health of communities can improve. Rational comprehensive planning, communicative planning, advocacy planning, grassroots organizing, visioning, development of best-practice examples, sustainability indicators, ecological footprint analysis, GIS systems and mapping, and last but not least frameworks of intergovernmental incentives and mandates can all help. Other more general factors such as leadership, inspiration, and insightful analysis are also important. Which strategies are most worth focusing on will vary from time to time and place to place.

A strategic perspective

What is essential, then, is a strategic perspective in which all of us as planners or citizens look for ways to bring about more sustainable urban development by incremental improvements in institutions, knowledge, values, society, or the physical environment. Mutually reinforcing initiatives in all these areas can help bring about change occur, hence sustainability planning's emphasis on coordinated action at different scales. International agreements such as Local Agenda 21 or climate change treaties can help national policies change; state policy, incentives, and leadership can help regional and local action occur; and so forth. No single level of initiative is enough by itself – a mutually reinforcing set of actions at all levels is necessary.

A strategic perspective on sustainability planning must be a long-term one, because change often does not happen rapidly and it may be most useful to develop contexts in one generation that can support positive action many decades in the future. Setbacks occur, and unfortunately there will be a great deal of permanent loss along the way (of species, ecosystems, cultures, and individuals). None of this should be countenanced without a fight. But the long-term sustainability, evolution, and health of life on this planet are the ultimate criteria of our success.

It is crucial to be aware of how small, incremental steps can lead to larger long-term goals. Each new, appropriately designed building in a city, each small piece of a natural ecosystem restored to health, each human being empowered to contribute to society in a meaningful way represents a major step forward. Over even five or ten years, such incremental steps can make a big difference in a city or town. Over 20 or 30 years, they can be revolutionary. It is up to us to figure out how best to contribute to the small as well as the large changes taking place.

The pace of change in urban environments and human values since the beginning of the Industrial Revolution, barely two hundred years ago, has been phenomenally rapid and is still accelerating. Widespread acceptance of goals of civil rights and environmental protection has occurred only in the last generation or two. Modern environmentalism dawned in the 1960s, and sustainable development first emerged as a theme in the 1970s. Environmental justice concerns came on the scene in the 1980s. The New Urbanism, smart growth, and livable communities became strong movements in the 1990s. What has been happening, in other words, is a rapid growth in awareness of the problems related to creation of a global industrial society and in consciousness of alternative ways of designing and managing communities. Though events around us are often discouraging, the emergence of movements such as those above should give us optimism for the future.

Dealing with power and improving democracy

Action towards more sustainable communities will depend in large part on having a political system that can respond to current situations and lead sustainable development activities. Democracy is widely accepted as the political ideal of our age and appears to offer many sustainability benefits itself in that it can be resilient, adaptable, equitable, and inclusive. However, democracy is often subverted by power, especially economic power. What we have in the United States and many other nations is more plutocracy – government by wealth and powerful economic constituencies – than democracy.[8] One of the Three Es, economy, has dominated the other two and has profoundly shaped the institutions and value structure of the culture. With such a flawed system, only imperfect and partial progress towards sustainable development is possible. What seems needed is a democratic system in which social and environmental values really do balance economic ones, and in which the public sector can play a more substantial role in asserting these values. The tradition of social democracy offers one such path, as Giddens and others have argued,[9] but the label is less important than the concept of a more balanced democratic system.

A truly functional democracy would rest on a foundation of three components: a clean political system, an enlightened electorate, and real choices within elections. Sadly, none of these elements is strongly present in the United States currently. Our political system is notoriously open to the influence of money and large corporations; this wealth allows certain interests to distort the political agenda and sets up a merry-go-round in which officials shuttle back and forth between corporate boardrooms and public office. The need for constant fundraising discourages many good people from running for office, from city council seats to the presidency. Americans often lack the most basic knowledge about the world and about history, and are more influenced by television and commercial culture than any other source. For its part the media fails to provide information about many of the most pressing isues of the day – including many sustainability planning topics – preferring to focus instead on crime, disasters, personalities, and scandals. Finally, we lack the full spectrum of political views that could give us real choice in elections. The two-party system, the enormous financial requirements for serious campaigns, and the "winner-take-all" structure of our representative democracy prevent third parties and alternative points of view from emerging. In contrast, within European parliamentary democracies (which are not perfect either) Green parties and social democratic view-points have been able to gain some degree of power. Our media also tends not to cover alternative points of view, and our national mind is uniquely closed to these, illustrated most famously by the anti-communist witchhunts of the 1950s.

Progress on all three of these fronts will be required if nations are to progress toward the ideal of democracy and create an environment in which sustainability planning can most effectively take place. Many authors have suggested ways in which a healthier politics can be created. The communitarian philosophy promoted by Amitai Etzioni,[10] the discursive democracy of John Dryzek,[11] and the "politics of meaning" promoted by Michael Lerner[12] represent several of these approaches. A variety of nuts-and-bolts political organizing, coalition-building, and educational strategies will be needed as well.

If equity and environment are to have equal standing with economic goals in our political system, we will need to reduce the influence of economic interests. We will need to place limits on the power of wealth and large corporations to capture government institutions to serve their own ends. This will mean taking steps such as reforming election financing, increasing participation in elections, adopting a more progressive tax system, and instituting extensive programs designed to assist lower-income individuals and small businesses (that is, those with relatively little power). Doing these things will depend on a fundamental growth in public consciousness about the nature of power in our society which may seem far-fetched in the current environment. Yet such a change is likely to come sooner or later, and in future eras the present time may be seen as an age of excessive inequality, materialism, and corporate greed. Giddens also urges a redefinition of rights and responsibilities, away from the present condition in which individuals and corporations are far more interested in receiving benefits from social institutions than in contributing to the common good. As he puts it, "One might suggest as a prime motto for the new politics, *no rights without responsibilities*" (emphasis in the original).[13] Or as John F. Kennedy urged Americans, "Ask what you can do for your country . . . ." Perhaps future leaders can reawaken this civic spirit.

In the United States we are reluctant to acknowledge the realities of power within our society. But we must. Sustainability planning will necessarily include efforts to build alternative forms of political power through grassroots organizing, coalitions, urban social movements, lobbying, occasional litigation, and political leadership. Hard-headed political organizing has almost always been necessary to bring about progressive change, whether in civil rights, environmental, women's, or peace movements. A whole set of strategies devoted to nonviolent social change has arisen to assist such movements, spearheaded historically by figures such as Ghandi and Martin Luther King Jr. Current organizing against economic globalization represents a recent phase of such activity. Such political organizing includes founding groups (in particular nonprofit advocacy organizations), planning demonstrations and conferences, building coalitions, lobbying elected officials, and contributing to such efforts through donations, labor, and professional creativity.

One question that often arises is how sustainability-oriented improvements are to be financed. Federal, state, and local governments complain that they are strapped for cash. Anti-tax crusaders seem constantly to be launching ballot referenda or legislative initiatives to lower taxes further. Public interest groups and even private philanthropic foundations have relatively limited resources. Each of us individually often feels financially strapped and unable to afford green products or additional charity.

For one thing, it should be clear by now that much sustainability planning does not require expensive new investment. It simply involves rethinking existing spending and regulation, and making small strategic amounts of funding available to catalyze new programs. Changing zoning codes, for example, requires no new expense. Designing new communities well does not take much more money than designing them badly, and may actually save money. Increasing equity across an entire metropolitan region may be a question of distribution, not of new spending. On a personal level buying organic produce or adding green features to our houses does not cost much extra and brings many long-term benefits.

Yet in other areas new spending undoubtedly will be required. And here it is a question of priorities. The response that "we just don't have the money" is often an excuse for inaction. After all, the Rio Earth Summit Conference funding of the entire Agenda 21 program was expected to cost $128 billion, only one-tenth of the global arms budget.[14] Those of us in industrialized nations live in some of the richest societies the world has ever known. If we can't find the resources to make our public realm livable, to protect and restore our environment, and to care for our least well-off neighbors, it is because we choose not to. Confronting the reality of the social and political situation, in which various elite, politically powerful groups divert our collective resources for their own purposes and promote overly materialistic values, will be one of the main challenges ahead.

An agenda for the future

This book has essentially been about "how we get there from here." We do so through personal awareness, compassion, and action. But we also do so through strategic thinking, collective activity in professional fields such as urban planning, and through the development of institutions and social capital that can create a context for even more significant actions in the future.

Professions such as city and regional planning play a crucial role in the process of sustainable development in that they address the material conditions of our human environment. To a substantial extent, such as by helping create a more vibrant public realm, planners can help create community and literally bring people together. Planners can also play a more explicit role in promoting equity, public health, alternative local economies, and a less automobile-dependent society. Last but not least, they can both protect natural ecosystems, and lead the twenty-first-century process of restoring many of those elements of the environment that have been damaged in the past. These are some of the core planks of the new agenda of planning.

Other fields such as architecture, landscape architecture, economics, public health, and public administration also relate directly to questions of how to have more sustainable and livable communities. In the past all these professions have too often been sidetracked into narrow, conventional ways of doing business, or have been disempowered politically, institutionally, or economically. But it is possible both to be professional and to play more proactive roles in planning for sustainability. Within planning and design fields, a long line of pioneers has called for us all to take action on behalf of the human and ecological needs we can observe all around us. Now, more than ever, is the time to do this.

Notes

Chapter 1 Introduction

1 For background on the process of suburban sprawl, see Jackson (1985); Weiss (1987); Schuyler (1986); and Daniels (1999).

2 This process of land conversion is well described in the planning literature. See for example Mark Gottdiener's classic description of the collusion of development interests on Long Island in Gottdiener (1977). The seminal notion of "growth coalitions" was developed by Harvey Molotch and others, for example in Logan and Molotch (1987).

3 Lewis (1922).

4 See for example the *1997 Natural Resources Inventory*, available at www.nhq.nrcs.usda.gov/land/pubs/97highlights.html.

5 For further background on problems related to sprawl, see Benfield *et al.* (1999); US Environmental Protection Agency (2001, p. 6); Daniels (1999); Kunstler (1993); and Burchell *et al.* (1997), available at http://nationalacademies.org/trb/publications/tcrp/tcrp_rpt_39-a.pdf.

6 Benfield *et al.* (1999, p. 5).

7 US Environmental Protection Agency (2001, p. 6).

8 US Bureau of the Census (2001).

9 See Stephen Marshall, "The Challenge of Sustainable Transport," in Layard *et al.* (2001, p. 133).

10 See Simin Davoudi, "Planning and Sustainable Waste Management," in Layard *et al.* (2001, p. 194).

11 McGinn (2001, p. 7).

12 Cited in McGinn (2001, p. 31).

13 Blakely and Snyder (1997).

14 For example, a 1995 study found that between 1970 and 1990 the isolation of the poor in the 100 largest US metropolitan areas rose by 9 percent (Abramson *et al.*, 1995). Similarly, Myron Orfield (1997) has mapped growing metropolitan inequality.

15 Kunstler (1993).

16 See for example Hayden (1984).

17 Putnam (2000).

18 Hall (1996, pp. 14ff.).

19 Isard (1975, p. 2).

20 See for example Krizek and Power (1996).

21 See for example Beatley and Manning (1997); Mazmanian and Kraft (1999); Roseland (1998).

22 See for example Congress for the New Urbanism (1999); Calthorpe (1993); Duany *et al.* (2000).

23 See for example Lennard *et al.* (1997); Lennard and Lennard (1995); Langdon (1994); Corbett and Zykofsky (1996); Kunstler (1993, 1996).

24   See www.healthycities.org.
25   See for example Bullard (1990).
26   See for example Bullard *et al.* (2000).
27   See for example Forman and Godron (1986); Dramstad *et al.* (1996); Farina (2000).
28   See for example Spirn (1984); Hough (1984, 1990); Lyle (1999); Pratt *et al.* (1994).
29   See for example Riley (1998); Thompson and Sorvig (2000); Todd and Todd (1993); Peck (1998).
30   See for example Innes (1995); Forester (1989).
31   Davidoff (1968).

Chapter 2 Sustainable development

1   Good historical discussions of the emergence of the concept are provided by Kidd (1992, pp. 1–26); Mitlin (1992); and Beatley (1995).
2   See for example Ostrom (1990).
3   Marsh (1864).
4   Wurster (1993, pp. 144–5).
5   See particularly Mumford (1934).
6   See for example Mazmanian and Kraft (1999); Gottlieb (1993).
7   Carson (1962); Commoner (1971); Roszak (1972).
8   Ward and Dubos (1972, pp. 192–3).
9   Ibid, p. 126.
10   See for example Teilhard de Chardin (1959).
11   See for example Gilligan (1982); Belenky *et al.* (1986).
12   Robert Gottlieb's 1993 book *Forcing the Spring* provides an excellent history of the evolution of environmentalism during this time.
13   Meadows *et al.* (1972).
14   Simon (1981).
15   Meadows *et al.* 1992.
16   Goldsmith *et al.* (1972).
17   Ibid, pp. 3–4.
18   Ibid, pp. 21–2.
19   See especially Hayes (1987); Brown (1981); and Lowe (1990, 1991).
20   Callenbach (1975, 1981).
21   World Commission on Environment and Development (1987, p. 8).
22   See for example Adams (1990).
23   Pearce *et al.* (1990, p. 1).
24   Trainer (1986, p. 5).
25   In Norberg-Hodge and Goering (1986, p. 127).
26   Ibid, p. 20.
27   Mander (1991).
28   For a more detailed discussion of key characteristics of the modernist worldview, see Chapter 6 of Norgaard (1994).
29   See for example Le Corbusier (1929 [1987]).
30   Ellin (1999 [1996], pp. 154ff.).
31   Dear (2000).
32   Harvey (1990).
33   See for example Capra (1996).
34   Layard *et al.* (2001, p. 14).

Chapter 3 Theory of sustainability planning

1 Owens and Cowel (2002, p. 7).
2 Cronon (1991).
3 See Webber *et al.* (1964).
4 Friedmann and Weaver (1979).
5 See for example Castells (1997).
6 Lefebvre (1974 [1991]); Soja (1989).
7 Beatley and Manning (1997).
8 The AICP Code of Ethics and Professional Conduct is available online at www.planning. org/ethics/conduct.html.
9 Blanco (1994, p. 3).
10 See Charles J. Hoch, "Making Plans," in Hoch *et al.* (2000, p. 23). Blanco demonstrates (1994, p. 68) that this process is similar to John Dewey's five-stage process of inquiry within his philosophy of instrumentalism.
11 For classic descriptions of this model, see Kent (1964) and Kaiser *et al.* (1995).
12 Lindblom (1959).
13 Etzioni (1967, 1971).
14 See for example Harvey (1973, 1990); Massey (1984); Lefebvre (1974); Castells (1977, 1978); Friedman (1987); Fainstein and Fainstein (1979); Beauregard (1989); Forester (1989, 1993).
15 Logan and Harvey (1987).
16 Stone (1989).
17 Mollenkopf (1983).
18 Friedman (1987).
19 Judge *et al.* (1995).
20 Healey (1997, pp. 28–9).
21 Arnstein (1969).
22 See for example Innes (1995).
23 See for example Habermas (1995 [1983], pp. 43ff.).
24 Warburton (1998, p. 2).
25 Helling (1998).
26 Warburton (1998, pp. 28–34); Friere (1970).
27 Davidoff (1968).
28 Zukin (1991).
29 See for example Bartholomew (1999).
30 See for example Castells (1983); Susan S. Fainstein and Clifford Hirst, "Urban Social Movements," in Judge *et al.* (1995).
31 Castells (1997).
32 Especially Giddens (1984).
33 Berger and Luckmann (1966).
34 Friedma (1987, pp. 181ff.).
35 Healey (1997, p. 61).
36 Putnam, (2000, pp. 183–284).
37 Ibid, p. 403.
38 Ibid, pp. 404–14.
39 Lindblom (1959).
40 A number of authors have written on the subject of how planners actively facilitate inequitable or unsustainable forms of development. See for example Harvey (1973), Beauregard (1959), Gottdiener (1977), Kunstler (1993), and Benfield *et al.* (1999).
41 See for example Howe (1994).

Chapter 4   Planning and the Three Es

1   See Mazmanian and Kraft (1999). Leslie Paul Thiele takes a somewhat longer viewpoint, discerning four waves of environmentalism, the first starting in the late nineteenth century with the rise of preservationists and resource conservationists, the second in the 1960s and 1970s with the rapid growth of the modern movement, the third in the 1980s with a period of cooptation and mainstreaming, and the fourth beginning in the 1990s with growing understanding of global interdependence and the rise of widespread concern about sustainable development. See Thiele (1999).
2   See for example Mazmanian and Kraft (1999); Gottlieb (1993); and Shabecoff (2000).
3   See for example Simon (1981).
4   See for example Rees (1992).
5   We have seen how capital subverts democratic institutions particularly in the United States, which according to observers such as Kevin Phillips has teetered historically between democracy and plutocracy, with the latter being in ascendance at the beginning of the twenty-first century. See Phillips (2002).
6   See for example Pearce and Markandya (1989); Pearce and Warford (1993); and Turner (1988).
7   See Costanza (1991); and Repetto (1985).
8   Hawken (1993); Hawken et al. (1999).
9   See for example Morris (1983); Sale (1985).
10 Nader et al. (1976).
11 Redclift (1987, p. 4).
12 See for example Massey (1994); Harvey (1973, 1990); Martin Khor, "Development, Trade, and the Environment: A Third World Perspective," in Goldsmith et al. (1992); Escobar (1995); Stavrianos (1981).
13 Frank (1969).
14 Orfield (1997).
15 Wilson (1987, 1996).
16 Gottlieb, Robert (1993 pp. 235 ff.).
17 US General Accounting Office (1983).
18 United Church of Christ, Commission for Racial Justice (1987).
19 Howard (1902 [1898]).

Chapter 5   Issues central to sustainability planning

1   See Gordon and Richardson (1997).
2   Planners in Portland, Oregon, for example, have been exploring the possibility of setting targets of around 1/3 infill.
3   Cullingworth and Nadin (2002, p. 172).
4   Rudlin and Falk (1999, p. 128).
5   Nicholson-Lord (2003).
6   See Peter Hall "Sustainable Cities or Town Cramming?," in Layard et al. (2001, pp. 101–14).
7   See Cervero (1998).
8   See for example Newman and Kenworthy (1999).
9   See Vale and Vale (1975); Van der Ryn and Cowan (1996).
10 Phillips (2002).
11 Daly and Cobb (1989).
12 See www.rprogress.org.
13 Waring (1988).
14 Hawken (1993).

15 Lovins (1977).
16 Schumacher (1973).
17 Henderson (1991).
18 See Kenneth Boulding's work in particular.
19 For more background on coops, see for example Co-op America (www.coopamerica.org).
20 See for example International Forum on Globalization (2002).

Chapter 6   Tools for sustainability planning

1   See Charles J. Hoch, "Making Plans," in Hoch *et al.* (2000, p. 23).
2   Phil Berke and Maria Manta Conroy (2000) published an interesting study comparing high-quality local plans with and without an explicit sustainability focus. They found little difference in whether these plans actually met sustainability goals, implying that an overt orientation toward "sustainability" may not be essential. However, this could simply mean that much progressive planning is headed in these directions these days anyway, whether under the guise of "sustainability," "livability," "environmental protection," "quality of life," or other catch-phrases.
3   Neuman (1998).
4   The most recent of these follows the "Three Es" format common to much sustainability planning. See Yaro and Hiss (1996).
5   Register (1987).
6   See www.iaia.org.
7   Counsell and Haughton (2002).
8   See for example Benfield *et al.* (2001); Ewing (1996); Bragado *et al.* (1995); Corbett and Corbett (2000); Corbett (1996); Fader (2000); Jacobs (1993); Lennard and Lennard (1995); Condon and Moriarty (1999); Roseland (1998); Wheeler (2002a), available at www.greenbelt.org. See also websites of the UN-related Best Practices database (www. bestpractices.org or www.sustainabledevelopment.org); the International Council for Local Environmental Initiatives (www.iclei.org); the Sustainable Communities Network (www.sustainable.org); the Resource Renewal Institute (www.rri.org); the Local Government Commission (www.lgc.org); the Rocky Mountain Institute (www.rmi.org); the International City/County Management Association (www.icma.org); the American Planning Association (www.planning.org); and the US Department of Energy's Center of Excellence for Sustainable Development (www.sustainable.doe.gov).
9   See www.dubai-award.dm.gov.ae.
10 Portnoy (2003, p. ix).
11 See for example Maclaren (1996).
12 Sustainable Seattle Coalition (1995).
13 http://www.sustainableseattle.org/AboutUs/History.shtml.
14 One good source on the subject of indicators is Maureen Hart's *Guide to Sustainable Community Indicators* (1999), available from www.sustainablemeasures.com.
15 Keough (2003). See also www.sustainablecalgary.ca.
16 See Southworth and Ben-Joseph (1997).
17 More information on LEED is available from the US Green Building Council, www. usgbc.org.
18 See for example Duany and Plater-Zyberk (1991).
19 Wackernagel and Rees (1996); Wackernagel and Yount (1998).
20 For one version, see www.rprogress.org.
21 Girardet (1999, p. 29).
22 Roy and Caird (2001, p. 277).
23 Shen (2001).
24 Moudon and Hess (2000).

25 See Kent and Klosterman (2000).
26 For further information on this field, see the International Association for Impact Assessment at www.iaia.org, the European Community's EIA home page at www.europa. eu.int/comm/environment/eia/home.htm, and a helpful resource list from the Canadian International Development Agency at www.iaia.org/eialist.html.
27 See http://cfpub.epa.gov/ncea/, www.ceaa.gc.ca/index_e.htm, and www.environment-agency.gov.uk.
28 See www.unece.org/env/eia/.
29 Wheeler (2000).
30 See Abbott (1997); Abbott *et al*. (1994).
31 Friedman (1987, pp. 181ff.).
32 Putnam (1993. p. 167).
33 See for example Gittell and Vidal (1998); Serageldin (1996).

Chapter 7 International planning

1 Morris (1994, pp. 305–6).
2 Gertler (1978).
3 For more detailed information on these conferences, see www.lib.msu.edu/publ_ser/docs/igos/unconfs.htm.
4 Section 28.3. Text of Agenda 21 available at www.un.org/esa/sustdev/agenda21.htm.
5 International Council on Local Environmental Initiatives (ICLEI) (1997), available at www.iclei.org/la21/la21rep.htm.
6 Ashley (2002, p. 460).
7 For a concise evaluation of this conference, see Ashley (2002).
8 United Nations Human Settlements Programme (United Nations Center on Human Settlements) (1996), available through www.unchs.org.
9 For more information see http://whc.unesco.org/nwhc/pages/sites/main.htm.
10 Frost (2001, p. 213).
11 FERN, The EU's Fifth Environmental Action Plan, September 1998, available at www. fern.org/pubs/archive/5eap.html.
12 Colwell (2001, pp. 220–1).
13 For further information, see http://europa.eu.int/scadplus/leg/en/s15001.htm.
14 Cullingworth and Nadin (2002, p. 34).
15 Rydin (1998, pp. 3, 129ff.).
16 Available at http://www.nature.coe.int/english/main/planning/guidingp.pdf.
17 Osadcha (2001). Section 3.2 available at www.rri.org/envatlas/europe/euprofile.html.
18 For more information, see www.sustainable-cities.org or www.iclei.org.
19 Chaibva (2000).
20 For further information, see www.earthcharter.org.
21 See for example Korten (1995); Gross (1980).
22 US Environmental Protection Agency, *Executive Summary of the Inventory of U.S. Greenhouse Gas Emissions and Sinks, 1990–2000*.
23 BBC wire, "Germany Close to Meeting Kyoto Protocol Target," November 5, 2002.
24 United Nations Framework Convention on Climate Change (2003).
25 Scientific Assessment Panel of the Montreal Protocol on Substances that Deplete the Ozone Layer, *Scientific Assessment of Ozone Depletion: 2002*, available at http://www. unep.org/ozone/pdf/execsumm-sap2002.pdf.
26 For more information, see www.biodiv.org.
27 See Barry G. Rabe, "Sustainability in a Regional Context: The Case of the Great Lakes Basin," in Mazmanian and Kraft (1999).
28 Rabinovich and Leitman (1996).

29 See for example Braidotti *et al.* (1994).

Chapter 8 National planning

 1 See for example Jackson (1985).
 2 See for example Markusen *et al.* (1991).
 3 Macey (2001).
 4 Southern Burlington County NAACP v. Mount Laurel, 67 N.J. 151, 336 A.2d 713, appeal dismissed and cert. Denied, 423 U.S. 808 (1975).
 5 Hoch *et al.* (2000, pp. 250–1).
 6 President's Council on Sustainable Development (1996).
 7 Urban Task Force (1999).
 8 For further information, see www.nrtee-trnee.ca/eng/main_e.htm.
 9 Urban Task Force (1999, p. 11).
10 Parkinson and Roseland (2002).
11 Extensive information about national Green Plans is available from the Resource Renewal Institute in San Francisco, an NGO devoted to promoting these national environmental planning policies, at www.rri.org.
12 Resource Renewal Institute, Case Study: The Netherlands' National Environmental Policy Plan, San Francisco, 1999. Available at www.rri.org/envatlas/europe/netherlands/nl-index.html.
13 See for example Van Eeten and Roe (2000).
14 See Alexander (2000, pp. 91–2); Beatley (2000b, pp. 98–100).
15 Rowe and Fudge (2003).
16 Eckerberg and Försberg (1998).
17 See Keil (1999), available online at www.rri.org/envatlas/europe/germany/de-index.html.
18 Ravetz (2000, p. 146).
19 See Peter Hall, "The View from London Centre: London Docklands Development Corporation," in Blowers and Evans (1997, p. 131).
20 Cullingworth and Nadin (2002, p. 208).
21 Department of Transport, Local Government and the Regions (2001b), available at www.dtlr.gov.uk/.
22 *Planning Green Paper* (2001b), Section 4.7.
23 Warburton (2002, pp. 82–4).
24 Department of Transport, Local Government and the Regions (2000a), available at www.dtlr.gov.uk/.
25 Department of Transport, Local Government and the Regions (2001a,) available at www.dtlr.gov.uk/.
26 *Sustainable Communities: Building for the Future* (2003).
27 Resource Renewal Institute, Case Study: New Zealand's Resource Management Act, San Francisco, 2000. Available at www.rri.org/envatlas/oceania/new-zealand/nz-index.html.
28 For more information, see www.deh.gov.au/esd/.
29 Barrett and Usui (2002).
30 Resource Renewal Institute, Development of Japan's Environmental Policy. San Francisco, 2000. Available at www.rri.org/envatlas/asia/japan/jp-dev.html.
31 Japan for Sustainability Information Center (2002), available through www.japanfs.org.
32 See Japan for Sustainability at www.japanfs.org/en/jfs/background.html.
33 The President's Council on Sustainable Development (1996).
34 Hall (1992, p. 223).
35 Rydin (1998, p. 110).
36 Lusk (2003).

Chapter 9   State and provincial planning

1   See http://www.csbr.umn.edu/B3/index.html.
2   Bollens (1992).
3   Available at www.njstateplan.com/plan2/main.htm.
4   See www.njfuture.org/.
5   Abbott (1997).
6   Abbott (1997).
7   Available at www.mnplan.state.mn.us/pdf/2000/eqb/ModelOrdWhole.pdf.
8   US EPA, Balancing Growth and Quality of Life: US EPA's National Smart Growth Achievement Award Recipients Announced, November 19, 2003.
9   Nadel and Geller (2001, p. vi), available at www.aceee.org/pubs/.
10 California Integrated Waste Management Board, press release available at www.ciwmb.ca.gov/lgcentral/Rates/Diversion/2000.htm.
11 See House Bill 3978, available at www.leg.state.or.us/01reg/measures/hb3900.dir/hb3948.en.html.
12 See www.scotland.gov.uk/about/ERADEN/SCU/00017108/home.aspx.
13 See http://mediambient.gencat.es/eng/aindex.htm.
14 See Parayil (2000).

Chapter 10   Regional planning

1   Calthorpe and Fulton (2001).
2   Wheeler (2002b).
3   See e.g. Greenstein and Wiewel (2000); Altshuler et al. (1999); Katz (2000); Pastor et al. (2001).
4   See e.g. Calthorpe and Fulton (2001); Daniels (1999).
5   See e.g. Peirce (1993).
6   Geddes (1915, p. 18).
7   Ibid, p. 16.
8   Howard (1902 [1898], pp.51–2).
9   See Luccarelli (1995); Susman (1976).
10 Odum and Estill Moore (1938).
11 Friedmann and Alonso (1964, p. 1).
12 Ibid, p. 497.
13 See Webber et al. (1964).
14 For more information including an updated Chesapeake 2000 agreement, see www.chesapeakebay.net.
15 Dramstad et al. (1996); Peck (1998).
16 Bartholomew (1999).
17 For more information, see www.cbf.org/.
18 For more information, see www.keeptahoeblue.com/.
19 For more information, see www.atlantaregional.com/.
20 For more information, see www.mtc.ca.gov/.
21 For an overview of growth management techniques, see Porter (1997).
22 Bae (2003).
23 Neuman (2000).
24 See Calthorpe (1993); Calthorpe and Fulton (2001); Downs (1994).
25 Lynch (1981, p. 235).
26 Brand (1994).
27 Alexander et al. (1977). Not all "patterns" in this framework are appropriate for sustainable community development as we now conceive of it. For example, the authors promote

"looped local roads" to prevent through-traffic, similar to suburbs in Canada. However, such loops, like cul-de-sacs, also make walking and bicycling difficult unless specific bike and pedestrian cut-through paths are also provided. Most current New Urbanist designers emphasize a more tightly connected grid-like pattern instead, with traffic calming measures to slow or reduce traffic.

28 Alexander *et al.* (1977); Alexander (1979).
29 For more information, see www.ceres.ca.gov/snep/.
30 Yaro and Hiss (1996).
31 Box *et al.* (2001, pp. 210–12).
32 For a good overview of LA's air quality planning struggles, see Daniel A. Mazmanian, "Los Angeles' Transition from Command-and-Control to Market-based Clear Air Policy Strategies and Implementation," in Mazmanian and Kraft (1999, pp. 77–112).
33 For more information, see www.chesapeakebay.net/.
34 For a history of Great Lakes cleanup efforts, see Barry G. Rabe, "Sustainability in a Regional Context: The Case of the Great Lakes Basin," in Mazmanian and Kraft (1999, pp. 247–81).
35 For more information, see www.calfed.water.ca.gov.
36 See Barry G. Rabe in Mazmanian and Kraft (1999).
37 See for example Orfield (1997); and john a. powell, "Addressing Regional Dilemmas for Minority Communities," in Katz (2000).
38 Orfield (1997, p. 87).
39 See Robert D. Bullard and coauthors, "Dismantling Transportation Apartheid: The Quest for Equity," in Bullard *et al.* (2000, pp. 54–5).
40 See for example Renner (1991).
41 Gottlieb (2002), available at www.brook.edu/dybdocroot/es/urban/publications/ gottlieb. pdf.

Chapter 11 Local government planning

1 See for example Hom (2002).
2 Yaro and Hiss (1996, p. 197).
3 Cullingworth and Nadin (2002, p. 1).
4 Jackson (1985, p. 242).
5 See Hall (1992, pp. 223ff.). Urban renewal and redevelopment programs in the United States have indeed represented a proactive public role, but unfortunately often had negative results in that older urban neighborhoods were bulldozed and replaced with sterile, modernist urban environments. Partly as a result, cities have been cautious about activist planning ever since.
6 See for example Duany and Plater-Zyberk (1991).
7 Winter (2001a, p. 27).
8 Warner (1962).
9 Juri Pill, quoted in Cervero (1998, p. 89).
10 See Southworth and Owens (1993).
11 See Beatley (1994).
12 See Burchell *et al.* (1997, p. 49).
13 Speir and Stephenson (2002, p. 67).
14 Friedman (2003, pp. 5–9).
15 Cullingworth and Nadin (2002, p. 217).
16 See Lyle (1994, Chapter 8).
17 Peck and Associates (2000, p. 52).
18 Ravetz (2000, p. 165).
19 See for example Kinsley and Lovins (n.d. [1996?]).

20 Logan and Molotch (1987).
21 C. Wright Mills, cited in Green (1973).
22 Sirkis (2000).
23 See Newman and Kenworthy (1999).
24 For more information on this and other BRT projects, see www.fta.dot.gov/brt/.
25 See Cervero (1998).
26 For further information, see www.glasgow.gov.uk/healthycities.
27 See Caroline Brown and Stefanie Dohr, "Understanding Sustainability and Planning in England: An Exploration of the Sustainability Content of Planning Policy at the National, Regional and Local Levels," in Rydin and Thornley (2000).
28 Portnoy (2003, pp. 64–71).
29 Livingstone (2002).
30 Munt (2001, pp. 292–3).

Chapter 12   Neighborhood planning

1 See Perry (1939).
2 Riesman (1953).
3 Bellah *et al.* (1985).
4 Putnam (2000).
5 Some nodes of suburban development may actually be relatively dense and contain apartments or other traditionally urban building forms, but the overall design of the neighborhood will not be particularly urban or livable. See for example Moudon and Hess (2000).
6 Schuyler (1996).
7 Schuyler (1986, p. 41).
8 In Nolan (1916, pp. 94–5).
9 See for example Mumford (1938, 1961).
10 Whyte (1968).
11 Jacobs (1961).
12 See for example Marcus Sarkissian (1986); Marcus and Francis (1990).
13 Former Toronto Mayor John Sewell (1993) provides a fascinating description of how different eras of neighborhood design philosophy have shaped one city in his book *The Shape of the City: Toronto Struggles with Modern Planning*.
14 Nelessen (1994, pp. 83ff.).
15 See Peter Hall, "Sustainable Cities or Town Cramming?", in Layard *et al.* (2001, pp. 101–13).
16 Rapoport (1975).
17 Rudlin and Falk (1999, p. 125).
18 These and other strategies are detailed in a guidebook I wrote entitled *Smart Infill: Creating More Livable Communities in the Bay Area*, published in 2002 by the San Francisco-based Greenbelt Alliance (Wheeler, 2002a).
19 See for example Grant (2002).
20 See www.princes-foundation.org/foundation/projdir-uep-poundbury.html.
21 See Jacobs (1993); Jacobs *et al.* (2001).
22 Jacobs (1961, pp. 363ff.).
23 Rudofsky (1969).
24 Appleyard (1981).
25 Engwicht (1993).
26 See Riley (1998).

Chapter 13 Site planning and architecture

1 Historically of course the process was simpler, and many people built houses and workplaces for themselves with little regulation or oversight. Self-built housing was relatively common on the outskirts of North American cities and towns until the middle of the twentieth century, and is still the norm in much of the developing world. For a historical description of do-it-yourself housing in one metropolitan area, see Harris (1996).
2 For an excellent description of the rise of large-scale homebuilding, see Weiss (1987).
3 For examples of well-designed, context-sensitive affordable housing, see Jones *et al.* (1997).
4 Eisenberg and Yost (2001).
5 McHarg (1969).
6 See for example Van der Ryn and Cowan (1996); Lyle (1994, 1999); Todd and Todd (1993); and architect William McDonough's Hannover Principles.
7 Condon and Moriarty (1999).
8 Van der Ryn and Cowan (1996).
9 Birkeland (2002, p. 13).
10 Winter (2001b, p. 52).
11 See for example Van der Ryn and Cowan (1996); Lyle (1994, 1999); Kelbaugh (1997).
12 Kelbaugh (1997, pp. 52ff.).
13 For more information, contact AsiaRain at www.asiarainhardwoods.com; 1104 Firenze Street, McCloud, California 96057; 530–964–2131 or 800–220–9062.
14 "Paving Without Asphalt or Concrete," *Environmental Building News*, 8 (11), 1999, available at www.buildinggreen.com/products/road_oyl.html.
15 Easton (1996).
16 Lipke (2001).
17 Eisenberg and Yost (2001).
18 See McCamant and Durrett (1994).
19 See Norwood and Smith (1995).
20 Alexander *et al.* (1977); Alexander (1979).

Chapter 14 How do we get there from here?

1 Owens Cowell (2001, p. 170).
2 Tuchman (1984).
3 See Jacobs (1985).
4 See for example Hiss (1990); Seamon (1993).
5 Seamon (1993, p. 16).
6 Inglehart (1977, 1990, 1997).
7 A substantial literature relates to the various ways we structure our own realities, either internally within ourselves through cognitive processes or externally within society through institutions and culture. Particularly useful works include Berger and Luckmann (1966); Giddens (1984, 1994); and Manuel Castells' trilogy (1996, 1997, 1998).
8 See among others Phillips (2002).
9 Giddens (1998).
10 Etzioni (1993).
11 Dryzek (1990, 2000).
12 Lerner (1996).
13 Giddens (1998, p. 65).
14 Ravetz (2000, p. 6).

Bibliography

1997 Natural Resources Inventory, Washington, DC: US Department of Agriculture, Natural Resources Conservation Service, 1997.

Abbott, Carl, The Portland Region: Where Cities and Suburbs Talk to Each Other – and Often Agree, *Housing Policy Debate*, Winter, 1997, pp. 11–51.

Abbott, Carl, Deborah Howe, and Sy Adler, *Planning the Oregon Way: A Twenty-Year Evaluation*, Corvallis: Oregon State University Press, 1994.

Abramson, Alan U., Mitchell S. Tobin, and Matthew R. VanderGoot, The Changing Geography of Metropolitan Opportunity: The Segregation of the Poor in U.S. Metropolitan Areas, 1970 to 1990, *Housing Policy Debate*, 6 (1), 1995, pp. 45–72.

Adams, W.M., *Green Development: Environment and Sustainability in the Third World*, New York: Routledge, 1990.

Alexander, Christopher, *The Timeless Way of Building*, New York: Oxford University Press, 1979.

Alexander, Christopher *et al.*, *A Pattern Language,* New York: Oxford University Press, 1977.

Alexander, Ernest R., Netherlands' Planning: The Higher Truth, *Journal of the American Planning Association*, 67 (1), 2000, pp. 91–2.

Altshuler, Alan *et al.* (eds.), *Governance and Opportunity in Metropolitan America*, Washington, DC: National Academy Press, 1999.

Appleyard, Donald, *Livable Streets*, Berkeley: University of California Press, 1981.

Arnstein, Sherry, A Ladder of Citizen Participation, *Journal of the American Institute of Planners*, 8 (3), 1969.

Ashley, Mike, Local Government and the WSSD, *Local Environment* (7) 4, 2002, pp. 459–63.

Audirac, Ivonne, *Rural Sustainable Development in America*, New York: John Wiley & Sons, 1997.

Bae, Chang-Hee Christine and Myung-Jin Jun, Counterfactual Planning: What If There Had Been No Greenbelt in Seoul? *Journal of Planning Education and Research*, 22, 2003, pp. 374–83.

Barrett, Brendan and Mikoto Usui, Local Agenda 21 in Japan: Transforming Local Environmental Governance, *Local Environment*, (7) 1, 2002, pp. 49–67.

Bartholomew, Keith, The Evolution of American Nongovernmental Land Use Planning Organizations, *Journal of the American Planning Association*, (65) 4, 1999, pp. 357–63.

Beatley, Timothy, *Habitat Conservation Planning*, Austin: University of Texas Press, 1994.

Beatley, Timothy, The Many Meanings of Sustainability. Planning and Sustainability: The Elements of a New (Improved?) Paradigm, *Journal of Planning Literature*, 9 (4), 1995, pp. 339–42; 383–95.

Beatley, Timothy, *Green Urbanism: Learning from European Cities*, Washington, DC: Island Press, 2000a.

Beatley, Timothy, Dutch Green Planning More Reality Than Fiction. *Journal of the American Planning Association*, 67 (1), 2000b, pp. 98–100.

Beatley, Timothy and Kristy Manning, *The Ecology of Place: Planning for Environment, Economy, and Community*, Washington, DC: Island Press, 1997.

Beauregard, Robert, Between Modernity and Postmodernity: The Ambiguous Position of U.S. Planning, *Environment and Planning D: Society and Space*, 7, 1989, pp. 381–95.

Becker, Egon and Thomas Jahn (eds.), *Sustainability and the Social Sciences: A Cross-disciplinary Approach to Integrating Environmental Considerations into Theoretical Reorientation*, London: Zed Books, 1999.

Belenky, Mary Field *et al.*, *Women's Ways of Knowing: The Development of Self, Voice, and Mind*, New York: Basic Books, 1986.

Bellah, Robert *et al.*, *Habits of the Heart: Individualism and Commitment in American Life*, Berkeley: University of California Press, 1985.

Benfield, F. Kaid, Matthew D. Raimi, and Donald D.T. Chen, *Once There Were Greenfields*, Washington, DC: Natural Resources Defense Council, 1999.

Benfield, F. Kaid, Jutka Terris, and Nancy Vorsanger, *Solving Sprawl: Models of Smart Growth in Communities Across America*, New York: Natural Resources Defense Council, 2001.

Berger, Peter L. and Thomas Luckmann, *The Social Construction of Reality*, New York: Doubleday, 1966.

Berke, Philip R. and Maria Manta Conroy, Are We Planning for Sustainable Development?, *Journal of the American Planning Association*, 66 (1), pp. 21–34.

Birkeland, Janis, *Design for Sustainability, A Sourcebook of Integrated Eco-logical Solutions*, London: Earthscan, 2002.

Blakely, Edward J. and Mary Gail Snyder, *Fortress America: Gated and Walled Communities in the United States*, Washington, DC: Brookings Institution Press, 1997.

Blanco, Hilda, *How to Think about Social Problems: American Pragmatism and the Idea of Planning*, Westport, CT: Greenwood Press, 1994.

Blowers, Andrew (ed.), *Planning for a Sustainable Environment: A Report by the Town and County Planning Association*, London: Earthscan, 1993.

Blowers, Andrew and Bob Evans (eds.), Town Planning into the 21st century, London: Routledge 1997.

Bollens, Scott A., State Growth Management: Intergovernmental Frameworks and Policy Objectives, *Journal of the American Planning Association*, 58 (4), 1992, pp. 454–66.

Boulding, Kenneth, *The Meaning of the Twentieth Century*, 1964.

Box, John, Veronica Cossons, and Jan McKelvey, Sustainability, Biodiversity, and Land Use Planning, *Town & Country Planning*, July/August 2001, pp. 210–12.

Bragado, Nancy, Judy Corbett, and Sharon Sprowls, *Building Livable Communities: A Policymaker's Guide to Infill Development*, Sacramento: Local Government Commission, 1995.

Braidotti, Rosi *et al.*, *Women, the Environment and Sustainable Development: Towards a Theoretical Synthesis*, London: Zed Books, 1994.

Brand, Stewart, *How Buildings Learn: What Happens After They're Built*, New York: Viking, 1994.

Brown, Lester R., *Building a Sustainable Society*, New York: Norton, 1981.

Brown, Lester R., Christopher Flavin, and Sandra Postel, *Saving the Planet: How to Shape an Environmentally Sustainable Global Economy*, New York: Norton, 1991.

Bullard, Robert D., *Dumping in Dixie: Race, Class, and Environmental Quality*, Boulder: Westview Press, 1990.

Bullard, Robert D., Glenn S. Johnson and Angel O. Torres, *Sprawl City: Race, Politics, and Planning in Atlanta*, Washington, DC: Island Press, 2000.

Burchell, Robert W. *et al. Costs of Sprawl Revisited: The Evidence of Sprawl's Negative and Positive Impacts*, Washington, DC: Transportation Research Board, 1997.

Burnham, Daniel, *Plan of Chicago*, Chicago: Commercial Club, 1909.

Callenbach, Ernest, *Ecotopia*, New York: Bantam Books, 1975.

Callenbach, Ernest, *Ectopia Emerging*, Berkeley: Banyan Tree Books, 1981.

Calthorpe, Peter, *The Next American Metropolis: Ecology, Community and the American Dream*, New York: Princeton Architectural Press, 1993.

Calthorpe, Peter and William Fulton, *The Regional City: Planning for the End of Sprawl*, Washington, DC: Island Press, 2001.

Capra, Fritjof, *The Web of Life: A New Scientific Understanding of Living Systems*, New York: Anchor Books, 1996.

Carson, Rachel, *Silent Spring*, New York: Fawcett Crest, 1962.

Castells, Manuel, *The Urban Question: A Marxist Approach*, London: Edward Arnold, 1977.

Castells, Manuel, *City, Class and Power*, New York: St Martin's Press, 1978.

Castells, Manuel, *The City and the Grassroots*, Berkeley: University of California Press, 1983.

Castells, Manuel, *The Rise of the Network Society* (Vol. 1 of *The Information Age*), Cambridge, MA: Blackwell, 1996.

Castells, Manuel, *The Power of Identity* (Vol. 2 of *The Information Age*), Cambridge, MA: Blackwell, 1997.

Castells, Manuel, *End of Millennium* (Vol. 3 of *The Information Age*), Cambridge, MA: Blackwell, 1998.

Cervero, Robert, *The Transit Metropolis: A Global Inquiry*, Washington, DC: Island Press, 1998.

Chaibva, Shem, EIA Training for Urban Local Authorities in Africa, *Local Environment* (5) 1, 2000, pp. 97–106.

Colwell, Adrian, Setting the Direction for EU Environmental Policy, *Town & Country Planning*, July/August 2001, pp. 220–1.

Commoner, Barry, *The Closing Circle: Nature, Man and Technology*, New York: Knopf, 1971.

Condon, Patrick and Stacy Moriarty (eds.), *Second Nature: Adapting LA's Landscape for Sustainable Living*, Los Angeles: Treepeople, 1999.

Congress for the New Urbanism, *Charter of the New Urbanism,* New York: McGraw-Hill, 1999.

Corbett, Judy, *A Policy-maker's Guide to Transit-oriented Development*, Sacramento: Local Government Commission, 1996.

Corbett, Judy and Michael Corbett, *Designing Sustainable Communities: Learning from Village Homes*, Washington, DC: Island Press, 2000.

Corbett, Judy and Paul Zykofsky, *Building Livable Communities: A Policymaker's Guide to Transit-oriented Development*, Sacramento: Center for Livable Communities/Local Government Commission, 1996.

Costanza, Robert, Herman E. Daly, and Joy A. Bartholomew, Goals, Agenda, and Policy Recommendations for Ecological Economics in Costanza, Robert (ed.), *Ecological Economics: The Science and Management of Sustainability*, New York: Columbia University Press, 1991.

Counsell, David and Graham Haughton, Sustainability Appraisal – Delivering More Sustainable Regional Planning Guidance?, *Town & Country Planning*, January 2002, pp. 14–17.

Cronon, William, *Nature's Metropolis: Chicago and the Great West*, New York: Norton, 1991.

Cullingworth, J.B. and Vincent Nadin, *Town and Country Planning in the UK*, 13th edn, London and New York: Routledge, 2002.

Daly, Herman and John B. Cobb, Jr, *For the Common Good: Redirecting the Economy Toward Community, the Environment, and a Sustainable Future*, Boston: Beacon Press, 1989.

Daniels, Thomas L., *When City and Country Collide: Managing Growth in the Metropolitan Fringe*, Washington, DC: Island Press, 1999.

Davidoff, Paul, Advocacy Planning, *Journal of the American Institute of Planners*, 21 (4), 1968.

Dear, Michael J., *The Postmodern Urban Condition*, Malden, MA: Blackwell, 2000.

Department of Transport, Local Government and the Regions, *Planning Policy Guidance 3: Housing,* London: The Stationery Office, 2000a.

Department of Transport, Local Government and the Regions, *Good Practice Guide on Sustainability Appraisal of Regional Planning Guidance*, London: The Stationery Office, 2000b.

Department of Transport, Local Government and the Regions, *Planning Policy Guidance 13: Transport*, London: The Stationery Office, 2001a.

Department of Transport, Local Government and the Regions, *Planning Green Paper Planning: Delivering a Fundamental Change*, London: The Stationery Office, 2001b.

Downs, Anthony, *New Visions for Metropolitan America*, Washington, DC: Brookings Institution, 1994.

Dramstad, Wenche E., James D. Olson, and Richard T.T. Forman, *Landscape Ecology Principles in Landscape Architecture and Land-use Planning*, Cambridge, MA: Harvard University GSD and Washington, DC: Island Press, 1996.

Dryzek, John S., *Discursive Democracy: Politics, Policy, and Political Science*, New York: Cambridge University Press, 1990.

Dryzek, John S., *Deliberative Democracy and Beyond: Liberals, Critics, Contestations*, New York: Oxford University Press, 2000.

Duany, Andres and Elizabeth Plater-Zyberk, *City and Placemaking Principles*, New York: Rizzoli, 1991.

Duany, Andres, Elizabeth Plater-Zyberk, and Jeff Speck, *Suburban Nation: The Rise of Sprawl and the Decline of the American Dream*, New York: North Point, 2000.

Easton, David, *The Rammed Earth House*, White River Junction, VT: Chelsea Green Publishers, 1996.

Eckerberg, Katerina and Bjorn Försberg, Implementing Agenda 21 in Local Government: The Swedish Experience, *Local Environment*, 3 (1), 1998, pp. 333–47.

Eisenberg, David and Peter Yost, Sustainability and Building Codes, *Environmental Building News*, 10 (9), 2001, pp. 1; 8–15.

Elkin, Tim *et al.*, *Reviving the City: Towards Sustainable Urban Development*, London: Friends of the Earth, 1991.

Ellin, Nan, *Postmodern Urbanism*, revised edn, New York: Princeton Architectural Press, 1999 [1996].

Engwicht, David, *Reclaiming Our Cities and Towns: Better Living with Less Traffic*, Philadelphia: New Society Publishers, 1993.

Escobar, Arturo, *Encountering Development: The Making and Unmaking of the Third World*, Princeton: Princeton University Press, 1995.

Etzioni, Amitai, Mixed Scanning: A "Third" Approach to Decision-making, *Public Administration Review*, 27, 1967, pp. 153–69.

Etzioni, Amitai, *The Active Society: A Theory of Societal and Political Processes*, New York: Free Press, 1971.

Etzioni, Amitai, *The Spirit of Community: Rights, Responsibilities, and the Communitarian Agenda*, New York: Crown Publishers, 1993.

Evans, Peter (ed.), *Livable Cities?: Urban Struggles for Livelihood and Sustainability*, Berkeley: University of California Press, 2002.

Ewing, Reid, *Best Development Practices*, Chicago: American Planning Association, 1996.

Ewing, Reid, Is Los Angeles-style Sprawl Desirable?, *Journal of the American Planning Association*, 63 (1), 1997, pp. 107–27.

Fader, Steven, *Density by Design: New Directions in Residential Development*, Washington, DC: Urban Land Institute, 2000.

Fainstein, N.I. and S.S. Fainstein, New Debates in Urban Planning: The Impact of Marxist Theory, *International Journal of Urban and Regional Research* 3, 1979, pp. 381–401.

Farina, Almo, *Landscape Ecology in Action*, Boston: Kluwer Academic Publishers, 2000.

FERN, The EU's Fifth Environmental Action Plan, September 1998, available at www.fern.org/pubs/archive/5eap.html.

Fernandez, Edesio (ed.), *Environmental Strategies for Sustainable Development in Urban Areas: Lessons from Africa and Latin America*, Brookfield, VT: Ashgate/Aldershot, 1998.

Forester, John, *Planning in the Face of Power*, Berkeley: University of California Press, 1989.

Forester, John, *Critical Theory, Public Policy, and Planning Practice: Toward a Critical Pragmatism*, Albany, NY: SUNY Press, 1993.

Forman, Richard T.T. and Michel Godron, *Landscape Ecology*, New York: John Wiley & Sons, 1986.

Frank, Andre Gunder, *Capitalism and Underdevelopment in Latin America*, New York: Monthly Review Press, 1969.

Friedman, Andrew, Recycling Redux: The Ups and Downs of the Waste-not Movement, *Planning*, 69 (10), 2003, pp. 5–9.

Friedmann, John, *Planning in the Public Domain: From Knowledge to Action*, Princeton: Princeton University Press, 1987.

Friedmann, John and William Alonso (eds.), *Regional Development and Planning: A Reader*, Cambridge, MA: MIT Press, 1964.

Friedmann, John and Clyde Weaver, *Territory and Function: The Evolution of Regional Planning*, London: Edward Arnold, 1979.

Freire, Paolo, *Pedagogy of the Oppressed*, trans. Myra Bergman Ramos, New York: Seabury Press, 1970.

Frost, Peter, Urban Biosphere Reserves: Reintegrating People with the Natural Environment, *Town & Country Planning*, July/August 2001, pp. 213–15.

Geddes, Patrick, *Cities In Evolution: An Introduction to the Town Planning Movement and to the Study of Civics*, London: Williams & Norgate, 1915.

Gertler, Len, *Habitat and Land*, Vancouver: University of British Columbia Press, 1978.

Giddens, Anthony, *The Constitution of Society: Outline of the Theory of Structuration*, Berkeley: University of California Press, 1984.

Giddens, Anthony, *Beyond Left and Right: The Future of Radical Politics*, Palo Alto, CA: Stanford University Press, 1994.

Giddens, Anthony, *The Third Way: The Renewal of Social Democracy*, Cambridge, UK: Polity Press, 1998.

Gilliard, Geoff, Surrey Shifts Sustainability Status Quo with Neighborhood Concept Plan, *Plan*, 43 (1), Spring 2003, pp. 13–14.

Gilligan, Carol, *In a Different Voice: Psychological Theory and Women's Development*, Cambridge, MA: Harvard University Press, 1982.

Girardet, Herbert, *Creating Sustainable Cities*, Devon, UK: Green Books, 1999.

Gittell, Ross and Avis Vidal, *Community Organizing: Building Social Capital as a Development Strategy*, Thousand Oaks, CA: Sage Publications, 1998.

Global Cities Project, *Building Sustainable Communities: An Environmental Guide for Local Government*, San Francisco: The Center for the Study of Law and Politics, 1991.

Goldsmith, Edward, *The Way: An Ecological World View*, Boston: Shambhala, 1993.

Goldsmith, Edward *et al.*, *Blueprint for Survival*, Boston: Houghton Mifflin, 1972.

Goldsmith, Edward *et al.*, *The Future of Progress: Reflections on Environment & Development*, Bristol, UK: The International Society for Ecology and Culture, 1992.

Gordon, Peter and Harry Richardson, Are Compact Cities a Desirable Planning Goal?, *Journal of the American Planning Association*, 63 (1), 1997, pp. 95–107.

Gottdiener, Mark, *Planned Sprawl: Private and Public Interests in Suburbia*, Beverly Hills: Sage Publications, 1977.

Gottlieb, Paul. D., *Growth Without Growth: An Alternative Economic Development Goal for Metropolitan Areas*, Washington, DC: The Brookings Institution, 2002.

Gottlieb, Robert, *Forcing the Spring: The Transformation of the American Environmental Movement*, Washington, DC: Island Press, 1993.

Grant, Jill, Mixed Use in Theory and Practice: Canadian Experience with Implementing a Planning Principle, *Journal of the American Planning Association*, 68 (1), 2002, pp. 71–84.

Green, Mark, The Corporation and the Community, in Nader, Ralph and Mark J. Green (eds.), *Corporate Power in America*, New York: Grossman Publishers, 1973.

Greenbelt Alliance, *Reviving the Sustainable Metropolis: Guiding Bay Area Conservation and Development into the 21st Century*, San Francisco, 1989.

Greenstein, Rosalind and Wim Wiewel (eds.), *Urban–Suburban Interdependencies*, Cambridge, MA: Lincoln Institute of Land Policy, 2000.

Gross, Bertram, *Friendly Facism: The New Face of Power in America*, New York: M. Evans, 1980.

Habermas, Jurgen, *Moral Consciousness and Communicative Action*, trans. Christian Lenhardt and Shierry Weber Nicholsen, Cambridge, MA: MIT Press, 1995 [1983], pp. 43ff.

Hall, Peter, *City and Regional Planning*, 3rd edn, New York: Routledge, 1992.

Hall, Peter, *Cities of Tomorrow: An Intellectual History of Urban Planning and Design in the Twentieth Century*, Cambridge, MA: Blackwell, 1996.

Hamm, Bernd and Pandurang K. Muttagi (eds.), *Sustainable Development and the Future of Cities*, London: Centre for European Studies, 1998.

Hardoy, Jorge E. *et al.*, *Environmental Problems in Third World Cities*, London: EarthScan, 1992.

Harris, Jonathan M., Timothy A. Wise, Kevin P. Gallagher, and Neva R. Goodwin, *A Survey of Sustainable Development: Social and Economic Dimensions*, Washington, DC: Island Press, 2001.

Harris, Richard, *Unplanned Suburbs: Toronto's American Tragedy 1900–1950*, Baltimore: Johns Hopkins University Press, 1996.

Hart, Maureen, *Guide to Sustainable Community Indicators*, Ipswich, MA: QLF/Atlantic Center for the Environment, 1995.

Hart, Maureen, *Guide to Sustainable Community Indicators*, 2nd edn, North Andover, MA: Sustainable Measures, 1999.

Harvey, David, *Social Justice and the City*, London: Edward Arnold, 1973.

Harvey, David, *The Condition of Postmodernity: An Enquiry into the Origins of Cultural Change*, Cambridge, MA: Blackwell, 1990.

Hawken, Paul, *The Ecology of Commerce: A Declaration of Sustainability*, New York: HarperCollins, 1993.

Hawken, Paul, Amory Lovins, and L. Hunter Lovins, *Natural Capitalism: Creating the Next Industrial Revolution*, London: Earthscan, 1999.

Hayden, Dolores, *Redesigning the American Dream: The Future of Housing, Work, and Family Life*, New York: W.W. Norton & Company, 1984.

Hayes, Denis, *Repairs, Reuse, Recycling (First Steps Toward a Sustainable Society*, Washington, DC: Worldwatch Institute, 1987.

Healey, Patsy, *Collaborative Planning: Shaping Places in Fragmented Societies*, Vancouver: University of British Columbia Press, 1997.

Helling, Amy, Collaborative Visioning: Proceed With Caution!: Results from Evaluating Atlanta's Vision 2020 Project, *Journal of the American Planning Association*, 64 (3), 1998, pp. 335–49.

Henderson, Hazel, *Paradigms in Progress: Life Beyond Economics*, Indianapolis: Knowledge Systems, Inc, 1991.

Hiss, Tony, *The Experience of Place*, New York: Vintage Books, 1990.

Hoch, Charles, J., Linda C. Dalton, and Frank S. So (eds.), *The Practice of Local Government Planning*, 3rd edn, Washington, DC: International City/County Management Association, 2000.

Holmberg, Johan (ed.), *Making Development Sustainable: Redefining Institutions, Policy, and Economics*, Washington, DC: Island Press, 1992.

Hom, Leslie, The Making of Local Agenda 21: An Interview with Jeb Brugmann, *Local Environment*, (7) 3, 2002, pp. 251–6.

Hough, Michael, *City Form and Natural Process: Towards a New Urban Vernacular*, New York: Routledge, 1984.

Hough, Michael, *Out of Place: Restoring Identity to the Regional Landscape*, New Haven, CT: Yale University Press, 1990.

Howe, Elizabeth, *Acting on Ethics in Planning*, New Brunswick: Centre for Urban Policy Research, 1994.

Howard, Ebenezer, *Garden Cities of To-morrow*, London: S. Sonnenschein & Co, 1902 [1898].

Inglehart, Ronald, *The Silent Revolution: Changing Values and Political Styles among Western Publics*, Princeton, NJ: Princeton University Press, 1977.

Inglehart, Ronald, *Culture Shift in Advanced Industrial Society*, Princeton, NJ: Princeton University Press, 1990.

Inglehart, Ronald, *Modernization and Postmodernization: Cultural, Economic, and Political Change in 43 Societies*, Princeton, NJ: Princeton University Press, 1997.

Innes, Judith, Planning Theory's Emerging Paradigm: Communicative Action and Interactive Practice, *Journal of Planning Education and Research*, 14 (3), 1995, pp. 128–35.

International Council on Local Environmental Initiatives (ICLEI), *Local Agenda 21 Survey*, Toronto, 1997.

International Forum on Globalization, *Alternatives to Economic Globalization: A Better World is Possible*, San Francisco: Berrett-Koehler, 2002.

International Union for the Conservation of Nature (IUCN), *World Conservation Strategy*, 1980.

Isard, Walter, *Introduction to Regional Science*, Englewood Cliffs: Prentice Hall, 1975.

Jackson, Kenneth T., *Crabgrass Frontier: The Suburbanization of the United States*, New York: Oxford University Press, 1985.

Jacobs, Allan, *Looking at Cities*, Cambridge, MA: Harvard University Press, 1985.

Jacobs, Allan, *Great Streets*, Cambridge, MA: MIT Press, 1993.

Jacobs, Allan, Elizabeth Macdonald, and Yodan Roffe, *The Boulevard Book*, Cambridge, MA: MIT Press, 2001.

Jacobs, Jane, *The Death and Life of Great American Cities*, New York: Random House, 1961.

Jacobs, Jane, *The Economy of Cities*, New York: Random House, 1969.

Japan for Sustainability Information Center, Tomamae Is "Japan's Denmark" and a Mecca for Wind Farms, 2002. Available through www.japanfs.org.

Jones, Tom, William Pettus, and Michael Pyatok, *Good Neighbors: Affordable Family Housing*, Melbourne, Australia: The Images Publishing Group, 1997.

Judge, David, Gerry Stoker, and Harold Wolman (eds.), *Theories of Urban Politics*, Thousand Oaks, CA: Sage Publications, 1995.

Kaiser, Edward J., David R. Godschalk, and F. Stuart Chapin Jr, *Urban Land Use Planning*, 4th edn, Urbana: University of Illinois, 1995.

Katz, Bruce (ed.), *Reflections on Regionalism*, Washington, DC: Brookings Institution Press, 2000.

Keil, Claudia, *Environmental Policy in Germany*, San Francisco: Resource Renewal Institute, 1999.

Kelbaugh, Douglas, *Common Place: Toward Neighborhood and Regional Design*, Seattle: University of Washington Press, 1997.

Kent, Robert B. and Richard E. Klosterman, GIS and Mapping: Pitfalls for Planners, *Journal of the American Planning Association*, 66 (2), 2000, pp. 189–98.

Kent, T.J., *The Urban General Plan*, San Francisco: Chandler Publishing Co, 1964.

Keough, Noel, Calgary's Citizen-led Community Sustainability Indicators Project, *Plan*, 43 (1), Spring 2003, pp. 35–6.

Kidd, Charles V., The Evolution of Sustainability, *Journal of Agricultural and Environmental Ethics*, 5 (1), 1992, pp. 1–26.

Kinsley, Michael J. and L. Hunter Lovins, Paying for Growth, Prospering from Development, Snowmass, CO: Rocky Mountain Institute, n.d. [1996?].

Korten, David C., *When Corporations Rule the World*, West Hartford, CT: Kumarian Press, 1995.

Krizek, Kevin J. and Joe Power, *A Planners Guide to Sustainable Development*, Chicago: American Planning Association, 1996.

Kunstler, James Howard, *The Geography of Nowhere: The Rise and Decline of America's Man-made Landscape*, New York: Simon & Schuster, 1993.

Kunstler, James Howard, *Home from Nowhere: Remaking Our Everyday World the the 21st Century*, New York: Simon & Schuster, 1996.

Langdon, Philip, *A Better Place to Live: Reshaping the American Suburb*, New York: HarperCollins, 1994.

Layard, Antonia, Simin Davoidi, and Susan Batty (eds.), *Planning for a Sustainable Future*, New York: Spon Press, 2001.

Le Corbusier, *The City of Tomorrow and its Planning*, New York: Dover Publications, 1929 [1987].

Lefebvre, Henri, *The Production of Space*, trans. Donald Nicholson-Smith, Cambridge, MA: Blackwell, 1974 [1991].

Lennard, Suzanne H. and Henry L. Lennard, *Livable Communities Observed*, Carmel, CA: Gondolier Press, 1995.

Lennard, Suzanne H., Henry L. Lennard, and Sven von Ungern-Sternberg (eds.), *Making Cities Livable*, Carmel, CA: Goldolier Press, 1997.

Leopold, Aldo, *A Sand County Almanac*, New York: Ballantine Books, 1949 [1970].

Lerner, Michael, *The Politics of Meaning: Restoring Hope and Possibility in an Age of Cynicism*, New York: Addison-Wesley, 1996.

Lindblom, Charles E., The Science of Muddling-through, *Public Administration Review*, 19, 1959, pp. 79–88.

Lipke, David J., Green Homes: Eco-friendly Building Trends, *American Demographics*, January 2001.

Lewis, Sinclair, *Babbitt*, New York: Harcourt, Brace & Co, 1922.

Livingstone, Ken, *The Draft London Plan: A Summary*, London: City Hall, 2002.

Logan, John R. and Harvey L. Molotch, *Urban Fortunes: The Political Economy of Place*, Berkeley: University of California Press, 1987.

Lovins, Amory, *Soft Energy Paths: Toward a Durable Peace*, San Francisco: Friends of the Earth, 1977.

Lowe, Marcia D., *Alternatives to the Automobile: Transport for Livable Cities*, Worldwatch Paper 98, Washington, DC: The Worldwatch Institute, 1990.

Lowe, Marcia D., *Shaping Cities: The Environmental and Human Dimensions*, Worldwatch Paper 105, Washington, DC: The Worldwatch Institute, 1991.

Luccarelli, Mark, *Lewis Mumford and the Ecological Region*, New York: The Guilford Press, 1995.

Lusk, Ann, Designing the Active City: The Case for Multi-use Paths, *Planners Network*, Fall 2003.

Lyle, John Tillman, *Regenerative Design for Sustainable Development*, New York: John Wiley & Sons, 1994.

Lyle, John Tillman, *Design for Human Ecosystems: Landscape, Land Use, and Natural Resources*, Washington, DC: Island Press, 1999.

Lynch, Kevin, *A Theory of Good City Form*, Cambridge, MA: MIT Press, 1981.

McCamant, Kathryn and Charles Durrett, *Co-housing: A Contemporary Approach to Housing Ourselves*, 2nd edn, Berkeley: Ten Speed Press, 1994.

Macey, Jonathan R., *Property Rights in Sweden: A Law and Economics Perspective*, Stockholm: Swedish Center for Business and Policy Studies, SNS Occasional Paper No. 85, January 2001.

McGinn, Anne Platt, *Why Poison Ourselves?: A Precautionary Approach to Synthetic Chemicals*, Worldwatch Paper 153, Washington, DC: Worldwatch Institute, 2001.

McHarg, Ian L., *Design With Nature*, Garden City, NY: The Natural History Press, 1969.

McHarg, Ian L. and Frederick Steiner, *To Heal the Earth: Selected Writings of Ian L. McHarg*, Washington, DC: Island Press, 1998.

Maclaren, Virginia, Urban Sustainability Reporting, *Journal of the American Planning Association*, 62 (2), 1996, pp. 184–202.

Mander, Jerry, *In the Absence of the Sacred: The Failure of Technology and the Survival of the Indian Nations*, San Francisco: Sierra Club Books, 1991.

Marcus, Clare Cooper and Wendy Sarkissian, *Housing as if People Mattered: Site Design Guidelines for Medium-density Family Housing*, Berkeley: University of California Press, 1986.

Marcus, Clare Cooper and Carolyn Francis (eds.), *People Places: Design Guidelines for Urban Open Space*, New York: Van Nostrand Reinhold, 1990.

Markusen, Ann *et al.*, *The Rise of the Gunbelt: The Military Remapping of Industrial America*, New York: Oxford University Press, 1991.

Marsh, George Perkins, *Man and Nature, or Physical Geography as Modified by Human Action*, New York: C. Scribner, 1864.

Massey, Doreen, *Spatial Divisions of Labour: Social Structures and the Geography of Production*, London: Macmillan, 1984.

Massey, Doreen, *Space, Place and Gender*, Cambridge, UK: Polity Press, 1994.

Mazmanian, Daniel A. and Michael E. Kraft (eds.), *Toward Sustainable Communities: Transition and Transformations in Environmental Policy,* Cambridge, MA: MIT Press, 1999.

Meadows, Donella, Dennis L. Meadows, Jörgen Randers, and William W. Behrens III, *The Limits to Growth*, New York: Universe Books, 1972.

Meadows, Donella, Dennis L. Meadows, and Jörgen Randers, *Beyond the Limits: Confronting Global Collapse, Envisioning a Sustainable Future*, Post Mills, VT: Chelsea Green, 1992.

Merchant, Carolyn, *Radical Ecology*, New York: Routledge, 1992.

Mitlin, Diana, Sustainable Development: A Guide to the Literature, *Environment and Urbanization*, 4(1), 1992.

Mollenkopf, John H., *The Contested City*, Princeton, NJ: Princeton University Press, 1983.

Morris, A.E.J., *History of Urban Form Before the Industrial Revolutions*, 3rd edn, New York: John Wiley & Sons, 1994, pp. 305–6.

Morris, David, *Self-reliant Cities: Energy and the Transformation of Urban America*, San Francisco: Sierra Club Books, 1983.

Moudon, Anne Vernez and Paul Hess, Suburban Clusters: The Nucleation of Multifamily Housing in Suburban Areas of the Central Puget Sound, *Journal of the American Planning Association*, 66 (3), 2000, pp. 243–64.

Mumford, Lewis, *Technics and Civilization*, New York: Harcourt, Brace & Co, 1934.

Mumford, Lewis, *The Culture of Cities*, New York: Harcourt, Brace & Co, 1938.

Mumford, Lewis, *The City in History: Its Origins, Its Transformations, and Its Prospects*, New York: Harcourt, Brace & World, 1961.

Mumford, Lewis, *The Urban Prospect*, New York: Harcourt, Brace & World, 1968.

Munt, Ian, An Exemplary Sustainable World City?, *Town & Country Planning*, November 2001, pp. 292–3.

Nadel, Steve and Howard Geller, *Smart Energy Policies: Saving Money and Reducing Pollutant Emissions through Greater Energy Efficiency*, Washington, DC: Tellus Institute, 2001, p. vi.

Nader, Ralph and Mark J. Green (eds.), *Corporate Power in America*, New York: Grossman Publishers, 1973.

Nader, Ralph, Mark Green, and Joel Seligman, *Taming the Giant Corporation*, New York: W.W. Norton & Co, 1976.

Natural Resources Defense Council and US Environmental Protection Agency, *Environmental Characteristics of Smart Growth Neighborhoods: An Exploratory Case Study*, Washington, DC, 2000.

Nelessen, Anton, *Visions for a New American Dream*, Chicago: Planners Press, 1994.

Neuman, Michael, Does Planning Need the Plan?, *Journal of the American Planning Association*, 64 (2), 1998, pp. 215ff.

Neuman, Michael, Regional Design: Recovering a Great Landscape Architecture and Urban Planning Tradition, *Landscape and Urban Planning*, 47, 2000, pp. 115–28.

Newman, Peter and Jeffrey Kenworthy, *Cities and Automobile Dependence: An International Sourcebook*, Aldershot, UK: Gower, 1989.

Newman, Peter and Jeffrey Kenworthy, *Sustainability and Cities: Overcoming Automobile Dependence*, Washington, DC: Island Press, 1999.

Nicholson-Lord, David, Myopia, *Town & Country Planning*, March/April 2003, 100–3.

Nolan, John (ed.), *City Planning: A Series of Papers Presenting the Essential Elements of a City Plan*, New York: D. Appleton and Co, 1916.

Norberg-Hodge, Helena and Peter Goering (eds.), *The Future of Progress: Reflections on Environment and Development*, Berkeley: International Society for Ecology and Culture, 1986.

Norgaard, Richard, *Development Betrayed: The End of Progress and a Coevolutionary Revisioning of the Future*, New York: Routledge, 1994.

Norwood, Ken and Kathleen Smith, *Rebuilding Community in America: Housing for Ecological Living, Personal Empowerment, and the New Extended Family*, Berkeley: Shared Living Resource Center, 1995.

O'Connor, Martin (ed.), *Is Capitalism Sustainable? Political Economy and the Politics of Ecology*, New York: The Guilford Press, 1994.

Odum, Howard W. and Harry Estill Moore, *American Regionalism: A Cultural-historical Approach to National Integration*, New York: Henry Holt & Co, 1938.

Orfield, Myron, *Metropolitics: A Regional Agenda for Community and Stability*, Washington and Cambridge, MA: Brookings Institution Press and Lincoln Institute of Land Policy, 1997.

Osadcha, Oksana, *Sustainable Development Policy in the European Union*, San Francisco: Resource Renewal Institute, 2001.

Osborn, Fairfield, *Our Plundered Planet*, New York: Grosset & Dunlap, 1948.

Ostrom, Elinor, *Governing the Commons: The Evolution of Institutions for Collective Action*, New York: Cambridge University Press, 1990.

Owens, Susan and Richard Cowell, Planning for Sustainability: New Orthodoxy or Radical Challenge, *Town & Country Planning*, June 2001, pp. 170–2.

Owens, Susan and Richard Cowell, *Land and Limits: Interpreting Sustainability in the Planning Process*, London: Routledge, 2002.

Parayil, Govindan (ed.), *Kerala: The Development Experience: Reflections on Sustainability and Replicability*, London: Zed Books, 2000.

Parkinson, Sarah and Mark Roseland, Leaders of the Pack: An Analysis of the Canadian "Sustainable Communities" 2000 Municipal Competition, *Local Environment* (7) 4, 2002, pp. 422–9.

Pastor, Manuel Jr. *et al.*, *Regions that Work: How Cities and Suburbs Can Grow Together*, Minneapolis: University of Minnesota Press, 2001.

Pearce, David *et al.*, *Sustainable Development: Economics and Environment in the Third World*, London: Edward Elgar, 1990.

Pearce, David and Edward B. Barbier, *Blueprint for a Sustainable Economy*, London: Earthscan, 2000.

Pearce, David, Edward Barbier, and Anil Markandya, *Blueprint for a Green Economy*, London: Earthscan, 1989.

Pearce, David and Jeremy J. Warford, *World Without End: Economics, Environment and Sustainable Development*, New York: Oxford University Press, 1993.

Peck and Associates, *Implementing Sustainable Community Development: Charting a Federal Role for the 21st Century*, Toronto: Canada Mortgage and Housing Corporation, 2000.

Peck, Sheila, *Planning for Biodiversity: Issues and Examples*, Washington, DC: Island Press, 1998.

Peirce, Neal R., *Citistates: How Urban America Can Prosper in a Competitive World*, Washington, DC: Seven Locks Press, 1993.

Perry, Clarence Arthur, *Housing for the Machine Age*, New York: Russell Sage Foundation, 1939.

Phillips, Kevin, *Wealth and Democracy: A Political History of the American Rich*, New York: Broadway Books, 2002.

Platt, Rutherford H., Rowan A. Rountree, and Pamela C. Muick, *The Ecological City: Preserving and Restoring Urban Biodiversity*, Amherst: University of Massachusetts Press, 1994.

Porter, Douglas R., *Managing Growth in America's Communities*, Washington, DC: Island Press, 1997.

Portnoy, Kent E., *Taking Sustainable Cities Seriously: Economic Development, the Environment, and Quality of Life in American Cities*, Cambridge, MA: MIT Press, 2003.

President's Council on Environmental Quality, *Global 2000 Report*, Washington, DC: US Government Printing Office, 1981.

President's Council on Sustainable Development, *Sustainable America: A New Consensus for Prosperity, Opportunity, and a Healthy Environment for the Future*, Washington, DC: US Government Printing Office, 1996.

Putnam, Robert B., *Making Democracy Work: Civic Traditions in Modern Italy*, Princeton, NJ: Princeton University Press, 1993.

Putnam, Robert B., *Bowling Alone: The Collapse and Revival of American Community*, New York: Simon & Schuster, 2000.

Rabinovich, Jonas and Josef Leitman, Urban Planning in Curitiba, *Scientific American*, March 1996, pp. 46–53.

Rapoport, Amos, Toward a Redefinition of Density, *Environment and Behavior*, 7 (2), 1975, pp. 133–58.

Ravetz, Joe, *City-Region 2020: Integrated Planning for a Sustainable Environment*, London: Earthscan, 2000.

Redclift, Michael, *Sustainable Development: Exploring the Contradictions*, London: Methuen, 1987.

Rees, William E., Ecological Footprints and Appropriated Carrying Capacity: What Urban Economics Leaves Out, *Environment and Urbanization*, 4 (2), 1992, pp. 121–30.

Register, Richard, *Ecocity Berkeley: Building Cities for a Healthy Future*, Berkeley: North Atlantic Books, 1987.

Register, Richard, *Ecocities: Building Cities in Balance with Nature*, Berkeley: Berkeley Hills Books, 2002.

Renner, Michael, *Jobs in a Sustainable Economy*, Worldwatch Paper 104, Washington, DC: The Worldwatch Institute, 1991.

Repetto, Robert (ed.), *The Global Possible: Resources, Development, and the New Century*, New Haven, CT: Yale University Press, 1985.

Riesman, David, *The Lonely Crowd*, Garden City, NY: Doubleday, 1953.

Riley, Ann L., *Restoring Streams in Cities*, Washington, DC: Island Press, 1998.

Roseland, Mark, *Toward Sustainable Communities: Resources for Citizens and their Governments*, Stony Creek, CT: New Society Publishers, 1998.

Roszak, Theodore, *Where the Wasteland Ends: Politics and Transcendence in Post-industrial Society*, Garden City, NY: Doubleday, 1972.

Rowe, Janet and Colin Fudge, Linking National Sustainable Development Strategy and Local Implementation: A Case Study in Sweden, *Local Environment*, 8 (2), 2003, pp. 125–40.

Roy, Robin and Sally Caird, Household Ecological Footprints: Moving Towards Sustainability?, *Town & Country Planning*, October 2001, pp. 277–9.

Rudlin, David and Nicholas Falk, *Building the 21st Century Home: The Sustainable Urban Neighborhood*, Oxford: Architectural Press, 1999.

Rudofsky, Bernard, *Streets for People: A Primer for Americans*, Garden City, NY: Doubleday, 1969.

Rydin, Yvonne, *Urban and Environmental Planning in the UK*, London: Macmillan, 1998.

Rydin, Yvonne and Andy Thornley (eds.), *Planning in the UK: Agendas for the New Millennium*, Aldershot: Ashgate, 2000.

Sale, Kirkpatrick, *Dwellers in the Land: The Bioregional Vision*, San Francisco: Sierra Club Books, 1985.

Satterthwaite, David (ed.), *The Earthscan Reader in Sustainable Cities*, London: Earthscan, 1999.

Schumacher, E.F., *Small Is Beautiful: A Study of Economics as if People Mattered*, London: Blond and Briggs, 1973.

Schuyler, David, *The New Urban Landscape: The Redefinition of City Form in Nineteenth-century America*, Baltimore: Johns Hopkins University Press, 1986.

Schuyler, David, *Apostle of Taste: Andrew Jackson Downing, 1815–1852,* Baltimore: Johns Hopkins University Press, 1996.

Scientific Assessment Panel of the Montreal Protocol on Substances that Deplete the Ozone Layer, *Scientific Assessment of Ozone Depletion: 2002*, New York: United Nations Environment Program. Available at http://www.unep.org/ozone/pdf/execsumm-sap2002.pdf.

Seamon, David (ed.), *Dwelling, Seeing, and Designing: Toward a Phenomenological Ecology*, Albany: SUNY Press, 1993.

Serageldin, Ismail, Sustainability as Opportunity and the Problem of Social Capital, *Brown Journal of World Affairs*, 3 (2), 1996, pp. 187–203.

Sewell, John, *The Shape of the City: Toronto Struggles with Modern Planning*, Toronto: University of Toronto Press, 1993.

Shabecoff, Philip, *Earth Rising: American Environmentalism in the 21st Century*, Washington, DC: Island Press, 2000.

Shen, Quin, A Spatial Analysis of Job Openings and Access in a U.S. Metropolitan Area, *Journal of the American Planning Association*, 67 (1), 2001, pp. 53–68.

Simon, Julian L., *The Ultimate Resource*, Princeton, NJ: Princeton University Press, 1981.

Sirkis, Alfredo, Bike Networking in Rio: The Challenges for Non-motorised Transport in an Automobile-dominated Government Culture, *Local Environment*, 5 (1), 2000, pp. 83–95.

Soja, Edward W., *Postmodern Geographies: The Reassertion of Space in Critical Social Theory*, New York: Verso, 1989.

Southworth, Michael and Eran Ben-Joseph, *Streets and the Shaping of Towns and Cities*, New York: McGraw-Hill, 1997.

Southworth, Michael and Peter M. Owens, The Evolving Metropolis: Studies of Community, Neighborhood, and Street Form at the Urban Edge, *Journal of the American Planning Association*, 59 (3), 1993, pp. 271–87.

Speir, Cameron and Kurt Stephenson, Does Sprawl Cost Us All?: Isolating the Effects of Housing Patterns on Public Water and Sewer Costs, *Journal of the American Planning Association*, 68 (1), 2002, pp. 56–70.

Spirn, Ann Whiston, *The Granite Garden: Urban Nature and Human Design*, New York: Basic Books, 1984.

Stavrianos, L.S., *Global Rift: The Third World Comes of Age*, New York: Morrow, 1981.

Stivers, Robert L., *The Sustainable Society: Ethics and Economic Growth*, Philadelphia: Westminster Press, 1976.

Stone, Clarence, *Regime Politics: Governing Atlanta, 1946–1988*, Lawrence: University Press of Kansas, 1989.

Stren, Richard *et al.* (eds.), *Sustainable Cities: Urbanization and the Environment in International Perspective*, Boulder: Westview Press, 1992.

Susman, Carl (ed.), *Planning the Fourth Migration: The Neglected Vision of the Regional Planning Association of America*, Cambridge, MA: MIT Press, 1976.

Sustainable Communities: Building for the Future, London: Office of the Deputy Prime Minister, 2003. Available through www.odpm.gov.uk.

Sustainable Seattle Coalition, *Indicators of Sustainable Community*, Seattle, 1995.

Teilhard de Chardin, Pierre, *The Phenomenon of Man*, New York: Harper, 1959.

Thiele, Leslie Paul, *Environmentalism for a New Millenium: The Challenge of Coevolution*, New York: Oxford University Press, 1999.

Thompson, J. William and Kim Sorvig, *Sustainable Landscape Construction: A Guide to Green Building Outdoors*, Washington, DC: Island Press, 2000.

Todd, Nancy Jack and John Todd, *From Eco-cities to Living Machines*, Berkeley: North Atlantic Books, 1993.

Tomalty, Ray, Growth Management in the Vancouver Region, *Local Environment*, 7 (4), 2002, pp. 431–45.

Trainer, Ted, *Abandon Affluence!*, London: Zed Books, 1986.

Tuchman, Barbara, *The March of Folly: From Troy to Vietnam*, New York: Knopf, 1984.

Turner, R. Kerry (ed.), *Sustainable Environmental Management: Principles and Practice*, Boulder: Westview Press, 1988.

United Church of Christ, Commission for Racial Justice, *Toxic Wastes and Race in the United States: A National Report on the Racial and Socioeconomic Characteristics of Communities with Hazardous Waste Sites*, Washington, DC: United Church of Christ, 1987.

United Nations Framework Convention on Climate Change, *Report on the In-depth Review of the Third National Communication of the United Kingdom*, New York, 2003.

United Nations Human Settlements Programme (United Nations Center on Human Settlements), *Habitat Declaration*, Nairobi, 1996.

Urban Ecology, Inc, *Blueprint for a Sustainable Bay Area*, Oakland, 1996.

Urban Task Force (Lord Rogers, Chair), *Towards an Urban Renaissance*, London: Department of the Environment, Transport, and the Regions, 1999.

US Bureau of the Census, *Statistical Abstract of the United States*, Washington, DC: US Government Printing Office, 2001.

US Environmental Protection Agency, *Executive Summary of the Inventory of U.S. Greenhouse Gas Emissions and Sinks, 1990–2000*, Washington, DC.

US Environmental Protection Agency, *Our Built and Natural Environments: A Technical Review of the Interactions between Land Use, Transportation, and Environmental Quality*, Washington, DC, 2001, p. 6.

US General Accounting Office, *Siting of Hazardous Waste Landfills and their Correlation with Racial and Economic Status of Surrounding Communities*, Washington, DC: GAO/RCED-83-168, B-211461, 1983.

Vale, Brenda and Robert Vale, *The Autonomous House: Design and Planning for Self-sufficiency*, London: Thames and Hudson, 1975.

Van der Ryn, Sim and Peter Calthorpe (eds.), *Sustainable Communities*, San Francisco: Sierra Club Books, 1984.

Van der Ryn, Sim and Stuart Cowan, *Ecological Design*, Washington, DC: Island Press, 1996.

Van Eeten, Michel and Emery Roe, When Fiction Conveys Truth and Authority: The Netherlands Green Heart Planning Controversy, *Journal of the American Planning Association*, 66 (1), 2000, pp. 58–67.

Vogt, William, *Road to Survival*, New York: W. Sloane Associates, 1948.

Wackernagel, Mathis and William Rees, *Our Ecological Footprint: Reducing Human Impact on the Earth*, Gabriola Island, BC: New Society Publishers, 1996.

Wackernagel, Mathis and J. David Yount, The Ecological Footprint: An Indicator of Progress Toward Regional Sustainability, *Environmental Monitoring and Assessment*, 51, 1998, pp. 511–29.

Warburton, Diane (ed.), *Community and Sustainable Development: Participation in the Future*, London: Earthscan, 1998.

Warburton, Diane, Participation – Delivering Fundamental Change, *Town & Country Planning*, March 2002, pp. 82–4.

Ward, Barbara and René Dubos, *Only One Earth: The Care and Maintenance of a Small Planet*, New York, W.W. Norton & Co, 1972.

Waring, Marilyn, *If Women Counted: A New Feminist Economics*, San Francisco: Harper & Row, 1988.

Warner, Sam Bass, *Streetcar Suburbs: The Process of Growth in Boston 1870–1900*, Cambridge, MA: Harvard University Press, 1962.

Webber, Melvin M. *et al.* (eds.), *Explorations into Urban Structure*, Philadelphia: University of Pennsylvania Press, 1964.

Weiss, Marc, *The Rise of the Community Builders: The American Real Estate Industry and Urban Land Planning*, New York: Columbia University Press, 1987.

Wheeler, Stephen M., Planning for Metropolitan Sustainability, *Journal of Planning Education and Research*, 20, 2000, pp. 133–45.

Wheeler, Stephen M., *Smart Infill: Creating More Livable Communities in the Bay Area*, San Francisco: Greenbelt Alliance, 2002a.

Wheeler, Stephen M., The New Regionalism: Key Characteristics of an Emerging Movement, *Journal of the American Planning Association*, 68 (3), 2002b, pp. 267–78.

Wheeler, Stephen M. and Timothy Beatly (eds.), *The Sustainable Urban Development Reader*, London and New York: Routledge, 2004.

Whyte, William H., *The Organization Man*, Garden City, NY: Doubleday, 1956.

Whyte, William H., *The Last Landscape*, Garden City, NY: Doubleday, 1968.

Wilson, William Julius, *The Truly Disadvantaged: The Inner City, the Underclass, and Public Policy*, Chicago: University of Chicago Press, 1987.

Wilson, William Julius, *When Work Disappears: The World of the New Urban Poor*, New York: Knopf, 1996.

Winter, Paul, Sustainable Housing – The Legal Context, Part I, *Town & Country Planning*, January 2001a, pp. 26–8.

Winter, Paul, Sustainable Housing – The Legal Context, Part II, *Town & Country Planning*, February 2001b, pp. 50–3.

World Commission on Environment and Development (Brundtland Commission), *Our Common Future*, New York: Oxford University Press, 1987.

Wurster, Donald, *The Wealth of Nature: Environmental History and the Ecological Imagination*, New York: Oxford University Press, 1993.

Yaro, Robert D. and Tony Hiss, *A Region at Risk: The Third Regional Plan for the New York-New Jersey-Connecticut Metropolitan Area*, Washington, DC: Island Press, 1996.

Zukin, Sharon, *Landscapes of Power: From Detroit to Disney World*, Berkeley: University of California Press, 1991.

Index